Water Resources Development and Management

Indexed by Scopus

Each book of this multidisciplinary series covers a critical or emerging water issue. Authors and contributors are leading experts of international repute. The readers of the series will be professionals from different disciplines and development sectors from different parts of the world. They will include civil engineers, economists, geographers, geoscientists, sociologists, lawyers, environmental scientists and biologists. The books will be of direct interest to universities, research institutions, private and public sector institutions, international organisations and NGOs. In addition, all the books will be standard reference books for the water and the associated resource sectors.

Dipankar Saha · Karen G. Villholth ·
Mohamed Shamrukh
Editors

Managed Groundwater Recharge and Rainwater Harvesting

Outlook from Developing Countries

Springer

Editors
Dipankar Saha
Formerly with Central Ground Water Board (CGWB)
Faridabad, India

Manav Rachna International Institute of Research and Studies
Faridabad, Haryana, India

Karen G. Villholth
Formerly with International Water Management Institute (IWMI)
Pretoria, South Africa

Water Cycle Innovation (Pty) Ltd.
Limpopo, South Africa

Mohamed Shamrukh
Formerly with Minia University
Minia, Egypt

Ministry of Environment and Climate Change
Doha, Qatar

ISSN 1614-810X ISSN 2198-316X (electronic)
Water Resources Development and Management
ISBN 978-981-99-8756-6 ISBN 978-981-99-8757-3 (eBook)
https://doi.org/10.1007/978-981-99-8757-3

This Springer imprint is published by the registered company Springer Nature Singapore Pte Ltd.
The registered company address is: 152 Beach Road, #21-01/04 Gateway East, Singapore 189721, Singapore

Paper in this product is recyclable.

Foreword

Water policies and programmes all over the world, throughout the twentieth century, were formulated ignoring the profound inter-connectedness of different elements of the water cycle. Many parts of the world today face water crises, almost entirely engendered by the way we have gone about managing, or rather mismanaging the groundwater. Groundwater is a shared, common pool resource that needs to be shared and managed collectively, with careful attention being paid to maintaining a balance between the rates of extraction and rates of recharge. Degraded environments and climate change underscore the need and challenge of restoring these natural balances in the wider ecosystem. The very future of the planet is at stake in this era of the Anthropocene.

'Command-and-control over nature', the dominant paradigm of the twentieth century, is no longer an option. We need to urgently move towards a paradigm of 'working with, not against nature', weaving our interventions into the contours of nature, to leverage beneficial natural and ecosystem-based functions and processes that may help to support the protection and regeneration of natural resources, including water, as part of integrated human-natural systems.

This volume critically examines a set of nature-based solutions, namely managed groundwater recharge and rainwater harvesting, which hinges on the purposeful use of the subsurface formations (soil and aquifers, the natural storage 'container' for groundwater) as a workable space for water capture and storage. Water storage and its judicious management are becoming increasingly important parameters in climate change adaptations to address the growing variability, extremes, and unpredictable nature of water resource availability as evidenced globally.

The approaches span widely and get increasingly refined and adapted as per the bio-geo diversity, which this volume aims to convey. From relatively simple capture of rainwater in agricultural fields, overland flows, or on rooftops, to be diverted to underground storage, to advanced technologies to harvest and induce recharge into aquifers through the application of a new understanding of the system and technology, across a multitude of contexts, from rural to urban, from developing to developed countries, and across a variety of sectors, agriculture, water supply, and industry. These methods are also increasingly applied to improve groundwater quality through dilution. The source water for additional recharge is getting scarcer at places and allowing the use of treated wastewater for recharging under strict control can be adopted, underlining the circular approaches to water use. Importantly, such methods may also reduce flooding risks and erosion and can help counteract seawater intrusion. Using the systems as part of aesthetic and recreational amenities and ecosystem restoration, the micro-climate may be improved, environmental flows can be augmented, and biodiversity can be enhanced, creating a multitude of potential co-benefits of these approaches.

This volume is a testimony to the benefits accrued to a multitude of stakeholders engaged in managed groundwater recharge and rainwater harvesting, through real-life cases from varied terrain, geology, land use, and water use profiles. Contributions have been made from fifteen developing countries spanning the entire globe namely Afghanistan, Bangladesh, Chile, Colombia, Egypt, Guatemala, India, Jordan, Morocco, Nigeria, Palestine, Qatar, South Africa, Sri Lanka, and Tunisia. Historic developments, progress, outcomes, challenges, and emerging policies and regulations guiding and regulating these practices are investigated.

Much remains to be done further, especially on the management and regulatory side. Without addressing the demand side of water, a mere supply-side focus will remain endlessly condemned to playing catch-up! The supply-side requires integration of surface water, coastal waters, wastewater, and groundwater, as well as the subsurface space, land, and ecosystems, with the make-or-break key factor being public participation and equitable access to the benefits of these solutions. They should support, rather than being detrimental to the deprived communities.

In this regard, India has been somewhat of a pioneer when it comes to rainwater harvesting and managed aquifer recharge, often conceived and implemented as part of an integrated watershed management approach, backed by a scientific understanding of aquifers. India now holds a dedicated and ambitious Master Plan for Artificial Recharge to Groundwater from 2020 (revised from 2013), aiming at advancing and directing these approaches, which to a larger extent originally were initiated and pursued through a civil society movement in response to growing droughts and local groundwater overdraft.

Global institutions, like the World Bank, UNESCO, International Water Management Institute, the Global Environment Facility, and many others, are increasingly advocating and supporting developing countries in accomplishing these endeavours. For example, the summary of the proceedings of the 2023 Water Conference held in March 2023,[1] a global summit on water reinvigorating commitments from the only previous United Nations Water Conference held in 1977 in Mar Del Plata, Argentina, highlighted water storage and rainwater harvesting as innovative technologies needing accelerated investments.[2]

---

[1] United Nations Conference on the Midterm Comprehensive Review of the Implementation of the Objectives of the International Decade for Action "Water for Sustainable Development", 2018–2028.

[2] https://digitallibrary.un.org/record/4012972/files/PGA77SummaryforWaterConference2023.pdf.

I must compliment the Corresponding Editor, Dipankar Saha, and the Co-editors, Karen G. Villholth and Mohamed Shamrukh, for their tireless labour of love in putting together such an enriching volume. Particularly heartening is to see the contributions made by the developing countries across the world. I must also put on record the efforts made by the Centre for Science and Technology of the Non-Aligned and Other Developing Countries (NAM S&T Centre) in bringing out such a wonderful monograph, which is timely, and I daresay, will prove of immense value across multiple contexts and stakeholders.

Dr. Mihir Shah
Former Member
Planning Commission, Government
of India and Distinguished Professor
Shiv Nadar Institution of Eminence
New Delhi, India

Preface

Sustainable management of freshwater resources is a global issue of growing concern in both policy and practice. With an increasing gap between water demand and supply, compounding challenges and complexities of water resources management are negatively affecting people, nature, and economies across the globe.

Worldwide, in various geographic, geological, topographic, land use, and climatic conditions, aquifers are a major source of water for human development. The dependence on groundwater is increasing because of low availability and reliability of surface water and its widespread pollution. Groundwater provides an important insulator and buffer against climate change, especially droughts, though groundwater itself is also impacted by climate change and droughts, though in less predictable ways and with a certain delay.

Over-exploitation of groundwater resources is a global phenomenon, particularly in arid and semi-arid areas. Pollution of groundwater is also happening widely, while often going unnoticed until critical public health or environmental problems emerge. Sustainable management of groundwater is a great challenge as the resource is distributed, hidden, and mostly not subjected to monitoring and regulation. At the same time, as the water demand is increasing, with existing drilling and pumping technologies, a large volume of groundwater is getting extracted, often in great excess of natural replenishment. Finally, groundwater and surface water are connected as parts of the natural water cycle, making conjunctive and integrated management of the water resources imperative.

Solutions are sought to combat escalating water insecurity, which potentially leads to social unrest and political tension. A critical part of this is to adopt various water supplies and demand-side interventions tailored to local conditions.

To enhance groundwater supply, one of the most important and globally widely adopted techniques is to increase and manage recharge to groundwater. While ample literature covers interventions and experiences in developed countries, it is acknowledged that for less developed countries and communities, though a wealth of traditional knowledge and practice exists, there is a need to document current and evolving practices and further strengthen approaches through support from science, policy, practice, and partnerships.

The idea of developing a book addressing groundwater recharge management and rainwater harvesting emerged from Dr. Amitava Bandopadhyay, Director General, Centre for Science & Technology of the Non-Aligned and Other Developing Countries (NAM S&T Centre), New Delhi, India.

The book, through its 19 chapters, provides novel information, via status papers, case studies, and research outputs, on the manmade applications of groundwater recharge enhancement and rainwater harvesting across Africa, Asia, and Latin America. The contributions provide examples of how to resolve the water crises and address climate change impacts on available freshwater resources. The issues addressed through these approaches include supplementing potable water supply, reducing the intensity of groundwater overexploitation, stormwater management, intensifying greywater reuse, and improving groundwater quality, environmental flows, and other ecosystem services. The chapters deal with local-scale experimentation and management, adopted government schemes, community involvement, private sector engagement, and socio-economic, environmental, legal, institutional, and policy aspects.

The book collates contributions from 15 NAM and other developing countries, namely, Afghanistan, Bangladesh, Chile, Colombia, Egypt, Guatemala, India, Jordan, Morocco, Nigeria, Palestine, Qatar, South Africa, Sri Lanka, and Tunisia. The contributions have been made by researchers, government departments, civil societies, non-governmental organizations, policymakers, and water management practitioners.

The overall aim of the book is to place before the readers, the efforts being undertaken in developing countries to address water resource sustainability and climate change resilience through traditional and innovative groundwater recharge, rainwater harvesting, and water storage methods. It also takes a broader outlook on the prospects of these methods and techniques to enhance long-term socioeconomic development for millions of communities globally through enhancing water security.

The Editors would like to express our sincere gratitude to all the authors who kindly accepted to write the chapters, revise the content based on the reviewer's comments, and submit final versions well in time in spite of their busy schedules.

We are thankful to Dr. Amitava Bandopadhyay, Director General; Mr. Madhusudan Bandyopadhyay, Senior Adviser; Mr. Pankaj Buttan, Data Processing Manager; Ms. Abhirami Ramdas, Research Associate; and other staff members at the NAM S&T Centre for all the technical and administrative support rendered towards the publication of this Monograph.

We are grateful to Dr. Mihir Shah, Former Member, Planning Commission, Government of India and Distinguished Professor, Shiv Nadar Institution of

Eminence, India, for writing the 'Forewcrd' of this book. We also express our gratitude to Dr. Loyola D'Silva, Executive Editor, Springer Nature, Singapore, and his team for making this endeavour a reality and a success.

New Delhi, India — Dipankar Saha, Ph.D.
Bela-Bela, Limpopo, South Africa — Karen G. Villholth, Ph.D.
Doha, Qatar — Mohamed Shamrukh, Ph.D.

Introduction

The World Economic Forum has consistently identified water crises among the top five global challenges since the last few years. Seventy per cent of the world's megacities are already water-stressed and the problem is much more alarming in developing countries, where groundwater levels are depleting rapidly. Groundwater is a critical natural resource for efficient and cost-effective supply of domestic, agricultural, and industrial water in both urban and rural areas. However, population growth, expansion of agriculture, and urbanization coupled with climate uncertainties and unsustainable surface water storage have accelerated groundwater extraction. The storage in surface bodies and groundwater is gradually decreasing due to changes in land use patterns and the depleting trend of groundwater recharge. All these have caused over-extraction of water from ground storage and surface water bodies, thus making potable water, an increasingly scarce commodity.

Natural replenishment of aquifer is not able to balance the excessive and persistent exploitation of groundwater in numerous parts of the world. In order to increase the natural supply of groundwater, artificial recharge to groundwater has emerged as a vital management approach. The ever-increasing imbalance of water availability and requirement can be resolved either by capturing the overflowing water in the continental regions through dams or by storing the available rainwater as groundwater through artificial recharge techniques.

Despite the advantages of rainwater harvesting as a sustainable development tool to mitigate water-related issues, there still exists a significant knowledge gap among the water management professionals to deliver desirable results. It is abundantly clear that access to water is a fundamental issue that requires solutions from the active engagement of stakeholders as well as the mobilization of knowledge and emerging technologies.

In order to address the above issues, the Centre for Science and Technology of the Non-Aligned and Other Developing Countries (NAM S&T Centre), New Delhi, is bringing out this Monograph for dissemination of relevant knowledge and information to the scientists, researchers, managers, policymakers, and other stakeholders engaged in the groundwater management sector.

The book through its nineteen chapters underscores the concept and principles of water harvesting techniques for groundwater recharge and its management. It provides readers with an overview of the groundwater occurrence, availability, and recharge; effective water harvesting approaches for groundwater recharge; issues on water harvesting and water security; case studies of implementing rainwater harvesting and the scope of rainwater harvesting for groundwater management strategies. The book intends to provide information on sustainable groundwater recharge practices to the developing world. The methods discussed in various chapters contribute notably towards achieving water-related sustainable development goals (SDGs) globally, especially the SDG-6.

The book brings together scientific communities from 15 countries namely Afghanistan, Bangladesh, Chile, Colombia, Egypt, Guatemala, India, Jordan, Morocco, Nigeria, Palestine, Qatar, South Africa, Sri Lanka, and Tunisia to share their knowledge and expertise to provide a detailed insight into the benefits and gaps in existing knowledge, implementation, and funding strategies, and public participation related to the improvement of groundwater management.

I am immensely grateful to Dr. Mihir Shah, Former Member, Planning Commission, Government of India, and Distinguished Professor at Shiv Nadar Institution of Eminence, India, for kindly agreeing to our request to write the Foreword of this book in spite of his very busy schedule.

I am thankful to Dr. Loyola D'Silva, Executive Editor, Springer Nature, Singapore, for his kind support and guidance towards bringing out this Monograph, and Ms. Shalini Monica C. Selvam, Project Coordinator, Springer Nature, India, for managing all the technical and administrative tasks for the publication process.

I would like to express my sincere gratitude to the Editors of this book, Dr. Dipankar Saha, Former Member, Central Ground Water Board (CGWB), New Delhi, India, and Chair Professor at Manav Rachna International Institute of Research and Studies (MRIIRS), Faridabad, Haryana, India; Dr. Karen G. Villholth, Director, Water Cycle Innovation (Pty) Ltd., Negester Klein-Kariba, Limpopo, South Africa; and Dr. Mohamed Shamrukh, Emeritus Professor of Water Resources and Environmental Engineering, Minia University, Egypt, and External Consultant at United Nations Environment Programme and Environmental Consultant at Ministry of Environment and Climate Change, Qatar, for their initiatives and efforts and sparing their valuable time in reviewing the papers for this book and taking charge of this publication project.

I also acknowledge the valuable support of the entire team at the NAM S&T Centre and am especially thankful to Mr. Madhusudan Bandyopadhyay, Senior Adviser, for his support and guidance, and Ms. Abhirami Ramdas, Research Associate, for her contributions in taking this publication project forward and bringing it to a successful conclusion.

I also record my appreciation for the assistance and support rendered by my colleagues Mr. Rahul Kumra, Assistant Administrative Officer, and Mr. Pankaj Buttan, Data Processing Manager, towards bringing out this Monograph.

I believe that this Monograph would serve as an excellent resource material for scientists and researchers from various R&D institutions, hydrologists, environmentalists, administrators, professionals from the water industry, hydraulic engineers, government officials, policymakers, and students who are actively engaged in the areas of water science and technology to make decisions to allocate water resources and to develop innovative cost-effective measures and approaches to achieve sustainable groundwater management.

Amitava Bandopadhyay, Ph.D.
Director General
NAM S&T Centre
New Delhi, India

Contents

Chapter 1
Managed Groundwater Recharge and Rainwater Harvesting for Sustainable Development: Research, Practices, and Policies from Developing Countries

Dipankar Saha, Karen G. Villholth, and Mohamed Shamrukh

Abstract Groundwater is a critical and strategic natural resource. Globally, it plays a pivotal role in meeting societal demands and providing ecosystem services. Across large areas, particularly in arid and semi-arid regions, increasing extraction of this resource is causing over-exploitation and depletion. The main manifestations in such cases are: fall in groundwater levels and subsequent dwindling yield or drying up of water wells, rise in groundwater salinity, reduced environmental flow in rivers, shrinking inland wetlands, saline water intrusion in coastal aquifers, subsidence of land, and a host of associated implications. Such a situation potentially adversely impacts agricultural production and food security, widens inequity, and triggers social tension and conflicts. Climate change is further adding woes to the situation with unpredictable rainfall, increasing temperatures, and rising seawater levels. Rainwater harvesting has been practiced since the dawn of some of the oldest civilisations in different corners of the world to ameliorate water shortfall. Managed groundwater recharge , also referred to as artificial recharge or managed aquifer recharge, is a later

---

D. Saha
Chair Prof, Manav Rachna Institute of Research and Studies, Faridabad, India

D. Saha (✉)
Former Member, Central Ground Water Board, Govt of India, Faridabad, India
e-mail: dsaha002@gmail.com

K. G. Villholth
Director, Water Cycle Innovation, 8900 Randers C, Denmark

M. Shamrukh
Minia University, El-Minia, Egypt

Ministry of Environment and Climate Change, Doha, Qatar

K. G. Villholth
URL: https://watercycleinnovation.com/

D. Saha et al. (eds.), *Managed Groundwater Recharge and Rainwater Harvesting*, Water Resources Development and Management,
https://doi.org/10.1007/978-981-99-8757-3_19

idea, being practiced increasingly over the last six decades or so as a dominant supply-side intervention to augment groundwater availability and reliability of water supply in general, but importantly during droughts and dry periods. This chapter provides an introductory view on managed groundwater recharge and rainwater harvesting (MGR/RWR), their objectives and possible synergies, efficacy, impacts, policies, and practices, illustrating their diverse benefits in developing countries across a wide array of terrain, land use, geology, and climatic conditions. The chapter also gives a synopsis of eighteen ensuing chapters from a wide range of contributors representing fifteen countries, illustrating the multitude of applications of rainwater harvesting and groundwater recharge augmentation and management, providing useful lessons for a broader audience.

1.1 Introduction

A part of the rainfall that falls on the Earth's surface infiltrates the surface, percolates through the vadose zone, and finally joins the water of an aquifer below; thus, recharging the groundwater resource. Rainfall is a main source of recharge of groundwater in large parts of the world. Besides rainfall, there are other sources of recharge, like return seepage from irrigation, seepage from canals and tanks, infiltration from rivers and streams, etc. Nevertheless, natural recharge is generally a slow process, and it varies greatly depending on various factors like, soil, terrain, land use, depth to the water table, nature of the vadose zone and distribution of rainfall in time and space (Bouwer 2002; Scanlon et al. 2006).

Natural recharge is often not able to keep up with the increasing groundwater abstraction in various parts of the globe. If the groundwater extraction on an annual basis and for an extended period is more than the total recharge in an area, the area is demarcated as over-exploited (Wada et al. 2010; Gleeson et al. 2010). Over-exploited areas are adversely affected by various impacts such as reduction in well yields and drying up of shallow wells, which jeopardise the otherwise assured water supply for domestic and irrigation uses, especially by small-scale users. Groundwater level decline may also disrupt and dry rivers and wetlands and other aquatic and terrestrial ecosystems (Al-Kharabsheh 2000). Another significant adverse impact is the deterioration of water quality from rising salinity, especially in coastal areas and inland arid areas (Saha and Gor 2020; Saha et al. 2022a, b; Fatema et al. 2018). These impacts are expected to intensify with climate change.

The impact of over-exploitation can be surprising and far-reaching. For instance, significant air pollution in the Delhi National Capital Region of India during the winter season every year has been found to be linked to the Government's effort to reduce groundwater over-exploitation in the adjoining states of Punjab and Haryana (Saha et al. 2022a; Mukherjee et al. 2015; Ibkar et al. 2023).

Dependence on groundwater is increasing rapidly throughout the globe. As per an estimate by Wada et al. (2010), total groundwater depletion in sub-humid to arid areas in the world has increased from 126 (±32) km^3 per year in 1962 to 283 (±40)

km^3 per year in 2000. These numbers are consistent with, and supplemented by more recent estimates, indicating that the global groundwater depletion is intensifying (de Graf et al. 2017). The reasons are many, including expanding groundwater-based irrigation, domestic demand from a rising population, changing lifestyle in urban areas, and higher consumption by industries, energy production, and infrastructure. The hotspots are spread over the globe, mostly located in arid and semi-arid areas, well-known cases being Northeast Pakistan and Northwest India (Rodell et al. 2009), Central Valley of California (Reilly et al. 2008), Yemen (Al-Sakkaf et al. 1999), Iran (Kerami and Hayati 2005), southeastern parts of Spain (Custudio 2002), Qatar (Shamrukh et al. 2012), and northeast China (Konikow and Kendy 2005).

In a scenario where groundwater extraction is increasing, a supply-side measure to counteract depletion and ensure stable water storage and water supply is needed by purposefully enhancing the recharge to groundwater. Such efforts have been made since the ages through some interventions on the land surface. The interventions adopt two broad techniques —(i) retaining water on the surface so that the infiltration increases because of longer soil–water contact, and (ii) enhancing the rate of infiltration and/or percolation so that a higher volume of water reaches the aquifer. The first part is widely mentioned as water harvesting or rainwater harvesting (RWH), and the second part is early known as artificial recharge, or later as managed groundwater recharge (MGR).

1.2 Rainwater Harvesting and Managed Groundwater Recharge

Rainwater harvesting (RWH) is a process of collecting natural precipitation and runoff over land for societal uses (Bobrs and Ben-Ashar 1982). RWH is being practiced worldwide since historical times, particularly in the arid and semi-arid areas. There are differences in approaches and methodologies but three characteristics are common (Bobrs and Ben-Ashar 1982):

a. Predominantly applied in arid and semi-arid areas.
b. Dependent on local water such as rainwater, surface runoff, flood waters, springs, etc.
c. Relatively small-scale and simple stakeholder-friendly interventions.

Artificial recharge is defined as the recharge that occurs when the natural pattern of recharge is deliberately modified and augmented (ASCE 2001). The process of recharge is not artificial per se, as the same physical laws govern the process of recharge, but the mechanisms under which the recharge takes place are manipulated. Artificial recharge refers to a process by which the groundwater resource is augmented relative to the natural rate and conditions of replenishment through a man-made facility (CGWB 2007). It is good to keep in mind that when water is harvested in a catchment and stored through RWHstructures, a part of the stored water is recharged, which is in addition to the natural recharge from rainfall and overland flow. Thus, the distinction between artificial recharge and rainwater harvesting is often blurred, and the methods can work in conjunction.

The term "managed groundwater recharge" (MGR) is used preferentially here in place of the term "artificial recharge", which is commonly applied in publications, to better express the overarching objectives and benefits of the practices and to emphasise the needed component of management. There are many examples worldwide where interventions of MGR and RWH have greatly enhanced the groundwater resource and in turn improved the livelihoods of farmers as well as drinking water supplies. Besides, there are potential collateral benefits, like less evaporation losses, less runoff, especially if flood waters are captured, and also potential improvements in water quality and enhanced environmental flows in rivers with spillover effects on ecosystem functioning and biodiversity (Valhondo et al. 2020; Zheng et al. 2021). In the Saurashtra region of Gujarat, India, underlain by basaltic aquifers, a community-led movement to harvest rainwater (primarily through check dams) has resulted in better groundwater availability in the wells for crop production during the non-monsoon season (Patel et al. 2020, Alam et al. 2022).

RWH has a long and cherished history worldwide, particularly in arid and semi-arid regions. In northern Jordan, during the Byzantine era, RWH was adopted to quell water crises (AbdelKhaleg and Ahmed 2007). In Ethiopia, evidence of the existence of ponds used for irrigation is reported from the Axunit period around 560 BC (Fattovich 1990). Also in Ethiopia, evidence of rooftop water harvesting from the fifteenth–sixteenth century is still visible in places (Getachew 1999). Another example is the water harvesting system developed in 1500 BC in North America (Pandey et al. 2003). Efficient water harvesting systems developed by the Harappans of Dholavira in Kutch, India, around 1500 BC, for the conservation, harvesting, and storage of water is an excellent example of ~~their~~ advanced hydraulic engineering skills (Ghosh et al. 2020). In semi-arid western India, the Hamirsar Lake was created in Bhuj, Gujarat, an arid zone during the 1550 to harvest rainwater and to recharge the freshwater aquifer (Saha and Gor 2020). Storage facilities for rainwater harvesting have been built throughout the Mediterranean, including Egypt, since 3000 BC. The tanks were used to store rainwater runoff and to store aqueduct water for seasonal fluctuations (Mays et al. 2013).

Human interventions with a primary target to increase groundwater resources is a recent phenomenon, spanning over the last four to six decades (Ali et al. 2023; Saha et al. 2022a, b; Zheng et al. 2021; Saha and Roy 2018). The urgency of recharging aquifers through direct interventions has accelerated with the extent of over-exploitation of the resource. It is argued, and also noted in practice, that when recharging an aquifer, the focus traditionally remains on volumetric storage issues, ignoring other important factors like quality of the source water for recharge, etc. The prevailing thinking is now that the overall groundwater system is to be managed (Dillion et al. 2019). To get optimal long-lasting benefits, there is a need to consider factors beyond the general connotation of the local aquifer, like the broader groundwater flow regime in different seasons, annual replenishable and recoverable resource behaviour, resource deficit of an area, inter-aquifer interaction in the case of a multitiered aquifer systems, climate change, etc.

1.3 Contemporary Interventions

Worldwide, a lot of awareness has been created on recharging aquifers and harvesting rainwater to combat groundwater stress. They are often considered the primary supply-side interventions for taming the declining resource and attaining sustainability in groundwater use in water-stressed areas (CGWB 2007). Various government schemes have been introduced in different countries. In India, the major emphasis of the *Mahatma Gandhi National Rural Employment Guarantee Scheme* (MGNREGS), the largest poverty alleviation scheme in the world, is on widespread, labour-intensive construction and rehabilitation of RWH and recharge structures. The *Atal Bhujal Yojana* Scheme, with a $750 million investment plan, jointly funded by Govt of India and the World Bank, is being adopted in over-exploited areas of eight states of India, with an.[1] Objective to improve the acute groundwater resource depletion situation through various supply and demand side interventions. The community and stakeholders are involved at all stages in the planning and execution of interventions. Other examples include a major recharge scheme with large-scale adaptation of borehole injection and recovery scheme in hard rock fractured aquifers in Windhoek, Namibia, which prevented construction of about 700 km of pipeline to bring water from a perennial river (Murray and Tredoux 2002). A significant scheme was implemented in Egypt to befit from flash floods through engineered structures for rainwater harvesting and groundwater recharge in Sinai and desert areas adjacent to the Nile Valley (Gado and El-Agha 2020).

1.4 Challenges and Way Forward

It is often argued that the benefits of harvesting water and recharging aquifers often remained suboptimal as they are not backed by scientific understanding of the aquifer system, groundwater flow regime, source water availability, etc. The cost–benefit of such interventions should indicate positive economic benefits as well as social outcomes, like equity in benefit distribution, avoiding upstream–downstream conflicts etc. The following issues need to be highlighted when interventions are taken up:

a. Sustainable management of groundwater resources should accompany the interventions.
b. There should be enough space available in the aquifer to accommodate recharged water.
c. Hydraulic properties of the aquifer should be good enough to facilitate recharge.
d. The aquifer configuration and the hydraulic properties should be such that the recharged water does not quickly dissipate by lateral flow, depriving the user/community of the anticipated benefits.

[1] https://www.ataljal.mowr.gov.in/.

e. Recharged groundwater should be accessible (not too deep) to pump after recharge.
f. Source water with acceptable physical and chemical quality should be available at the time of recharge.
g. The community should be sensitised and involved in the planning, operation, and maintenance.
h. The cost–benefit factor should be favourable.

For points a to e, a holistic understanding of the aquifer geometry, hydraulics, and the behaviour of water levels is essential. In developing countries, site selection and design of the structures are often done on an ad-hoc basis without sufficient backup knowledge (Saha et al. 2022a, b). However, the policymakers are increasingly realising this and initiating research/pilot schemes to develop knowledge support for recharge. The Government of India has taken up a massive scheme, the *National Aquifer Management Programme* (NAQUIM) since 2017, to generate holistic and aquifer-specific understanding and information, including on chemical quality and groundwater and resource availability, for an area of 2.9 million km^2. The investigation is carried out upto 300 m below ground in soft rock and up to 200 m depth in hard rock areas (Saha et al. 2018). The outputs are of critical help for the planning of recharge activities (Singh et al. 2019). In Egypt, a national project was implemented to investigate the hydrogeology of the Nile Valley and the adjacent desert areas presented in maps and technical reports. The output helped in assessing the suitable places for groundwater recharge at the national level (RIGW 1988).

The understanding of aquifers is more necessary where complex hydrogeological situations exist, like in multilayered aquifer systems underlying large river floodplains. The hydraulic head, recharge path and mechanism, groundwater extraction, volume of recharge, and chemical quality of groundwater may be different for different aquifers at different depths. This is exemplified by Keesari et al. (2021) in the intensely exploited Gangetic plains.

Availability of source water for recharge is a critical issue. In arid and semi-arid areas, many interventions suffer from inadequate rainfall. Often, hardly any runoff is available for recharge, as commitments exist for dams, reservoirs, and other societal uses (Gee and Hellel 1988). In a study under NAQUIM, in the highly groundwater-stressed regions in north-western India, it is clearly revealed that MGR and RWH have limiting value for groundwater improvement, mainly due to limited source water available for recharge. In such cases, demand-side interventions, like crop diversification, efficient irrigation, better agronomic practices, etc., could have more significant impact (Saha et al. 2017). Reclaimed wastewater from the urban areas is being considered a source of water and is practiced in places. Such practices have been reviewed by Foster and Chilton (2004). A good example is the restoration of the Korba Aquifer in Tunisia by recharging treated wastewater through infiltration basins (Ouelhazi et al. 2014).

Besides the volumetric availability of water, the quality of the source water is also important. It is proven through research that the quality of water applied in infiltration schemes improve significantly by physical removal of suspended material, pathogens, and organic matter during the infiltration through the soil layers (BGS 2002). In some cases, water quality improvement is consciously sought by managing chemical and microbiological processes during filtration and further percolation through the vadose zone. This approach is also referred to as Soil Aquifer Treatment (Bouwer 2002). Improvement of recharged water quality owing to a decrease in pathogens is reported by Dillion et al. (2006). Improvement in chemical contaminants and organic matter is reported at various experimental sites (Hoppe-Jones et al. 2010; Valhondo et al. 2018; Bekele et al. 2011).

However, further research is needed to better understand the improvement of source water quality through various interactions as there is pressure to use overland flow from agricultural fields, often found to be charged with fertilisers, and pesticides, as well as reclaimed wastewater, which may pose threats to the quality of newly generated groundwater, and cause severe implications to human health, if not carefully implemented.

Globally, and in developing countries, no comprehensive guidelines on the quality of source water to be used for MGR exist. The World Health Organization (WHO) has recommended and many countries have adopted guidelines for the quality of water for potable, agricultural, and different industrial uses. Dillion et al. (2014) articulated a viability assessment before managed aquifer recharge (MAR) is put into use. In this assessment, the water quality issues are dealt with amply.

The enhanced MGR and RWH generally promote equity in water distribution and access. They are generally small structures in comparison to large water-capturing structures like dams, and the benefits can be harvested by small and marginal holders, through low-cost dug wells and shallow bore wells (Saha et al. 2022a, b). The cost per unit volume of captured water in a dam is much higher than water harvesting structures (Saha et al. 2022a, b), though it is also mentioned that the proponent of recharge underestimates the cost and inflates the benefits (Kumar et al. 2008). A study by Dashora et al. (2022) in a semi-arid area in India indicates that even after accounting for an average annual maintenance cost of 3%, the average benefit–cost ratio of a check dam comes to 400%.

Another analysis for a proposed recharge site at Baghmalek Mountain front in Iran, underlain by coarse alluvial deposits, reveals that the cost–benefit ratio based on the predicted increase in agricultural production is estimated to be 1:1.32. The analysis did not include the benefit of reduction of the fury of the flood because of the recharge interventions (Kalantari et al. 2010).

1.5 Book Contributions

The book contains 18 chapters from 15 developing and other countries in Africa, Asia, and Latin America. The contributions are from top-notch professionals, representing senior government officials, policymakers, water management practitioners, scientists, academicians, non-governmental organisations, and civil society. A brief description of the individual chapters is provided below.

In **Egypt**, the gap between renewable water resources and demands will grow by more than twofolds to about 50 billion cubic meters (BCM) by 2050. The chapter by Dawoud on "The Role of Groundwater Aquifers Artificial Recharge in Water Resources Management in Egypt" provides an assessment of the schemes of artificial recharge in the national water plan. Artificial recharge of aquifers using conventional water resources such as rainfall or surplus overland flow or unconventional water resources such as treated wastewater is considered an important tool for managing the scarce water resources in Egypt. Several projects on artificial recharge of aquifers were carried out in the country to enhance deteriorated groundwater quality and improve groundwater availability.

In **India**, rainwater harvesting and artificial recharge are considered the backbone of activities to counter water shortage in groundwater-dependent areas. As a major supply-side intervention, the Govt. of India and the state governments have put considerable resources and efforts into creating facilities for conserving water and recharging aquifers. In the chapter "Water Conservation and Artificial Recharge Efforts in India", Suresh et al. meticulously present various schemes adopted by the central and state govt., like the *Atal Bhujal Yojana, Pradhan Mantri Krishi Sinchai Yojana,* and *Jal Shakti Abhyan,* focussing on rural areas. These schemes also include groundwater recharge through rooftop rainwater harvesting in urban areas.

Nigeria is relatively blessed with water resources with an average annual precipitation of 1,150 mm. In the chapter "Groundwater Recharge, Rainwater Harvesting and Regulations for Sustainable Water Resources Development in Nigeria", Eduvie and Musilim reviewed the available surface water resources, present water consumption, and future water demand till 2030 in the country. In view of the increasing demand, there will be more pressure on groundwater resources. They have recommended that the National Water Resources Policy (NWRP) should address all challenges through effective Integrated Water Resources Management. A plan for rainwater harvesting and artificial recharge needs to be framed urgently for the country.

Farm ponds are traditional water harvesting structures in **India**, particularly in hard rock terrain. The main objective of the farm pond is to harvest rainwater that will support non-monsoon cultivation as well as store water that would recharge the aquifer below. However, in places in southern India, underlain by hard rock aquifers and suffering from over-exploitation of groundwater resources, the farm ponds are used for storing pumped-out groundwater for later use. The chapter "Farm Ponds in Semi-Arid Hard Rock Terrain in India: Are they Increasing Dependency on Groundwater?" presented by D'Souza et al. demonstrate how such farm ponds increase dependency on groundwater, while depriving the small and marginal farmers of groundwater access and also wasting this precious resource.

Qatar is suffering from limited natural freshwater resources. A comprehensive national project was carried out in 2009 to study the role of artificial recharge in sustaining the over-exploited groundwater in the country. In the chapter "Rainfall Recharge Wells for Groundwater Sustainability in Qatar", Shamrukh and Al-Muraikhi present a detailed description and main outputs of this project. The optimal design indicates that a total of 1,502 recharge wells (1,388 new and 114 existing) are required in hydrogeologically favourable locations across the country to infiltrate storm water, typically around 20 mm within a 5-days period, which will augment the groundwater by 33.5 Mm^3/year. Balanced groundwater use will require additional measures in the future.

Long-term droughts, combined with groundwater over-exploitation, cause serious concern about groundwater depletion and quality in **Morocco**. A managed aquifer recharge program was one of the main efforts to ensure the sustainability of water resources. The chapter "Managed Aquifer Recharge in a Semi-Arid Basin: A Case Study from the Souss Aquifer, Morocco" by Gouahi et al. presents the groundwater recharge estimations for the Souss River Basin. Maximum infiltration is taking place between the middle and the upstream part of the basin where the hydraulic conductivity is relatively high. The recharge occurs mainly along the bed of the Souss River and its tributaries, which is confirmed by the water level variations in piezometers and isotope analysis. However, enhanced groundwater recharge still remains insufficient to meet the water needs, therefore other strategies and alternative solutions have been suggested in the study.

With increasing water demand and recent prolonged droughts, **Chile** is embarking on national strategies and practices on MAR. The chapter "Managed Aquifer Recharge in Chile: A Promising Alternative to Enhance Water Security", by Bennison and Claro illustrates how Chile is working on not only supporting the technical solutions, but also developing national guidelines through structured dialogue with researchers, legal entities, and potential users and implementers.

India is extracting more groundwater than any other country in the world. Large geographical areas are experiencing unabated over-exploitation of groundwater. Sustainable management of the resource is being debated and trialled through policy and practice. The chapter "From Managed Aquifer Recharge to Managing Aquifer Recharge: Developing a Strategic Approach to Artificial Recharge in India" by Kulkarni et al. discusses how the concept of Managed Aquifer Recharge is being adopted in diverse aquifer conditions of India. They emphasise the importance of recharge planning within the larger framework of participatory groundwater management.

South Africa is experiencing more frequent and intense floods and droughts. In the semi-arid Western Karoo, MAR, as a part of the national Artificial Recharge Strategy, is actively pursued by the national authorities for small communities facing water stress. In the chapter "Managed Aquifer Recharge Projects in the Western Karoo, South Africa: Progress and Challenges", by Esterhuyse et al. give a detailed description of four cases in the Western Karoo, illustrating how relatively simple and low-maintenance MAR schemes building on flood water capture and infiltration close to existing or new abstraction wells can help close the increasing water demand gap.

The case from **Tunisia**, "Artificial Recharge of Groundwater in Tunisia: A Long and Fruitful Experience" by Jarraya-Horriche and Chaieb provides an outline of the artificial recharge that started back in the 1970s and has been considered as an important national strategy since the decade 1990–2000. The benefits of artificial recharge is variable in space and time, but its impact does not meet the original expectations. In order to quantify the AR impact, monitoring networks have been set up all around the sites to measure the groundwater level and quality. A significant positive impact has been recorded in many cases, which has encouraged the decision makers to continue promoting artificial recharge practices.

The chapter by Molano on "Artificial Groundwater Recharge in Santa Marta and Bogotá, **Columbia**" provides two examples of how MAR has been implemented and trialled in two widely different regions of Columbia, one in a large coastal city, Santa Marta, with more than half a million inhabitants and the other in the Bogotá Plains outside the capital city of Bogotá. Schemes encompassed in-stream modification and an aquifer storage and recovery pilot scheme under pressure, respectively. Interestingly, the successful Santa Marta scheme, after 10 years, had to be discontinued, as the source river flow reduced and water quality diminished with time, indicating a need to incorporate catchment management and protection in the scheme.

The chapter by Salameh and Abdallat entitled "Groundwater Artificial Recharge in **Jordan**—Case Studies and Potential Areas" provides a review of the previous works and makes an assessment of groundwater recharge potential in Jordan. Their work is based on the availability of recharge water and the potential use of the recharge water. Six high-potential sites are identified, which are located along the northern part of the Jordan Valley, along Wadi Araba, Jafr Depression, Sirhan Depression, Wadi Rajil, and in the south of Aqaba. Groundwater recharge in Jordan can help counter the impacts of climate change and the implications of depleting groundwater resources.

The shallow aquifer in the Marj Sanour Watershed in **Palestine** is experiencing excessive over-exploitation of groundwater. The water level of the shallow aquifer has dropped more than 80 m in the last decade. The chapter "Groundwater Artificial Recharge in the Marj Sanour Watershed, Palestine" authored by Wishahi presents the results of a pilot project where six recharge wells were drilled to replenish the depleted Marj Sanour Aquifer. The overland flow is filtered before reaching into recharge wells. The results are encouraging, showing that the estimated recharge rate through the six wells is about 20,000 m^3/day in days with sufficient rainfall.

In the chapter "Rainwater Harvesting for Groundwater Recharge: Experience from **Sri Lanka**", Ariyananda and de Silva show how groundwater levels drop in the dry season in many parts of Sri Lanka because of rising abstraction and how the quality of groundwater has ~~also~~ deteriorated over the years. Rainwater harvesting techniques are adopted to counter this. The effectiveness of different types of structures, like garden ponds, runoff tanks, and rooftop rainwater harvesting, has been studied. The improvement of the water level as well as the water quality has also been examined .

Kabul City, the capital of **Afghanistan**, gets its major water supply from the Kabul Plain Aquifers. In the chapter "Groundwater Recharge in the Kabul Plain (Afghanistan), through Rainwater Harvesting", Zaryab et al. highlight that groundwater extraction from the Kabul Plain Aquifers is taking place in an uncontrolled manner and is becoming unsustainable as the resources are getting depleted. A MAR pilot project was carried out during 2017–2020, which showed encouraging results. In addition, the residents of Kabul City are encouraged by the municipality to harvest rainwater and store it underground or in household tanks during the rainy seasons.

There is an urgent need to take up efforts to recharge the Kabul Plain Aquifers in a planned and coordinated manner.

In the chapter "Combating Urban Water logging with Support from Underlaying Overexploited Aquifer - A Case Study from **India**", Mukherjee et al. present a case study on how artificial recharge can help reduce urban water logging in Faridabad town, adjoining the Delhi National Capital Region of India. In India, many urban aquifers suffer from huge groundwater pumping and consequent over-exploitation. At the same time, such areas remain waterlogged often due to various reasons, like stormwater, drainage congestion, reduced infiltration owing to pavementing etc. The chapter illustrates how water logging can be reduced by diverting standing water into depleted aquifers through properly planned and designed recharge structures.

The case from **Guatemala**, covered by Iturbide-Chang et al.: "Guatemala: Water harvesting and Managed Aquifer Recharge to Combat Water Scarcity in a Country of Apparent Water Abundance" provides insights into a broad range of rainwater harvesting, managed aquifer recharge, stormwater management and water-sensitive landscaping options, that all work with the subsurface as critical water recipient and moderator of water extremes. These options span from implementation in schools, markets, private up-market homes, urban areas as well as coastal and mountainous regions. It is estimated that so far more than 100,000 people have benefitted.

Bangladesh, being located in the largest deltaic region of the world, is rich in water resources. Three of the largest rivers of the world, the Ganges, the Brahmaputra, and the Meghna, join the Bay of Bengal here. The thick alluvial deposits on the banks of the rivers hold potential aquifers. In the paper "Scope of Induced Recharge to Riverbank Aquifers in Bangladesh", Bashar presents case studies on how groundwater is being exploited through high-capacity deep tube wells located on the banks of the rivers. Because of good aquifer potential and hydraulic interconnectivity with the river, the well fields benefit from induced bank filtration. This can be replicated in other parts of the country after proper hydrogeological and geochemical studies.

Acknowledgements The Editors acknowledge the support extended by Dr. Amitava Bandopadhyay, Director General, Centre for Science and Technology of the Non-Aligned and Other Developing Countries (NAM S&T Centre), New Delhi, India during the preparation of the manuscript. The contributions by the Authors from different countries of the world, which have been included in the book as chapters and summarised in this chapter are thankfully acknowledged.

References

AbdelKhaleq RA, Ahmed IA (2007) Rainwater harvesting in ancient civilizations in Jordan. Water Supply 7(1):85–93. https://doi.org/10.2166/ws.2007.010

Al-Kharabsheh A (2000) Ground-water modelling and long-term management of the Azraq basin as an example of arid area conditions (Jordan). J Arid Environ 44(2):143–215. https://doi.org/10.1006/jare.1999.0580

Alam MF, Pavelic P, Villholth KG, Sikka A, Pande S (2022) Impact of high-density managed aquifer recharge implementation on groundwater storage, food production and resilience: a case from Gujarat, India. J Hydrol Reg Stud. https://doi.org/10.1016/j.ejrh.2022.101224

Ali S, Shekhar S, Kumar R, Brindha K, Li P (2023) Genesis and Mobilization of fluoride in groundwater of India: statistical evaluation, health impacts, and potential remedies. J Hazard Mater Adv 100352. https://doi.org/10.1016/j.hazadv.2023.100352

American Society of Civil Engineers (ASCE) (2001) Standard Guidelines for Artificial Recharge of Ground Water, EWRI/ASCE 34–01(ASCE Standard No. 34–01)

BGS (2002) The effectiveness of artificial recharge of groundwater, a review. British Geological Survey, Commercial Report. CR/02/108M. Gale I, Neumann I Carlow R, Monech I

Bekele E, Toze S, Patterson B, Higginson S (2011) Managed aquifer recharge of treated wastewater: water quality changes resulting from infiltration through the vadose zone. Water Res 45:5764–5772. https://doi.org/10.1016/j.watres.2011.08.058

Bores TM, Ben-Asher J (1982) A review of rainwater harvesting Agricultural Water Management. 5:145—158

Bouwer H (2002) Artificial recharge of groundwater: hydrogeology and engineering. Hydrogeol J 10:121–142

CGWB (2007) Manual of artificial recharge of groundwater, Central Ground Water Board, Govt of India. https://cgwb.gov.in/documents/Manual-Artificial-Recharge.pdf.

Dashora Y, Cresswell D, Dillon P, Maheshwari B, Clark R, Soni P, Singh PK (2022) Hydrologic and cost–benefit analysis of multiple check dams in catchments of ephemeral streams, Rajasthan. India. Water 14:2378. https://doi.org/10.3390/w14152378

de Graaf IEM, van Beek RLPH, Gleeson T, Moosdorf N, Schmitz O, Sutanudjaja EH, Bierkens MFP (2017) A global-scale two-layer transient groundwater model: Development and application to groundwater depletion. Adv Water Resour 102:53–67. https://doi.org/10.1016/j.advwatres.2017.01.011

Dillon P, Pavelic P, Toze S, Rinck-Pfeiffer S, Martin R, Knapton A, Pidsley D (2006) Role of aquifer storage in water reuse Desalination 188:123–134. https://doi.org/10.1016/j.desal.2005.04.109

Dillon D, Vanderzalm J, Sindu J, Page D, Chadha D (2014): A water quality guide to MAR in India. https://recharge.iah.org/files/2016/11/A-Water-Quality-Guide-to-MAR-in-India-2014.pdf

Dillon P, Stuyfzand P, Grischek T, Lluria M, Pyne RDG, Jain RC, Bear J, Schwarz J, Wang W, Fernandez E, Stefan C, Pettenati M, van der Gun J, Sprenger C, Massmann G, Scanlon BR, Xanke J, Jokela P, Zheng Y, Rossetto R, Shamrukh M, Pavelic P, Murray E, Ross A, Bonilla Valverde JP, Palma Nava A, Ansems N, Posavec K, Ha K, Martin R, M, (2019) Sixty years of global progress in managed aquifer recharge. Hydrogeol J 27(1):1–30

Fatema S, Marandi A, Zahid A, Hassan MQ, Hossain MA, Schüth C (2018) Seawater intrusion caused by unmanaged groundwater uses in a coastal tourist area, Cox's Bazar. Bangladesh. Environ Earth Sci 77:75. https://doi.org/10.1007/s12665-018-7260-6

Fattovich R (1990) Remarks on the Pre-Aksumite period in northern Ethiopia. J Ethiopian Stud 23:1–33

Foster SSD, Chilton PJ (2004) Downstream of downtown: urban wastewater as groundwater recharge. Hydrogeol J 12:115–120. https://doi.org/10.1007/s10040-003-0296-y

Gado T, El-Agha D (2020) Feasibility of rainwater harvesting for sustainable water management in urban areas of Egypt. Environ Sci Pollut Res 27:32304–32317

Gee GW, Hillel D (1988) Groundwater recharge in arid regions: review and critique of estimation methods. Hydrol Process 2(3):255–266

Getachew A (1999) Rainwater harvesting in Ethiopia: an overview. In: 25th WEDC conference on integrated development for water supply and sanitation, Addis ABabba, Ethoipia, file:///C:/Users/Dipankar%20saha/Downloads/387.pdf

Ghosh S, Umashankar M, Chowdhury S (2020) Modelling water resources in the ancient indus valley City of Dholavira and lessons learnt. In: Subramanian B, Chen SS, Reddy K (eds) Emerging technologies for agriculture and environment. Lecture notes on multidisciplinary industrial engineering. Springer, Singapore. https://doi.org/10.1007/978-981-13-7968-0_17

Gleeson T, VanderSteen J, Sophocleous AA, Taniguchi M, Alley WM, Allen DM, Zhou Y (2010) Commentary: Groundwater sustainability strategies. Nat Geosci 3:378–379. https://doi.org/10.1038/ngeo881

Hoppe-Jones C, Oldham G, Drewes JE (2010) Attenuation of total organic carbon and unregulated trace organic chemicals in u.s. riverbank filtration systems. Water Res 44:4643–4659. https://doi.org/10.1016/j.watres.2010.06.022

Ibkar A, Mukherjee A, Didwania N, Rai S (2023) Impact of urbanization on groundwater in changing climatic scenario: a case study. Chapter 17 In: Thambidurai P, Dikshit AK (eds) Impacts of urbanization on hydrological systems in India. Springer, Cham. https://doi.org/10.1007/978-3-031-21618-3_17

Kalantari N, Rangzan K, Shripad ST (2010) Rahimi M H (2010) Site selection and cost-benefit analysis for artificial recharge in the Baghmalek plain. Khuzestan Province, Southwest Iran, Hydrogeol J 18:761–773. https://doi.org/10.1007/s10040-009-0552-x

Keesari T, Sinha UK, Saha D, Dwivedi SN, Shukla RR, Mohokar H, Roy A (2021) Isotope and hydrochemical systematics of groundwater from a multi-tiered aquifer in the central parts of Indo-Gangetic Plains, India–Implications for groundwater sustainability and security. Sci Total Environ. https://doi.org/10.1016/j.scitotenv.2021.147860

Konikow LF, Kendy E (2005) Groundwater depletion: a global problem. Hydrogeol J 13:317–320. https://doi.org/10.1007/s10040-004-0411-8

Kumar MD, Patel A, Ravindranath R, Singh OP (2008) Chasing a mirage: water harvesting and artificial recharge in naturally water-scarce regions 43(35):61–71

Mays L, Antoniou G, Angelaki A (2013) History of water cisterns: legacies and lessons. Water 5:1916–1940

Mukherjee A, Saha D, Harvey CF, Taylor RG, Ahmed KM, Bhanja SN (2015) Groundwater systems of the IndianSub-Continent. J Hydrol: Regional Stud. https://doi.org/10.1016/j.ejrh.2015.03.005

Murray EC, Tredoux G (2002) Pilot artificial recharge scheme: testing sustainable water resource development in fractured aquifers. Water Research Commission Report No. 967/1/02, Pretoria. ISBN 1-86845-883-0

Ouelhazi H, Lachaal F, Charef A, Challouf B, Chaieb H, Horriche F (2014) Hydrogeological investigation of groundwater artificial recharge by treated wastewater in semi-arid regions: Korba aquifer (Cap-Bon Tunisia). Arab J Geosci 7:4407–4421. https://doi.org/10.1007/s12517-013-1090-3

Pandey DN, Gupta AK, Anderson DM (2003) Rainwater harvesting as an adaptation to climate change. Curr Sci 85(1):46–59

Patel P, Saha D, Shah T (2020) Sustainability of groundwater through community driven distributed recharge: an analyses of arguments for water scarce regions of semi-arid India. 2020, J Hydrol: Regional Stud. https://doi.org/10.1016/j.ejrh.2020.100680

RIGW (1988) Hydrogeological Maps for Egypt. Research Institute for Groundwater (RIGW), Egypt

Reilly TE, Dennehy KF, Alley WM, Cunningham WL (2008) Ground-water availability in the United States, U.S. Geol Surv Circ 1323. Available at http://pubs.usgs.gov/circ/1323/

Rodell M, Velicogna I, Famiglietti JS (2009) Satellite-based estimates of groundwater depletion in India. Nature 460:999–1002. https://doi.org/10.1038/nature08238

Saha D, Roy R (2018) Groundwater resources of India: Potential, challenges and management. In: Sikdar P (ed) Groundwater development and management, issues and challenges in South Asia. Springer, Berlin. https://doi.org/10.1007/978-3-319-75115-3_2

Saha D, Chakraborty M, Chowdhury A (2022a) Stubble burning in northwestern India is it related to groundwater overexploitation? In: Re V, Lilla R, Manzione TAA, Mukherji A, MacDonald A (eds) Groundwater for Sustainable livelihoods and equitable growth. https://doi.org/10.1201/9781003024101-7

Saha D, Gor N (2020) A prolific aquifer system is in peril in arid Kachchh region of India. Groundw Sustain Dev. https://doi.org/10.1016/j.gsd.2020.100394

Saha D, Marawaha S, Mukherjee A (2017) Groundwater Resources and Sustainable Management Issues in India. In: Saha D, Marwaha S, Mukherjee A (eds) Clean and sustainable groundwater in India

Saha D, Sikka AK, Goklani R (2022b) Artificial recharge endeavours in India: a review. Water Secur. https://doi.org/10.1016/j.wasec.2022.100121

Saha D, Marwaha S, Dwivedi SN (2018) Groundwater resources and sustainable management in India. https://doi.org/10.1007/978-981-10-4552-3_1

Scanlon BR, Keese KE, Flint AL, Flint LE, Gaye CB, Edmunds WM, Simmers I (2006) Global synthesis of groundwater recharge in semiarid and arid regions. Hydrol Process: Int J 20(15):3335–3370

Shamrukh M, Al-Muraikhi A, Al-Hamar Y (2012) Exploring of deep groundwater in the southwest aquifer of Qatar. In: The 10th Gulf Water Conference. Water Sciences and Technology Qatar, pp 1–13

Singh A, Saha D, Tyagi SAC (2019) A step towards water security in India. In: Singh A, Saha D, Tyagi AC (eds) Water governance: challenges and prospects. Springer, Berlin. https://doi.org/10.1007/978-981-13-2700-1_1

Valhondo C, Carrera J, Martínez-Landa L, Wang J, Amalfitano S, Levantesi C, Diaz-Cruz MS (2020) Reactive barriers for renaturalization of reclaimed water during soil aquifer treatment. Water 12(4):1012

Valhondo C, Martinez-Landa L, Carrera J, Ayora C, Nödler K, Licha T (2018) Evaluation of EOC removal processes during artificial recharge through a reactive barrier. Sci Total Environ 612:985–994

Wada Y, van Beek LPH, van Kempen CM, Josef WTM, Reckman JWM, Vasak S, Bierkens MFP (2010) Global depletion of groundwater resources. American Geophysical Union. Geophys Res Lett 37:L20402. https://doi.org/10.1029/2010GL044571

Zheng Y, Ross A, Villholth KG, Dillon P (eds) (2021) Managing aquifer recharge: a showcase for resilience and sustainability. UNESCO/IAH/GRIPP, p 379. ISBN: 978-92-3-100488-9. https://unesdoc.unesco.org/ark:/48223/pf0000379962

Chapter 2
The Role of Artificial Recharge of Aquifers in Water Resources Management in Egypt

Mohamed A. Dawoud

Abstract By 2050, the gap between Egypt's renewable water resources supplies and demands will grow by more than twofolds from 23 BCM in 2022 to approximately 50 BCM. Closing such a big water supply and demand gap will be challenging and expensive. The availability of water resources plays a crucial role in driving Egypt's economic development and facilitating rapid growth across multiple sectors. Currently, around 82% of the country's water demand is met through the utilization of surface water, while approximately 12% is fulfilled by groundwater sources. The remaining 6% comes from the reuse of agriculture drainage water and treated sewage effluent. Managing demand, particularly for agricultural water use, which uses more than 85% of the total water use in Egypt, must be given priority by planners and decision-makers. Failure to reduce the water demand in various sectors and curb uneconomic agricultural use will have severe socioeconomic repercussions. Once renewable freshwater resources are depleted, the most significant source of new water will be desalination of seawater and brackish groundwater or reusing of treated sewage effluent. Artificial recharge of aquifers using conventional water resources such as rainfall or surplus overland flow or treated wastewater is an important tool for managing the scarce water resources in Egypt. Several projects on artificial recharge of aquifers were carried out in Egypt using surface water, rainfall, and treated sewage effluent. Artificial recharge is an essential element of the National Water Plan for building up strategic water reserves, controlling declining groundwater levels, enhancing deteriorated groundwater quality, and improving water management.

Keywords ASR · Artificial recharge · MAR · Water management

M. A. Dawoud (✉)
National Water Research Center, Egypt, and Senior Water Advisor at Environment Agency, Abu Dhabi, United Arab Emirates
e-mail: mdawoud@ead.gov.ae

D. Saha et al. (eds.), *Managed Groundwater Recharge and Rainwater Harvesting*, Water Resources Development and Management,
https://doi.org/10.1007/978-981-99-8757-3_1

2.1 Introduction

As an arid and hyper-arid country, Egypt is facing a growing challenge due to its limited freshwater resources. The country's water supply is under threat due to increasing demand, high population growth, pollution, and climate change. In 2022, the annual per capita renewable freshwater availability decreased to about 580 m^3, which is significantly below the 1,000 m^3 water scarcity threshold (Fig. 2.1). About 98% of Egypt's freshwater resources originate outside of its borders, such as the Nile River and from the aquifers. The Nile River provides the country with 93% of its water requirements, posing a serious challenge for water policy and decision-makers (ref?). In 2022, the total annual renewable freshwater resources were about 58.40 BCM and the estimated virtual water was about 35 BCM. The total projected water demand was about 114 BCM, indicating a deficit of about 20.6 BCM annually (Dawoud 2019). Non-conventional water resources that include agricultural drainage water, desalinated brackish groundwater and/or seawater, non-renewable groundwater, and treated municipal wastewater represent about 22% of the total available water resources are generally used for agriculture, landscaping, and industry through specialized processes (Dawoud 2019).

Egypt remains the most downstream country in the Nile basin with an annual share of 55.5 BCM (as per 1959 Treaty between Egypt and Sudan). The average annual volume of rainfall that is effectively utilized for agricultural purposes in Egypt is estimated as 1.0 BCM, while the annual groundwater abstraction from both renewable Quaternary aquifer and non-renewable deeper aquifer systems is about 6.5 BCM. The annually produced wastewater volume is 8.75 BCM, of which 3.15 BCM is secondary treated, 0.9 BCM is primary treated and only 0.068 BCM is tertiary treated in 2022 (Ref?). The total capacity of the desalination plants in Egypt at present (2022) is 685,060 m^3/day, and the Govt. plans to increase the daily capacity to about 1,559,330 m^3/day by 2030. Table 2.1 shows the present water resources in Egypt (2022). Out of all the 4.05 BCM of treated wastewater, only 0.50 BCM can be reused directly for irrigation all over Egypt, while the remaining amount is

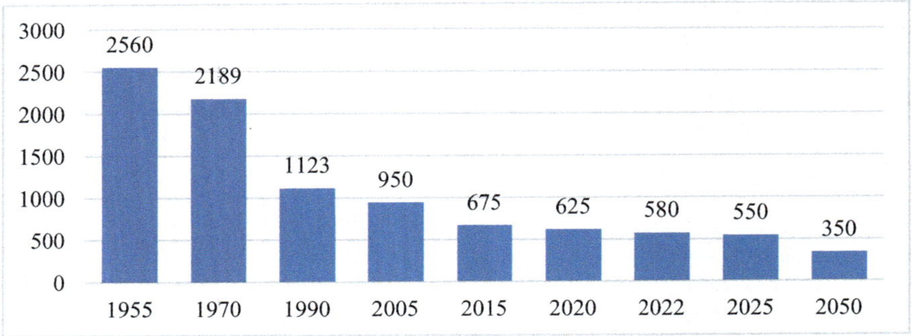

Fig. 2.1 Renewable water resources per capita share in Egypt (m^3/capita/year). *Source* Dawoud (2019)

Table 2.1 Water resources availability in Egypt (2022)

Resource	Quantity (BCM)
Nile water	55.5
Wastewater	8.75
Agriculture drainage	10.566
GW nile aquifer	6.5
GW desert	2.85
Rain harvesting	1.01
Desalination	0.25
Total	85.426

Source MWRI (2022)

discharged back to environment. This volume can be artificially recharged to aquifer system to enhance the groundwater reserve.

The high floods of the Nile River in the last three years resulted in the gradual formation of four new inland lakes in the Toshkah Depression. Total water storage in the dams was 40.2 km^3, but a component escaped due to evaporation. The stored water in these four lakes can be used as a potential source for artificial recharge. Assessing Toshkah Depression lake's water storage is important to develop better management scenarios for these four lakes (Abdel-Latif and Yacoub 2011). Evaporation from the water surface (2.7 m/y/km^2) is a major variable in analyzing the variations in a lake's level and its water storage (Elshemy and Meon 2011; Elsawwaf and Willems 2012; Abdel Wahab et al. 2020).

2.2 Aquifer Artificial Recharge Approach

Artificial recharge of aquifers is considered as a promising tool for sustainable groundwater management (also known as managed aquifer recharge, water storage, and enhanced recharge). In this technique, freshwater or treated wastewater can be injected into suitable aquifer systems to enhance/create freshwater storage for agricultural purposes, potable water supply, and other environmental benefits (Zhang et al. 2020; Saha et al. 2021). Literature review of most recent studies indicates that at present, over thirty countries all over the world have successfully utilized artificial recharge for enhancing aquifer reserves and water treatment (Vanderzalm et al. 2022). Treated wastewater can be injected through intermittent percolation in infiltration basins into the aquifer systems using the soil aquifer treatment (SAT) approach which provides nutrients to agricultural fields. In Egypt, excess surface water during the winter closure is injected into aquifer system for agricultural irrigation during peak seasons (El Arabi and Dawoud 2012). Figure 2.2 shows a schematic diagram of different types of artificial recharge techniques. Globally, there are about 1115 aquifer artificial recharge sites used for different purposes namely domestic

[55.8%], agricultural [27.4%], ecological and environmental protection [12%], and industrial and commercial [4.8%]. Aquifers' artificial recharge approach can also be used for specific applications and targets such as water resources balance adjustment and geothermal energy (Zhang et al. 2020).

The economic value of water arises when its supply becomes limited, in comparison to its demand, leading to competition among users. Aquifer recharge and recovery system offers several benefits, including the availability of additional water during periods of scarcity, improvement in water quality, or a combination of both. The process of water recharge can generate a new freshwater resource and can also

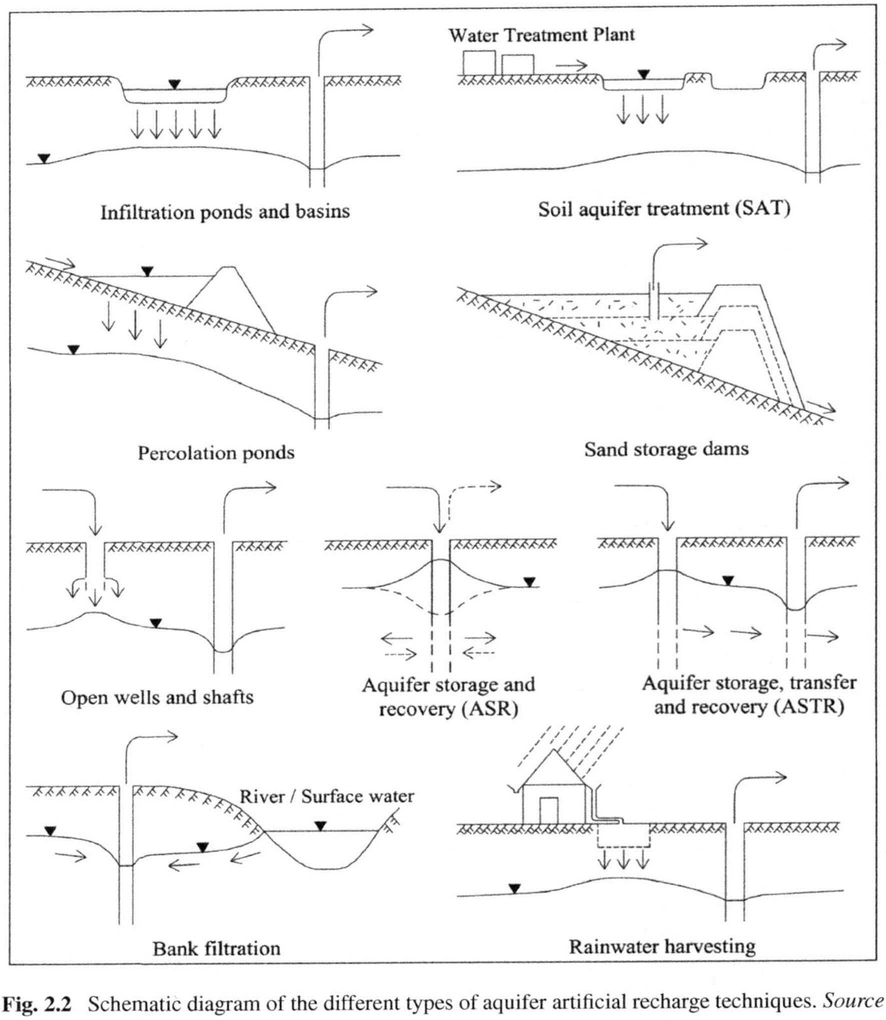

Fig. 2.2 Schematic diagram of the different types of aquifer artificial recharge techniques. *Source* Zhang et al. (2020)

augment water storage in an aquifer, thereby stabilize or even arise in water level. The cost of artificial aquifer recharge can vary depending on several factors such as the size of the project, the type of recharge method used, the location, and the specific requirements of the aquifer.

Capital Costs: Capital costs refer to the initial investment required to set up the infrastructure for artificial recharge. These costs typically include:

- Construction Costs: This includes the cost of building or modifying recharge facilities such as recharge basins, infiltration galleries, injection wells, or spreading grounds. The cost will depend on the size, complexity, and materials used in the construction;
- Permitting and Legal Costs: Obtaining necessary permits and complying with legal requirements can incur costs, including application fees, environmental impact assessments, and legal consultation;
- Engineering and Design Costs: These costs cover the expenses associated with the planning, designing, and engineering of the recharge system. It includes studies, surveys, geotechnical investigations, and design services;
- Monitoring and Control Systems: Aquifer recharge projects require monitoring and control systems to ensure proper operation and management. The cost includes installing sensors, data collection equipment, and control mechanisms; and
- Land Acquisition: If additional land is required for the recharge infrastructure, land acquisition costs may be incurred.

Operational Costs: Operational costs refer to the ongoing expenses associated with the maintenance and operation of the aquifer recharge system. These costs can include:

- Energy costs: Energy is required to pump water or operate other components of the recharge system. Operational costs include electricity or fuel expenses for powering pumps, valves, and control systems;
- Water treatment costs: Depending on the source water quality and the requirements of the aquifer, water treatment processes may be necessary before recharge. These costs cover the treatment processes such as filtration, disinfection, or pH adjustment;
- Maintenance and repairs: Regular maintenance, repair, and replacement of components such as pumps, pipelines, and control systems are necessary to ensure the smooth operation of the recharge system;
- Monitoring and testing: Ongoing monitoring and testing of water quality, recharge rates, and aquifer performance are essential. The costs include laboratory testing, data analysis, and personnel for monitoring activities;

It's important to note that the actual costs can vary significantly based on project-specific factors and local conditions. Therefore, it is advisable to consult with experts and conduct a detailed feasibility study to obtain accurate cost estimates for a specific aquifer artificial recharge project.

Recent studies indicate that the artificial recharge cost is estimated to be approximately 0.10–0.32 USD/m^3 for the projects with a capacity of one to five million m^3/year (ref?). Operation of such schemes may face several challenges as clogging, aquifer pollutions, and injected water movement calculation, 3-D design of the monitoring systems, the hydrogeochemical interaction between injected water and aquifer native water, and the need for recovered water purification. These challenges also impact the operational cost of such project.

According to the fundamental principle of benefit maximization, which aims to increase the overall value of limited resources, actions such as the implementation of artificial recharge of aquifer systems should be undertaken only if their total benefits outweigh the total costs (Ward and Michelsen 2002). Cost–benefit analysis (CBA) is extensively discussed in microeconomic textbooks as well as dedicated literature (Layard and Glaister 1994; Boardman et al. 1996). Environmental CBA, which encompasses factors like water quality and supply, represents a specialized area of investigation (Atkinson and Mourato 2008). The primary objective of CBA is to achieve allocative efficiency, where policies and investments are pursued only if they yield net positive benefits. The selection of a policy or investment should be based on maximizing net benefits. It is important to note that while CBA primarily focuses on economic efficiency, other considerations such as equity and national security may also be relevant to decision-making processes. CBAs are not conducted in a value-neutral context, and the societal desirability of costs and benefits should be considered. Even if decisions are not solely based on CBA, it is crucial to incorporate CBA as an integral input into the decision-making process (Pearce et al. 2006). Figure 2.3 shows the framework for the assessment of aquifer artificial recharge (Maliva et al. 2015).

2.3 The History of Aquifer Artificial Recharge in Egypt

Groundwater is a critical resource for Egypt and is considered a strategic water reserve and the second freshwater source for agriculture and domestic uses in rural areas. However, recently Egypt is facing severe water crisis due to rapid population growth, urbanization, lack of financial resources, and climate change. To address these challenges, artificial recharge of aquifers and recovery has been recommended as a sustainable tool for strategic water reserves. Abdel-Magid et al. (1991), assessed the feasibility of one of the early projects using treated sewage effluent. The study concluded that treated sewage effluent could be a sustainable source for aquifer recharge, especially in areas facing groundwater depletion and quality deterioration. Also, El-Gafy et al. (2012) evaluated the impact of aquifer recharge using surface water, on the groundwater quality in the Nile Delta region. It has been found that aquifer recharge with surface water can significantly improve the deteriorated groundwater quality by reducing the concentration of contaminants. El-Kadi et al. (2016) evaluated the suitability of artificial recharge techniques for the reuse of treated sewage effluents in Sinai Peninsula. The study found that aquifer recharge with treated sewage effluents could be an effective and sustainable solution. El-Didy et al. (2019) assessed the potential of aquifers' artificial recharge techniques

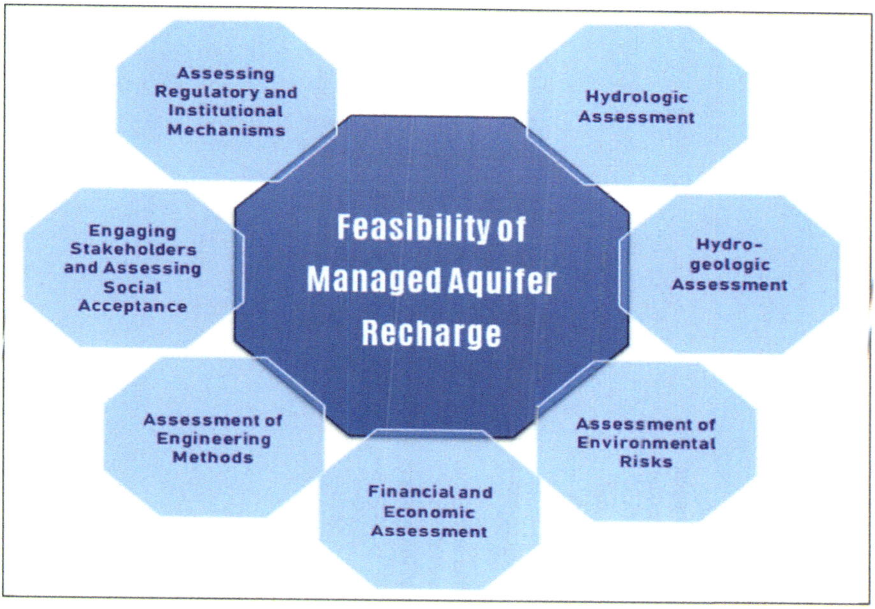

Fig. 2.3 The framework for the assessment of aquifer artificial recharge

for the replenishment of the depleted and deteriorated groundwater in the Western Desert. The study concluded that aquifer artificial recharge could be a viable technical, economic, and environmental solution for enhancing the groundwater quality and potentiality in Western Desert, but further in-depth hydrogeological evaluation is needed to optimize the technique and improve the system design. Several projects, either pilot projects or full scheme projects were implemented in Egypt over the last 20 years. Table 2.2 and Fig. 2.4 show examples of details of aquifer artificial recharge projects in Egypt. Darwish (1998) assessed the first trial for aquifer artificial recharge at El Bustan Extension Area in the Western Nile desert area using surface water. Three infiltration tests were carried out during (1995–1996) using Nile Water. The results showed that due to the specific hydrogeological conditions of this area, moderate infiltration rates of about 15–30 cm/day were obtained during the first 15–20 days of the infiltration process, and later if there is no clogging, the final infiltration rates are in the range of 10–12 cm/day (Darwish 1998). The technical feasibility and economic viability of artificial recharge using surface water in the western Nile Delta region are significant, particularly considering the suboptimal subsoil conditions due to the presence of clay, loam, and calcrete layers in some areas which limit the infiltration rate.

A comprehensive study was conducted to assess and select a potential site for conducting an experiment on artificial recharge using treated wastewater. Artificial recharge of aquifers with treated sewage water is an innovative approach that can be implemented in Egypt to utilize sewage water safely, as compared to its direct

Table 2.2 Aquifer artificial recharge projects in Egypt

No	Project name	Description	Location	Recharge water sources
1	Toshkah project	A canal system diverts surface water from Lake Nasser to be recharged in Toshkah Depression located in the Western Desert. The objective of this project was to annually recharge 5 BCM of water into aquifer system. However, the project has encountered several obstacles and challenges including, hydrogeological challenges, technical complexities, and high expenses. The research indicates that recharge can significantly enhance groundwater quality and has shown promise as an effective and sustainable solution for water shortage. Nonetheless, further in-depth research and assessments are required to optimize these techniques and address the challenges associated with their implementation	Toshkah, south Egypt	Surface water
2	El Bustan project	In 1995, a recharge basin was constructed with the dimension at the bottom of 155 × 135 m and at the top is 185 × 170 m, with a 2.5 m average depth The surface water was supplied to the recharge basin from a branch canal taken from El Bustan main canal. The agricultural lands surrounding the project site rely on surface water irrigation facilitated by modern irrigation techniques. The study area is characterized by a flat topography, with an average elevation of 30 m above mean sea level (AMSL)	El Bustan, Western Nile Delta (about 150 km Northwest of Cairo)	Surface water

(continued)

Table 2.2 (continued)

No	Project name	Description	Location	Recharge water sources
3	Abu Rawash project	Under this program, an extensive study was conducted, which involved the identification of potential sites for an artificial recharge experiment using treated wastewater. A pilot project was established at the Abu Rawash wastewater treatment plant. In 2001, comprehensive hydrogeological investigations and engineering designs were conducted. Multiple scenarios for aquifer storage and recovery were evaluated, taking into account technical, environmental, economic, and regulatory considerations. The results demonstrated the potential of artificial recharge of the groundwater aquifer using treated wastewater. However, further in-depth studies are necessary to evaluate how the characteristics of the aquifer influence the recharge process when utilizing treated wastewater. The health risks associated with wastewater recharge are contingent upon various factors, including the prevailing physical and chemical conditions in the aquifer, the limited adsorption capacities of the aquifer materials, and the rate at which microorganisms die off and toxic pollutants degrade into other harmful compounds	Abu Rawash, Western Nile	Treated sewage effluents

Fig. 2.4 Location of Aquifers artificial recharge projects in Egypt

use. By passing through the unsaturated zone of the aquifer, a significant portion of the biological load and a partial reduction in the concentration of certain chemicals takes place. This paper presents the initial hydrogeological investigation and various tested scenarios for aquifer storage, focusing on the Abu Rawash farm as a case study. Additionally, an environmental impact assessment (EIA) was conducted for the artificial recharge experiment to evaluate the technical and environmental considerations, providing guidelines for future recharge projects. The results indicate that the artificial recharge of groundwater aquifers using treated wastewater shows promise, but further detailed studies are necessary to assess how the aquifer characteristics influence the recharge mechanism with treated wastewater. It is crucial to evaluate the potential health risks associated with changes in the physical and chemical conditions within the aquifer, including limited adsorption capacities and the survival of microorganisms and degradation of toxic pollutants (El Arabi 2012). Table 2.2 shows the list of the present aquifer artificial recharge projects in Egypt.

To address the current water stress situation, the adoption of innovative technologies like managed aquifer recharge (MAR) is considered crucial for expanding the utilization of non-conventional water resources and ensuring sufficient water supplies (Ref?). Suitability maps for MAR are widely utilized as decision-making tools to balance water demand and supply. Conducting a sensitivity analysis to validate these suitability maps can provide a deeper understanding of the results and identify influential factors. The West Delta region was selected as a case study due to the presence of two MAR sites, allowing an assessment of the suitability of implementing MAR projects with reclaimed water. This study employs a spatially explicit sensitivity analysis on a newly developed framework for MAR suitability maps, which utilizes Multi-Criteria Decision Analysis (MCDA) and spreading methods techniques to determine appropriate locations for MAR implementation. The sensitivity analysis visually examines the impact of varying weighted criteria on the final outputs and identifies criteria that are particularly sensitive to changes in weights. The results of the sensitivity analysis demonstrate that the applied MCDA framework for suitability mapping in the West Delta generates robust outcomes in terms of identifying the most suitable sites for MAR. Furthermore, the results suggest that the proposed framework can be applied in arid environments with similar data availability. Moreover, the findings emphasize the potential utilization of suitability maps in sustainable groundwater management plans, supporting the practical implementation of MAR projects in the West Delta (Hani et al. 2023).

2.4 Site Selection for Aquifer Artificial Recharge

The selection of the potential locations for Artificial Recharge sites is controlled by a set of hydrogeological settings, availability of water resources, planning, economical, and environmental considerations.

2.4.1 Hydrogeological Settings

The hydrogeological characteristics are important factors in selecting the potential sites for artificial recharge, which include unsaturated thickness, hydraulic gradient, flow direction, aquifer layers, hydraulic conductivity (transmissivity), aquifer type, and native groundwater quality. Figures 2.5 and 2.6 show Egypt's hydrogeological map and general hydrogeological cross-section in the Nile Valley (RIGW/IWACO 1998). The hydrogeological setting of Egypt influences the choice of aquifer artificial recharge methods, site selection, and the long-term sustainability of the recharge process. There are five major groundwater aquifer systems in Egypt.

1. Nile Basin (Quaternary) Aquifer System which covers the Nile Valley, Delta, and the adjacent desert fringes, representing about 5% of Egypt's area. The Quaternary Aquifer System in Egypt holds major groundwater resource that covers vast areas of the country, including the Nile Delta, the Western Desert, and the Sinai Peninsula. It is the youngest and most extensive aquifer system in Egypt, consisting of a series of sedimentary layers of varying permeability and thickness. This aquifer system is primarily recharged by rainfall and irrigation return flows and is used for agricultural, industrial, and domestic purposes. However, the aquifer is facing significant challenges due to overexploitation, pollution, and

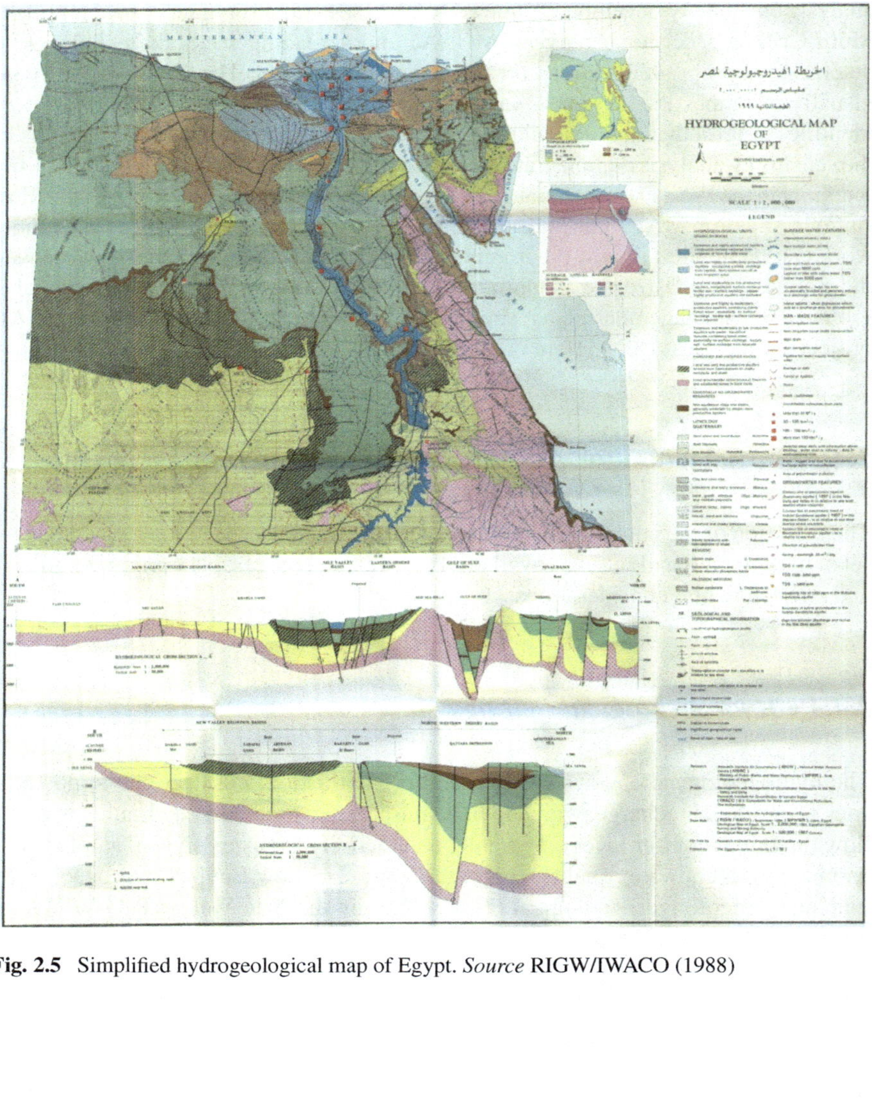

Fig. 2.5 Simplified hydrogeological map of Egypt. *Source* RIGW/IWACO (1988)

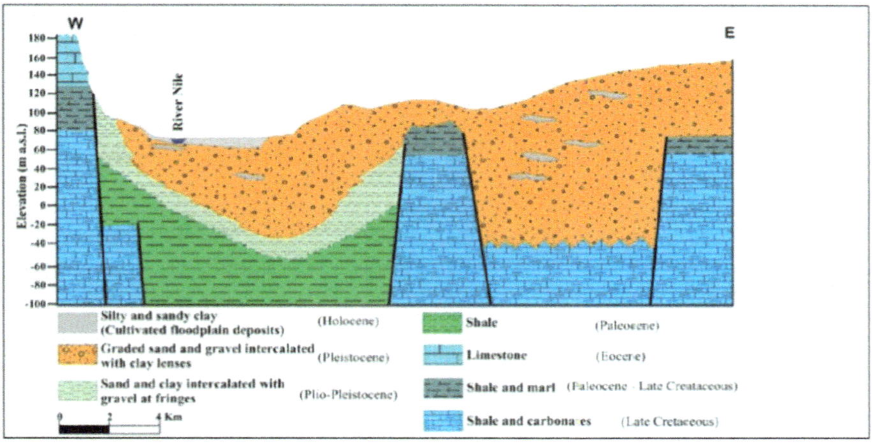

Fig. 2.6 General hydrogeological cross-section in nile valley area. *Source* RIGW/IWACO (1988)

climate change, which have led to declining water levels, deteriorating water quality, and land subsidence.

2. Nubian Sandstone Aquifer System (NSAS): This is the largest aquifer system in North Africa and covers an area of around two million km^2, of which approximately 540,000 km^2 are in Egypt. The NSAS consists of multiple layers of sandstone and is considered a fossil aquifer system as its recharge at present is minimal or non-existent. It is estimated that the NSAS contains around 150,000 km^3 of water, of which around 30,000 km^3 is located in Egypt (RIGW 1999).
3. The Moghrah Aquifer System: is a major groundwater aquifer located in the Western Desert of Egypt. It is situated in the Fayoum Depression, approximately 80 km southwest of Cairo. The aquifer consists mainly of sandstone and limestone formations and is estimated to contain about 2.7 BCM of water. The Moghrah Aquifer is an important source of water for agriculture and domestic use in the surrounding area. However, overuse and mismanagement have led to depletion of resources and declining water levels. This has also resulted in increased salinity and deterioration of water quality, making it unsuitable for certain uses. Efforts are being made to manage and conserve the Moghra Aquifer, including the development of alternative water sources and the implementation of better irrigation practices. In addition, the Egyptian government has launched a project to recharge the aquifer with treated wastewater. However, due to over-abstraction, groundwater quality is deteriorated and the salinity increased from 4000 ppm in 2015 to more than 12,000 ppm in 2022. This aquifer system is a non-renewable aquifer with very little natural recharge.
4. The Fracture Limestone (Carbonate) Aquifer System: In Egypt, it is a large, complex, and transboundary groundwater resource that extends across Egypt, Libya, Sudan, and Chad. It is one of the largest aquifer systems in the world, covering an area of approximately 2 million km^2 and is composed of carbonate

rocks with complex fissure networks that allow for the storage and flow of groundwater. This aquifer is an important source of water for the people living in the Western Desert of Egypt, where surface water is scarce. However, the aquifer is being depleted at a rate that is faster than it can be replenished, primarily due to over-extraction for agriculture and domestic use. The water quality in this aquifer system is brackish and the aquifer potentiality is low.

5. The Coastal Aquifer System: is a critical source of water for the country's densely populated coastal regions, including the cities of Alexandria and Port Said. It is a shallow, unconfined aquifer that is recharged by both rainfalls. The coastal aquifer faces numerous challenges, including overexploitation, pollution, and climate change. Overuse of groundwater has led to a decline in water levels, which in turn has caused seawater to intrude into the aquifer and contaminate groundwater. In addition, industrial and agricultural activities have caused pollution of the aquifer, making the water unsafe for consumption without treatment.

2.4.2 Unsaturated Thickness

The unsaturated thickness of an aquifer, also known as the vadose zone, has a significant impact on artificial recharge. The unsaturated zone refers to the portion of the subsurface above the water table, where the pore spaces are partially filled with air and partially with water. The thickness of this unsaturated zone affects the rate of recharge and effectiveness of the methods. A thicker unsaturated zone provides a greater buffer for water infiltration, allowing more time for natural processes such as filtration and attenuation to occur. This can enhance the quality of the recharged water by removing contaminants. Additionally, a thicker unsaturated zone provides more storage capacity for the recharged water, allowing for greater water availability during periods of drought. Conversely, a thin unsaturated zone may result in rapid percolation of recharge water, reducing the opportunity for natural filtration processes and potentially increasing the risk of contamination. Therefore, understanding the unsaturated thickness is crucial for designing and managing aquifer artificial recharge systems to ensure sustainable and high-quality water supply. It is important for the unsaturated zone to exhibit an infiltration rate of at least 0.25 m/day. This minimum rate ensures that water can effectively percolate through the unsaturated zone and recharge the underlying aquifer.

2.4.3 The Aquifer Transmissivity and Porosity

The transmissivity and porosity of an aquifer have significant impacts on aquifer artificial recharge. Transmissivity refers to the ability of the aquifer to transmit water horizontally, while porosity refers to the volume of pore space within the aquifer. Both of these characteristics influence the rate of recharge and efficiency of the method.

High aquifer transmissivity and porosity will help to prevent water mounding below the recharge basin bottom which can cause a decrease in infiltration rate and recharge capacity. Results from different aquifer artificial recharge sites indicate that effective porosity should be more than 0.1 and the transmissivity should be more than 500 m^2 /day.

2.4.4 The Water Flow Characteristics

The flow character within an aquifer has a profound impact on artificial recharge. Understanding the aquifer's water flow patterns is crucial for designing effective recharge strategies and ensuring successful replenishment of groundwater resources. The direction, velocity, and distribution of water flow within the aquifer determine how the recharged water will disperse and interact with the existing groundwater resource. By considering the aquifer's flow dynamics, recharge methods can be strategically placed to maximize the spread and penetration of recharged water throughout the aquifer. This helps in achieving a more uniform recharge and avoids the creation of localized preferential flow paths that might result in inadequate recharge in certain areas.

2.4.5 Groundwater Quality

Quality of groundwater in an aquifer plays an important role in artificial recharge. It is crucial to consider the existing groundwater quality before implementing recharge methods to ensure that recharged water does not negatively impact the overall aquifer quality. Recharge water should meet specific quality criteria and not introduce contaminants that may degrade the aquifer's water quality. In cases where the existing groundwater quality is already compromised, additional treatment processes may be required to improve the recharge water's suitability. By assessing the aquifer's groundwater quality, appropriate measures can be implemented to minimize the risk of water quality degradation during the recharge process. This may include pre-treatment of recharge water, selecting suitable recharge sites away from potential pollutant sources, or implementing natural filtration processes to improve water quality before recharge. Maintaining or enhancing groundwater quality through proper management of artificial recharge is essential for safeguarding water resources and ensuring their long-term usability.

2.5 Factors Affecting the Design of Aquifer Artificial Recharge

Several factors influence the design of aquifer artificial recharge in Egypt. These factors need to be considered to ensure the effectiveness and sustainability of recharge projects. Here are some key factors that impact the design of aquifer artificial recharge in Egypt:

- Hydrogeological Setting: The hydrogeological characteristics of the aquifer, including its type, transmissivity, porosity, and depth, play a significant role in determining the most suitable recharge methods. The availability of suitable recharge sites, the presence of confining layers, and the flow dynamics of groundwater all influence the design considerations.
- Water Availability: The quantity and quality of water available for recharge are essential factors. The availability of a reliable water source and its proximity to recharge sites are crucial considerations in the design process. The sustainable yield of the water source needs to be assessed to ensure long-term viability.
- Water Demand and Supply: Understanding the water demand and supply dynamics in the region is vital. The recharge design should align with the water needs and availability to effectively replenish the aquifer and address water deficits during periods of high demand or water scarcity.
- Regulatory Framework: Compliance with local regulations and policies related to water management and aquifer recharge is essential. The design should adhere to legal requirements and environmental standards to ensure proper governance and avoid potential conflicts.
- Water Quality: The quality of the recharge water and its compatibility with the aquifer framework and groundwater need to be considered. Pre-treatment processes may be necessary to address potential contamination or to improve water quality before recharge.
- Recharge Method Selection: The choice of recharge method depends on factors such as the hydrogeological setting, water availability, and local conditions. Common recharge methods include infiltration basins, injection wells, canal spreading, or recharge through existing watercourses. The method selected should align with the site-specific conditions and objectives of the recharge project.
- Monitoring and Evaluation: Establishing a robust monitoring and evaluation framework is crucial for assessing the performance and effectiveness of the recharge project. Regular monitoring of groundwater levels, water quality, and recharge efficiency helps in making informed decisions and adjusting the design as necessary.

These factors need to be comprehensively assessed and integrated into the design process of aquifer artificial recharge projects in Egypt. By considering these factors, the design can be optimized to ensure sustainable water management and maximize the benefits of artificial recharge in the country. In Egypt, broadly two methods were used:

Table 2.3 Factors affecting the selection of artificial recharge method

Parameter	Artificial recharge method		
	Surface basins and ditches	Vadose zone injection wells	Direct injection wells
Aquifer type	Phreatic	Phreatic	Confined/Phreatic
Pre-treatment requirements	Low technology	Removal of solids	High technology
Capacity	1000–20,000 m^3/ha/day	1000–3000 m^3/well/day	2000–6000 m^3/well/day
Maintenance requirements	Drying and scraping	Drying and disinfection	Disinfection and flow recovery
Soil-Aquifer treatment	Vadose zone and saturated zone	Vadose and saturated zones	Saturated zone

Source El Arabi and Dawoud (2012)

1. Surface basins, furrows, and ditches were used in El Bustan experimental station in the western Nile Delta region and Abu Rawash pilot project for aquifer recharge with treated wastewater.
2. Injection Wells were used in Toshkah pilot project for the injection of surface water into the groundwater aquifer system.

Open basins are inexpensive to construct and simple to operate. However, they can be used to recharge only aquifers that have a direct hydraulic connection with the ground surface (phreatic aquifers). Artificial recharge using surface basins requires permeable surface soils. Where these are not available, trenches or shafts in the unsaturated zone can be used, or water can be directly injected into aquifers through wells. The design of land application requires data on the infiltration rates of the soil, the hydraulic properties of the unsaturated zone and underlying aquifer, and the location of any soil or groundwater contamination that may be mobilized by the recharge (Oakford 1985). The aquifer should be sufficiently transmissive to avoid excessive build-up of groundwater mounds. Table 2.3 shows the major factors affecting the selection of artificial recharge method.

2.6 Source Water Availability for Artificial Recharge

Three types of water are available for aquifer artificial recharge in Egypt:

1. Rainfall: The rainfall pattern in Egypt is predominantly influenced by the Mediterranean climate, with most of the rainfall occurring during the winter months. The amount of rainfall gradually decreases as one moves southward within the country. Along the coastal areas, mean rainfall is around eight inches. However, as we move toward Cairo, the mean annual rainfall reduces to a little over an inch, and in Upper Egypt, it diminishes further to less than an inch. This

regional variation in rainfall distribution is a critical factor to consider in the design of aquifer artificial recharge projects in Egypt. With lower precipitation levels in certain areas, alternative water sources and recharge methods need to be implemented to meet the water demand and sustain groundwater resources throughout the country. The total estimated rainfall water used in Egypt is about 1.0 BCM/year. There is a potential for harvesting rainfall and surface runoff in the northern coast and Sinai for artificial recharge through building recharge dames.

2. Excess surface water from main or branch canals: During winter, there are excess of surface water in some canals which can be used for aquifer recharge. An example of the surface water available for aquifer recharge is the stored water at Toshkah depression in four lakes. The massive floods of the Nile River resulted in the gradual formation of four new inland lakes in the Toshkah depression area (first lake of 12.5 km^3, second of 0.4 km^3, third of 5.2 km^3, and fourth of 22.1 km^3 area) as shown in Fig. 2.6. Recent studies (Elshemy and Meon 2011; Elsawwaf and Willems 2012; Abdel Wahab et al. 2020) indicated the stored water in these four lakes are as follows:

   (1) The first lake storage is about 12.5 km^3 of water, with a maximum water depth of 46.1 m.
   (2) The second lake storage is about 0.4 km^3 of water and reached a water depth of 21.7 m.
   (3) The third lake storage is about 5.2 km^3 of water, and the water depth was 43.4 m.
   (4) The fourth lake storage is about 22.1 km^3 of water and has the highest water depth of 57.0 m.

The total storage is about 40.2 km^3. The groundwater aquifer system at the lakes area is hard ferruginous sandstone with extensive shale and clay intercalations with a thickness of about 140–230 m. Groundwater in this aquifer is of meteoric origin and belongs to the fresh to slightly brackish type with a salinity of about 240–2370 ppm. The recharge rate from these lakes to the main aquifer system is very small. The best way is to take this water into the Nile Valley and Delta for aquifer artificial recharge to the aquifers.

- Treated Wastewater: the total wastewater in Egypt is about 7.5 BCM/year (2022), out of which 4.5 BCM is treated and 3.0 BCM remains untreated. Currently, only 1.5 BCM/year of treated wastewater is being used, out of which 0.5 BCM is secondary treated and 0.44 BCM is primary treated for forest irrigation with an area of about 13,000 Feddans (about 5,460 ha). About 3.2 BCM/year is pumped to drains at Greater Cairo, and Nile Delta. Figure 2.7 shows the location of afforested areas irrigated with treated wastewater. However, it is also associated with certain risks that need to be carefully considered. These risks include water quality concerns, groundwater contamination, soil and crop contamination, public perception and acceptance, lack of regulation, and institutional framework. To mitigate these risks, it is important to implement proper treatment processes, monitoring systems, and regulatory frameworks. Regular monitoring of water quality, adherence to strict guidelines, and continual evaluation of the recharge system's performance can help minimize the potential risks associated with aquifer artificial recharge using treated wastewater.

Fig. 2.7 Location of afforested areas irrigated with treated wastewater

2.7 Monitoring and Assessment of Artificial Recharge of Aquifers

Monitoring and assessment for artificial recharge are crucial for ensuring the feasibility and sustainability of recharged water in terms of quantity and quality and the effectiveness of the technical methodology used. Some of the key reasons why monitoring plans are important include:

1. Assessing the impact of the structure: Monitoring the water level and quality of the aquifer can help determine whether the recharge structure is successful in replenishing the groundwater resource. Regular monitoring can also help in identifying any potential problems with the recharge process and allow for timely corrective action.
2. Ensuring water quality: Monitoring with proper planning can help identify potential water quality issues that may arise because of the recharge process. By regularly testing the water in the aquifer, it is possible to ensure that the water is safe for use and that contaminants are not entering the aquifer.
3. Understanding the impact on the environment: Monitoring can help assess the potential environmental impact of artificial recharge. By monitoring the surrounding environment, it is possible to identify any changes in water quality, vegetation, or wildlife that may result from the recharge process.
4. Compliance with regulations: Monitoring plans are typically required by regulatory agencies to ensure compliance with environmental and health standards.
5. Identifying opportunities for improvement: Regular monitoring can also help identify opportunities for improving the recharge process. By analyzing data collected through monitoring, it is possible to identify areas where changes can be made to improve the effectiveness and sustainability of the recharge process (Fig. 2.8).

2.8 Role of Aquifer Artificial Recharge in Integrated Water Management in Egypt

Artificial recharge of aquifers is an important technique in integrated water management that involves replenishing groundwater reserves by making more water infiltrating underground aquifers. This technique can help to sustainably manage water resources, particularly in regions where water is scarce or experiencing increased demand due to population growth, industrialization, and climate changes, like in Egypt. Figure 2.9 illustrates the synergistic relationship between managed aquifer recharge, demand management, and conjunctive use in restoring an over-exploited aquifer to hydrologic equilibrium. In a dry climate such as Egypt, where natural recharge is diminishing and irrigation demands are rising, managed aquifer recharge can serve as an adaptive strategy to aid in re-establishing hydrologic balance. By replenishing the aquifer through artificial recharge methods, water availability can

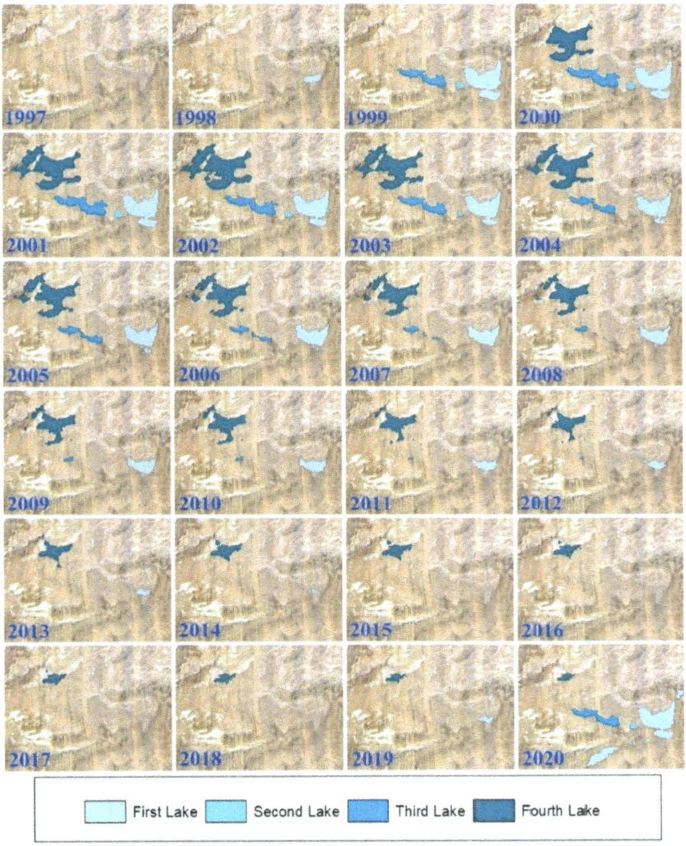

Fig. 2.8 The change in Toshkah depressions surface areas (1997–2020)

be enhanced, mitigating the impacts of declining natural recharge, and meeting the growing irrigation demands. Aquifer artificial recharge plays a crucial role in augmenting groundwater resources by utilizing available surface water, rainfall, and unused treated wastewater. It operates in conjunction with the practice of conjunctive use, which involves the coordinated management of both surface water and groundwater. This integrated approach ensures the sustainability of water supplies and helps achieve various management objectives, including the protection of ecosystems (Fig. 2.10). The conjunctive use of surface water and groundwater allows for optimized water allocation, improved water resource management, and enhanced ecosystem protection. Together, these strategies offer comprehensive and holistic approaches to effectively manage water resources, maintain ecological balance, and meet the diverse water needs of communities and the environment.

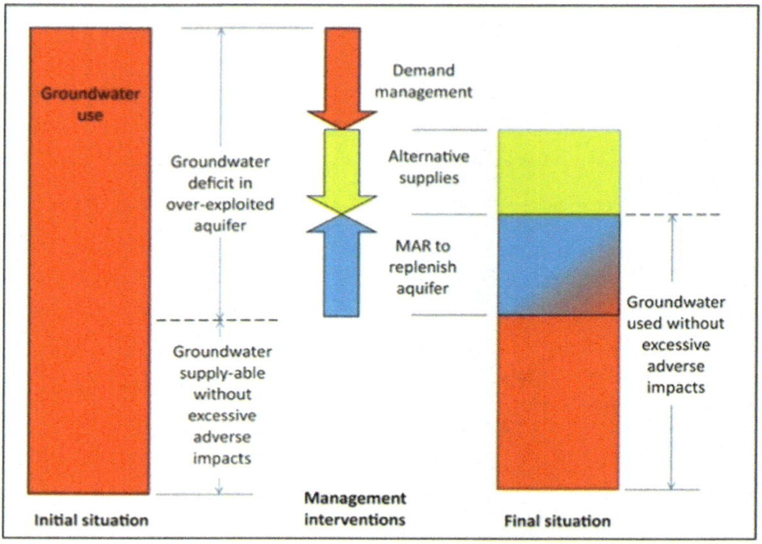

Fig. 2.9 The role of artificial recharge in enhancing groundwater aquifers. *Source* Dillon et al. (2012)

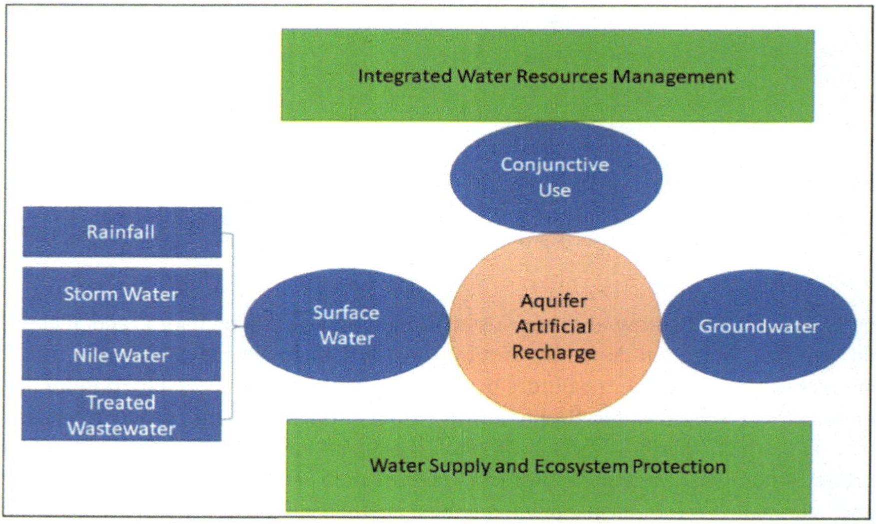

Fig. 2.10 Role of aquifer artificial recharge in integrated water resources management

2.9 Conclusions and Recommendations

Aquifer artificial recharge can play a crucial role in integrated water resource management in Egypt, offering wide-ranging benefits. Such recharge can contribute to conserving surface water resources, improving groundwater quality, minimizing evaporative losses, and increasing the volume of available groundwater, utilizing the unused treated wastewater. The economics of artificial recharge of aquifers depend on site-specific factors. Generally, recharge is the most cost-effective in areas with permeable soils, unconfined aquifers with fresh groundwater. Artificial recharge is particularly economically attractive in transmissive aquifers with fresh or mildly brackish groundwater such as Toshkah and Wester Nile Delta. It becomes especially beneficial in urban areas where the recovered water holds high value, requires minimal land area, and offers protection to the groundwater resource from overlying land uses. Aquifer artificial recharge offers economic advantages compared to alternative water sources in various contexts, ranging from urban to rural supplies, and holds potential for widespread implementation in floodwater recharge projects. Conducting site-specific assessments and validations can help determine the economic feasibility of aquifer artificial recharge and its suitability for different water management scenarios.

References

Abdel-Mageid H, Hago A, Abdel-Magid IM (1991) Analysis of pipe networks by the finite element method, Water International 16(2):96–101

Abdel-Latif M, Yacoub M (2011) Effect of change of discharges at Dongola station due to sedimentation on the water losses from Nasser Lake. Nile Water Sci Eng J 4(1):86–98

Abdel Wahab D, Adomako D, Abass G, Adotey DK, Anornu G, Ganyaglo S (2020) Hydrogeochemical and isotopic assessment for characterizing groundwater quality and recharge processes in the Lower Anayari catchment of the Upper East Region, Ghana, Environ, Dev Sustain 23(4):5297–5315

Atkinson G, Mourato S (2008) Environmental cost-benefit analysis. Annu Rev Environ Resour 33:317–344

Boardman A, Greenberg D, Vining A, Weimer D (1996) Cost-benefit analysis: concepts and practice. Prentice Hall, Upper Saddle River, NJ, USA, pp 1–135

Darwish M (1998) Artificial recharge experiments in bustan extension area, Egypt, At the book artificial recharge of groundwater, 1st edn. Taylor and Francis

Dawoud M (2019) Groundwater economics in arid regions: Abu Dhabi Emirate case study. In Proceedings of the 13th Gulf Water Conference Proceedings, Desalination and Water Treatment, Kuwait, Volume 176, pp 12–14

Dillon P, Fernandez EE, Tuinhof A (2012) Management of aquifer recharge and discharge processes and aquifer storage equilibrium. IAH contribution to GEF-FAO Groundwater Governance Thematic Paper 4. In Paper presented at Global Environment Facility (GEF)-Food and Agriculture Organization of the United Nations (FAO)

El Arabi N (2012) Environmental management of groundwater in Egypt via artificial recharge extending the practice to Soil Aquifer Treatment (SAT). Int J Environ Sustain 1(3):66–82

El Arabi N, Dawoud M (2012) Groundwater aquifer recharge with treated wastewater in Egypt: Technical, environmental, economical and regulatory considerations, Desalination and Water Treatment 47(1–3):266–278
El-Gafy I, El-Ganzori A, Mohamed A (2012) Decision support system to maximize economic value of irrigation water at the Egyptian governorates meanwhile reducing the national food gap, Water Science (27):1–18
Elshemy M, Meon G, (2011) Climate change impacts on water quality indices of the southern part of Aswan High Dam reservoir, Lake Nubia. Fifteenth Intern. Water Techn. Conf., IWTC-15, Alexandria, Egypt, p 17
Elsawwaf M, Willems P (2012) Analysis of the climate variability on Lake Nasser evaporation based on the Bowen ratio energy budget method, J Environ Biol 33(2):475–485
Hani HM, Nour El Din MM, Khalifa A, Elalfy E (2023) Sensitivity analysis for multi-criteria decision analysis framework for site selection of aquifer recharge with reclaimed water. Sustainability 15(6):1–21
Layard R, Glaister S (1994) Cost-benefit analysis, 2nd edn. Cambridge University Press, Cambridge, UK
Maliva RG, Herrmann R, Coulibaly K, Guo W (2015) Advanced aquifer characterization for optimization of managed aquifer recharge. Environ Earth Sci 73:7759–7767
Ministry of Water Resources and Irrigation (MWRI) (2022) Annual water resources statistical book. Egypt, pp 10–116
Oakford ET (1985) Artificial recharge: methods, hydraulics, and monitoring. In: Asamo T (ed) Artificial recharge of groundwater. Butterworth Publishers, Boston, pp 69–127
Pearce D, Atkinson G, Mourato S (2006) Cost-benefit analysis and the environment: recent developments; organization for economic co-operation and development. Paris, France
Pescod MB (1992) Wastewater treatment and use in agriculture, FAO Irrigation and Drainage paper No. 47, Rome, Italy
RIGW/IWACO (1999) Artificial recharge pilot project at el-bustan area, final report. Research Institute for Groundwater, Egypt, pp 1–56
RIGW/IWACO (1998) Interpretation of analysis results of the first monitoring network sampling round. Research Institute for Groundwater, Egypt, pp 1–110
RIGW (1988) Hydrogeological maps for Egypt. Research Institute for Groundwater, Egypt
Saha D, Sikka AK, Goklani R (2021) Artificial recharge endeavours in India: a review, water security, India. https://doi.org/10.1016/j.wasec.2022.100121
Van Puffelen J (1982) Artificial groundwater recharge in the netherlands. DVWK Bulletin 11 (Artificial Groundwater Recharge). Amsterdam, Netherland.
Vanderzalm J, Page D, Dillon P, Gonzalez D, Petheram C (2022) Assessing the costs of managed aquifer recharge options to support agricultural development. Agric Water Manag 263(107437):1–11
Ward FA, Michelsen A (2002) The economic value of water in agriculture: concepts and policy applications. Water Policy 4(5):423–466
Zhang H, Xu Y, Kanyerere T (2020) A review of the managed aquifer recharge: historical development, current situation, and perspectives, physics and chemistry of the earth, Parts A/B/C, vol 102887, pp 118–119

Chapter 3
Water Conservation and Artificial Recharge Efforts in India

S. Suresh, M. Senthilkumar, and S. N. Dwivedi

Abstract Groundwater resource in India has immensely contributed in achieving food and drinking water security of the country. Groundwater development has been the backbone of green revolution in India, and exponential growth in well construction has led to about 21.7 million wells presently in India. The easy access and extraction of groundwater to meet the demand has resulted in unplanned and skewed groundwater development in India and it has created pockets of over-exploitation, where recharge exceeds the annual replenishment. In order to combat the lopsided development, efforts have been made in India to augment the natural recharge through man made interventions under "Rain Water Harvesting and Artificial Recharge (RWH & AR)". In India, the various subjects of governance are vested with either the Union or the State Government as per the constitution of India. Water is primarily a State subject and only in cases of disputes between State Governments, the Union Government intervenes to resolve the issues. However, initiatives on RWH and AR are being taken up by both Central and State Governments through different schemes. The efforts made in India towards water conservation and groundwater resource augmentation have been compiled as a chapter to highlight the steps taken to meet the challenges in such a diverse hydrogeological environment.

S. Suresh (✉) · M. Senthilkumar · S. N. Dwivedi
Central Ground Water Board, Department of Water Resources, RD & GR, Ministry of Jal Shakti, Faridabad, India
e-mail: suranraj@gmail.com

M. Senthilkumar
e-mail: senthilkumar-cgwb@gov.in

S. N. Dwivedi
e-mail: shailendradwivedi-cgwb@gov.in

D. Saha et al. (eds.), *Managed Groundwater Recharge and Rainwater Harvesting*, Water Resources Development and Management,
https://doi.org/10.1007/978-981-99-8757-3_2

3.1 Introduction

Groundwater development in India has immensely contributed to achieving food and drinking water security in the country. During the past 5 decades, there has been a phenomenal increase in groundwater extraction in India and presently estimated to cater nearly 65% of the irrigation requirement, 85% of the rural water supply and nearly 50% of the urban water supply. There has been a spurt in the growth of the tubewells/borewells in the country during the past 5 decades and presently there are nearly **21.7 million** wells through which groundwater extraction for irrigation use is taking place. As per the latest assessment of dynamic groundwater resources (CGWB 2020a), the total Annual Ground Water Recharge is 436 Billion Cubic Meter (BCM) and the Annual Extractable Ground Water Resource is 398 BCM. The Annual Ground Water Extraction for all uses is 245 BCM. The Stage of Ground Water Extraction, which is a measure of Annual Ground Water withdrawal for all uses (irrigation, industrial and domestic uses), over Annual Extractable Ground Water Resource, is 62% for the country as a whole. However, as per the 2020 assessment, out of the total 6965 assessment units in the country, 1114 units in 15 States/UTs have been categorized as "Over-exploited" where the Annual Ground Water Extraction is more than the Annual Extractable Ground Water Resource. Unplanned and unsustainable extraction in some parts of the country has led to a severe decline in water levels in those areas. In this backdrop, rainwater harvesting and artificial recharge have assumed significant importance in the country and there has been a concerted effort on the part of the Central and State Governments in making water conservation and recharge a successful movement through people's participation.

Rainwater harvesting has been practised from time immemorial in different parts of the world. The civilization initially flourished along the river banks and when the population increased, they moved inland but fetched water from rivers for their existence. When the rivers were found insufficient to meet their need, tanks or ponds were made by men to harvest the rain and store water for future use. As the technology advanced, large dams were built across the rivers to harness rainwater for the non-rainy period. As the demand increased, a search was also directed beneath the ground for water and with time, advancement of technology has pushed it as the main source to meet the societal demands in many parts of the country. With increasing extraction, the withdrawal started exceeding the annual recharge and the water level started declining. In order to arrest the depletion of groundwater, rainwater harvesting has been used to augment the natural recharge of groundwater system.

Rainwater harvesting and artificial recharge can be categorized into three types, viz., storage of rainwater, recharge to groundwater system or a combination of both. In the areas of shallow groundwater level (<3 m below ground level) during post-monsoon or with poor-quality water, it is generally advocated to store the water for subsequent use. While in areas of deeper groundwater, the practice of recharge is advised. In case, the rate of aquifer intake is slower than the precipitation intensity, a combination of storage and recharge is suggested. The augmented recharge or artificial recharge to groundwater is defined as the recharge that occurs when the natural pattern of recharge is deliberately modified to increase recharge (ASCE 2001).

In case of recharge projects, the following points need to be considered while formulating the project.

Need—Defined by deeper water level and declining trend of water level

Scope—Defined by the availability of surplus source water (yield from the catchment—committed demand of the existing water bodies) and the ability of the system to accept the recharge.

Economic viability of the project—Defined by the benefit cost ratio.

The geomorphology, terrain and hydrogeology determine the type of structures and design to be used, while the available surplus runoff limits the number of structures.

In India, the various subjects of governance are vested with either the Union or the State Government as per the constitution of India. Water is primarily a state subject and only in case of disputes between State Governments, the Union Government intervenes to resolve the issues. However, initiatives on rainwater harvesting and artificial recharge (RWH & AR) are being taken up by both Central and State Governments through different schemes. Central Government initiatives are also taken up directly through different Ministries and Departments including Central Ground Water Board (CGWB), which is the apex organization dealing with groundwater governance in the country. The information for this chapter has been compiled from many reports of CGWB and State agencies and also from many internal meetings and discussions.

3.2 Initiatives of the Central Government

The Central Government implements groundwater-related activities mainly through the Central Ground Water Board; however, several water conservation-related efforts are also coordinated directly through the Ministry of Jal Shakti. The steps taken by the Union Government to control the depletion of water levels and promote rainwater harvesting/conservation are also hosted on the website and can be accessed at http://jalshakti-dowr.gov.in/sites/default/files/Steps_to_control_water_depletion_Feb2021.pdf. Some of the important points are listed below.

- Creation of a new Ministry of Jal Shakti (from earlier Ministry of Water Resources) for dealing with all matters relating to water at one place in an integrated manner.
- The National Water Policy (2012) has been formulated by the Ministry, which inter-alia advocates rainwater harvesting and conservation of water and highlights the need for augmenting the availability of water through direct use of rainfall.
- In compliance to the decision taken by the Committee of Secretaries, an "Inter-Ministerial Committee" under the Chairmanship of Secretary (Ministry of Jalshakti) has been constituted to take forward the subject of "Push on Water Conservation Related Activities for Optimum Utilization of Monsoon Rainfall".

- The Ministry has circulated a Model Bill to all the States/UTs to enable them to enact suitable groundwater legislation for its regulation and development, which includes provision of rainwater harvesting and artificial recharge. So far, 15 States/UTs have adopted and implemented the groundwater legislation along the lines of Model bill.
- Central Ground Water Authority (CGWA), the regulatory body to control groundwater extraction, has issued directions under Sect. 3.5 of "The Environment Protection Act, 1986" for mandatory Rain Water harvesting/Roof Top Rain Water Harvesting for all target areas in the Country including UTs. While granting "No Objection Certificate (NOC)" for drawing groundwater, CGWA insists on mandatory rainwater harvesting as per the guidelines issued.
- CGWB has also prepared a conceptual document entitled "Master Plan for Artificial Recharge to Ground Water in India" during the year 2020, which is a macro plan formulated to work out the feasibility of various structures for the different terrain conditions of the country and respective estimated costs. It envisages construction of 14.18 million structures for harvesting rainwater in the Country at an estimated cost of INR.1335 Billion to harness 185 BCM (Billion Cubic Metre) of water, in an area of 1.1 Million sq.km by harnessing surplus monsoon runoff to augment groundwater resources.
- CGWB has taken up Aquifer Mapping and Management programme during XII Plan (2012–2017), aiming to delineate aquifer disposition and their characterization for preparation of aquifer/area-specific groundwater management plans, that includes artificial recharge with community participation. The management plans are shared with the respective State Governments for taking appropriate measures.
- The Ministry has instituted National Water Awards to incentivize good practices in water conservation and groundwater recharge.
- Awareness programmes are conducted regularly, from time to time, under the Information, Education and Communication (IEC) Scheme of the Ministry in various parts of the country to promote rain water harvesting and artificial recharge to ground water.
- The Ministry of Rural Development in consultation and agreement with the Ministry of Jal Shakti and the Ministry of Agriculture and Farmers' Welfare has developed an actionable framework for Natural Resources Management (NRM), titled "Mission Water Conservation" to ensure gainful utilization of funds. The Framework strives to ensure synergies in *Mahatma Gandhi National Rural Employment Guarantee Scheme* (MGNREGS), *Pradhan Mantri Krishi Sinchayee Yojana* (PMKSY), erstwhile *Integrated Watershed Management Programme* (IWMP) now *PMKSY-Watershed Development Component and Command Area Development and Water Management* (CAD&WM), given their common objectives. Types of common works undertaken under these programmes/schemes are water conservation and management, water harvesting, soil and moisture conservation, groundwater recharge, flood protection, land development, and command area development. Department of Land Resources is currently implementing 8214 watershed development projects in 28 States covering an area of about 39.07 million ha. Under the *Watershed Development Component* (WDC) of the *Pradhan*

Mantri Krishi Sinchayee Yojana (PMKSY) principally for the development of rainfed portions of net cultivated area and culturable wastelands. The major activities taken up under the WDC-PMKSY, inter-alia, include ridge area treatment, drainage line afforestation, soil and moisture conservation, rainwater harvesting, horticulture, pasture development etc.

- Ministry of Housing and Urban Affairs has released Model Building Bye-laws, 2016, which recommends Rainwater Harvesting and recharge for all types of buildings with plot sizes 100 m^2 or more. Barring the States/UT of Sikkim, Mizoram and Lakshadweep, all the States have incorporated the provisions in their respective building bye-laws. The plans submitted to the local bodies shall indicate the system of stormwater drainage along with points of collection of rainwater in surface reservoirs or in recharge wells.
- The Government of India has initiated *Atal Bhujal Yojana*, a INR. 60,000 million Scheme (equally funded by the Govt of India and the World Bank), for sustainable management of groundwater resources with community participation in water-stressed blocks in the states of Gujarat, Haryana, Karnataka, Madhya Pradesh, Maharashtra, Rajasthan and Uttar Pradesh.

In addition to the above, a recent initiative of Union Government is *Jal Shakti Abhyan* (JSA), which commenced in 2019 for 256 water-stressed districts in the country with a focus on five intervention areas, viz., (1) water conservation and rainwater harvesting, (2) Renovation of traditional and other water bodies/tanks, (3) Reuse, borewell recharge structures, (4) watershed development and (5) intensive afforestation. Considering the success of the program, it was continued in 2021 as JSA-II, "*Catch the Rain Campaign*", extending the campaign to cover the whole country. The focussed interventions include (i) Water conservation and rainwater harvesting, (ii) enumerating, geo-tagging and making inventory of all water bodies; preparation of scientific plans for water conservation based on it, (iii) Setting up of *Jal Shakti Kendras* in all districts, (iv) intensive afforestation and (v) awareness generation. During JSA-2021, 1.622 million water conservation and rainwater harvesting structures were created across the country. Similarly, 0.3 million traditional water bodies were renovated during the period. INR 656.6 billion was spent against water conservation-related activities under MGNREGA during the period. As a sample, a study was taken up by CGWB in 2021 to assess the impact of JSA campaign in 15 select districts which revealed an overall improvement in 13 districts. No interventions have been carried out in the 2 districts due to COVID pandemic restrictions. However, the improvement in water levels is found to be a cumulative effect of various factors like increased rainfall and various interventions of both state and central government including JSA.

Ministry of Jal Shakti has also compiled information on the practices adopted by States, NGOs, Corporates and best practices and hosted them on its website under Citizen corner as "Best Practices" (http://jalshakti-dowr.gov.in/government-initiatives; http://jalshakti-dowr.gov.in/ngo-initiatives; http://jalshakti-dowr.gov.in/private-corporate-initiatives).

3.3 Initiatives of CGWB

Experiments on artificial recharge to aquifers started in India from 1970 onwards by Central and State Government Departments and individually by some NGOs in different parts of the country where early signs of overexploitation of groundwater were noticed. CGWB undertook artificial recharge experiments through injection well around Kamliwala in Central Mehsana, Gujarat, where sufficient water was available from Saraswati River during monsoon period. The injection well-accepted recharge at a rate of 225 m^3/day for 250 days. There was a build-up of 5 m piezometric head in the injection well and 0.6 to 1.0 m in wells in adjoining areas within 150 m from the injection well. Subsequently, several such studies were taken up in different states of the country. The details of demonstrative artificial recharge studies taken up by CGWB during different Five-Year Plans are furnished in Table 3.1.

Over-exploited, critical, coastal and urban aquifers as well as areas facing groundwater quality issues were given focus in the above schemes. A separate scheme on using dug wells for recharge was prepared for the states of Tamil Nadu, Gujarat, Madhya Pradesh, Karnataka, Maharashtra, Rajasthan and Andhra Pradesh at the estimated cost of INR 187.1 billion covering 4.455 million irrigation dug wells in 1155 blocks. Under this scheme, farmers were given funds directly for the construction of recharge pits near the dug well at an average cost of Rs 4000/-.

In the course of water conservation work, CGWB also brought out the following documentations in an attempt to disseminate the experiences gained during various groundwater augmentation projects implemented by the Board in the country. These

Table 3.1 Artificial recharge studies taken by CGWB during different five-year plans

Five-year plan	States	Cost in INR in million
VIII (1992–97)	Maharashtra, Karnataka, Andhra Pradesh, Delhi, Kerala, Madhya Pradesh, Tamil Nadu, West Bengal and Chandigarh (Total States/UT—9)	33.3
IX (1997–2002)	Andhra Pradesh, Arunachal Pradesh, Assam, Andaman and Nicobar, Bihar, Chandigarh, Gujarat, Haryana, Himachal Pradesh, Jammu and Kashmir, Jharkhand, Karnataka, Kerala, Lakshadweep, Madhya Pradesh, Maharashtra, Meghalaya, Mizoram, Nagaland, Delhi, Odisha, Punjab, Rajasthan, Tamil Nadu, Uttar Pradesh, Uttarakhand and West Bengal (Total States/UT—27)	331.0
X (2002–2007)	Andhra Pradesh, Karnataka, Madhya Pradesh and Tamil Nadu (Total States—4)	56.0
XI (2007–2012)	Andhra Pradesh, Arunachal Pradesh, Bihar, Chhattisgarh, Chandigarh, Delhi, Gujarat, Himachal Pradesh, Jammu and Kashmir, Jharkhand, Kerala, Karnataka, Maharashtra, Madhya Pradesh, Nagaland, Odisha, Punjab, Rajasthan, Tamil Nadu, Uttar Pradesh and West Bengal (Total States/UT—21)	998.7

documents are also hosted on the CGWB website. (http//cgwb.gov.in/Manuals-Guidelines.html.)

- Manual on Artificial Recharge of Ground Water
- Artificial Recharge Guide
- Rain Water Harvesting Guide
- Booklet on "Simple Ways to Save Water"
- Elixir of Life Water (In Urdu language)
- Standard Designs for Adoption of Roof Top Rainwater Harvesting In Delhi
- Rainwater Harvesting techniques to Augment Ground Water
- Interim Report on Project wise Impact Assessment of Completed Demonstative Artificial Recharge Projects of XI Plan (http://cgwb.gov.in/Ar-reports.html)
- Guidelines for Impact Assessment of Artificial Recharge Structures/Schemes (http://cgwb.gov.in/AR/Final-Guideline-IA-AR-Studies-submitted-website-hosting.pdf)

The manual on Artificial recharge provides detailed guidelines on investigative techniques for selection of sites, planning and design of artificial recharge structures, monitoring and economic evaluation of artificial recharge schemes. It also included elaborate case studies and field examples of artificial recharge schemes from different parts of the country. The manual has been used extensively for planning and implementation of schemes for the augmentation of groundwater resources by various agencies. Suitability of various structures considering the geomorphology, rainfall and terrain conditions has been identified and provided in the manual (CGWB 2007) and the same is also outlined in Table 3.2.

These apart, Central Ground Water Board has also published technical brochures on various aspects of artificial recharge through its regional offices in various states, in vernacular languages, which served as guidelines to various State Governments and non-governmental agencies and the general public. Some of the State Departments have also brought out manuals and guidelines on artificial recharge to groundwater, which dealt with specific areas in most cases.

3.4 Initiatives of State Governments

Several projects have been implemented with the State initiatives using people's participation in executing recharge projects. Some of the successful projects implemented are listed in Table 3.3 and discussed in brief subsequently.

Mukhya Mantri Jal Swawlamban Abhiyan in Rajasthan: The scheme aimed at making villages self-sufficient in water and creating "Islands of Excellence" with people's participation. It was a four-year programme, each phase of one year and it commenced in 2015–2016 and ended in 2019, covering 295 Blocks in 33 Districts. The scheme was launched with the objective of ensuring the effective implementation of water conservation and water harvesting-related activities in rural areas through

Table 3.2 Artificial recharge structures suitable under combination of different topographic slopes, hydrogeologic groups and rainfall distribution

Slope	Hydrogeologic group						Unconfined/ Confined aquifer
	Consolidated		Semi consolidated		Un-consolidated		
	Rainfall						
	Adequate	Limited	Adequate	Limited	Adequate	Limited	
Steep Slope (20–10%) Runoff zone	Bench Terrace Contour Trench	Gully Plug	Bench Terrace Contour Trench	Gully Plug	–	–	Unconfined
Moderate Slope (10–5%) Piedmont zone	Bench Terrace Contour Trench Gravity Head Recharge Well*	*Nalah* Bunds Contour Bunding Percolation Tanks *Nalah* Trench Gravity Head Recharge well* Bore Blasting	Bench Terrace Contour Trench Gravity Head Recharge Well*	Nalah Bund Contour Bund Percolation Tanks Nalah Trench Gravity Head Recharge Well	Ditch and Furrow Recharge Basin Pits and Shafts Contour Trench Gravity Head Recharge Well	Recharge Basin Pits* and Shafts* Contour Trench Gravity Head Recharge Well	Unconfined
	Deep Gravity Head Recharge Well Hydrofracturing Fracture Seal Cementation		Injection Well* Recharge Shafts*				Confined
Moderate to Gentle Slope (5–2%) Transition zone	Nalah Bunds Contour Bunding Percolation Tanks Recharge Pits Canal Irrigation* Induced Recharge Ground Water Dams Fracture Seal cementation	Nalah Bunds Contour Bunding Percolation Tanks Recharge Pits Ground Water Dams Canal Irrigation*	Recharge Basin Canal Irrigation* Induced recharge Stream Channel Modification Recharge Pits	Recharge Pits Stream Channel Modification	Flooding Recharge Basin Stream Channel Modification Induced Recharge Gravity Head Recharge Well* Canal Irrigation*	Stream channel Modification Gravity Head Recharge Well* Ditch and Furrow Recharge Basin* Recharge Shaft* Ground Water Dam (In shallow alluvium)	Unconfined

(continued)

Table 3.2 (continued)

Slope	Hydrogeologic group						Unconfined/ Confined aquifer
	Consolidated		Semi consolidated		Un-consolidated		
	Rainfall						
	Adequate	Limited	Adequate	Limited	Adequate	Limited	
	Gravity Head Recharge Well* Hydrofracturing Deep Fracture Seal Cementation		Recharge Shaft* Gravity Head Recharge Wells* Injection Wells* Hydrofracturing		Recharge Shafts* Gravity Head Recharge Wells* Injection Wells*		Confined
Gentle Slope (< 2%) Storage Zone	Surface Irrigation Recharge Basin Recharge Pits Gravity Head Recharge Wells	Induced Recharge Recharge Basin Recharge Pits Gravity Head Recharge Wells	Recharge Pits	Flooding Canal Irrigation* Induced Recharge Surface Spreading Infiltration Gallery	Flooding* Surface Spreading* Infiltration Gallery		Unconfined
	Gravity Head Recharge Wells (On Lineaments or their intersections)	Injection Wells	Injection Wells Connector Wells	Gravity Head Recharge Wells (On Lineaments or their intersections)	Injection Wells		

Note Rainfall is considered "adequate" if annual precipitation is more than 1000 mm
*Indicate availability of source water supply through canals, trans-basin transfer or treated wastewater (After CGWB 2007)

Table 3.3 Initiatives by the state governments

State	Name of scheme	Objectives
Rajasthan	Mukhyamantri Jal Swavlamban Abhiyan	To make villages self-sufficient in water and creating "Islands of Excellence"
Maharashtra	Jalyukt Shibar	Dovetailing all existing schemes under one roofto make "water for all-drought-free Maharashtra and to implement "Jalyukta Shivar" (water full surrounding) campaign to increase water availability"
Gujarat	Sujalam Sufalam Abhiyan	A conservation plan to deepen water bodies in the state before monsoon through desilting of lakes, check dams, rivers etc. people participation under MGNREGS
Telangana	Mission Kakatiya	Restoration of all existing 46,531 no. of Minor Irrigation sources/Tanks under "Mission Kakatiya"
Andhra Pradesh	Neeru Chettu	Interdepartmental convergence activity among "water conserving departments". The work includes repair and renovation of existing structures and construction of new structures, arresting soil erosion by following ridge-to-valley approach, development of cascade of tanks to divert the surplus, afforestation etc.
Tamil Nadu	Kudimaramathu	Storage of rainwater to overcome flood and drought by restoration of water bodies reviving the age-old practice of Kudimaramathu (or self-maintenance of tanks by community)
Bihar	Jal Jeevan Hariyali	Identification and rejuvenation of conventional water storage structures, construction of small structures across streams/rivers/Nala, Transfer from surplus to deficit areas, Afforestation, smart irrigation practices
Karnataka		Filling up of tanks with treated water to increase the soil moisture and improve the groundwater recharge

the convergence of schemes of various Govt Departments. The work was implemented through people's participation and based on the valley-to-ridge approach, starting from the treatment of the catchment to the renovation of existing structure and creation of new structures based on the available surplus runoff. The project included institutional arrangements from village level to State level. The outcome of the project included the creation of additional storage capacities to harvest 128 M.Cu.m (4516 M. Cu.ft) of water coupled with extensive and vigorous watershed development activities in an exhaustive and scientific manner which helped in intercepting an additional 11,170 M. Cu.ft. monsoon water.

Jalyukta Shivar Abhiyaan in Maharashtra was launched in the year 2014, by aggregation of all existing schemes under one roof in a bid to make Maharashtra a drought-free state by 2019. This flagship programme is aimed to permanently overcome drought situation and thus implement "Jalyukta Shivar" (water full surrounding) campaign to increase water availability in about 5000 villages to make them free of water scarcity. The scope of work includes widening and deepening of rivers, desilting of lakes, ponds and canals, catchment area treatment with the construction of contour trench, check and earthen stop dams, compartment bunds, farm ponds etc. The Scheme mainly envisages arresting rainwater within the village boundaries, increasing groundwater level and preparation of Village Action Plans including water budget. *Jalyukta Shivar Abhiyan* was launched during 2015–2016 in 6,202 villages, 5,288 villages during 2016–2017 and 5029 villages during 2017–2018. A total of 16,519 villages have been covered so far and in these villages till 27 July 2018, it is reported that 5,37,705 works are completed and 19,105 works are in progress. Water harvesting potential of 22.01 lakh TCM is created through the works completed.

Sujalam Suphalam Yojna in Gujarat is a project of using 332 km long *Sujalam Suphalam Spreading Canal*, traversing through seven districts from Mahi to Banas Rivers which includes diversion of surplus flood water of Kadana Reservoir and Narmada to the water deficit areas. The work included lift irrigation schemes, construction of tidal regulators/Bhandaras and canal to provide water to the water-scarce areas in Gujarat. The canal yojana has resulted in about 700 mcm of flood water being diverted to water-scarce areas.

Mission Kakatiya in Telangana is a project for the restoration of tanks in the State. The topography and rainfall pattern in Telangana have made tanks as a source of irrigation by storing water and regulating flow. During the Kakatiya period from the twelfth to fourteenth century, the construction of tanks was carried out with utmost technical expertise. Realizing the importance of protecting tanks in economic growth in the state, the Government of Telangana took up the programme and restored all existing 46,531 tanks from the year 2014–2019. The results include an increase in water retention capacity of the soil, thereby decreasing the number of waterless periods, improvement in groundwater recharge, application of silt to agricultural lands as nutrient/fertilizer, effective utilization of 255 TMC of water allocated for Minor irrigation sector under Godavari and Krishna River basins and brought around 0.1 million acres under irrigation command.

Mission Neeru Chettu in Andhra Pradesh was a project with a vision to make the state a drought-proof State and to eradicate poverty and reduce economic inequalities through better "water conservation and sustainable management". *Neeru Chettu* is an interdepartmental convergence activity among water conserving departments in the State, viz., Irrigation, Rural Development, Groundwater, Forest, Andhra Pradesh Space Application Centre (APSAC) and water utilizing departments, viz., Agriculture, Horticulture, Fisheries, Animal Husbandry, Rural Water Supply, Municipal Administration and Urban Development. The scope of work includes:

- Repairs and renovation of existing water harvesting structures and construction of new structures for groundwater recharge
- Arresting the soil erosion by following the ridge-to-valley approach.

- Repairs, renovation and desilting of tanks by convergence with Mahatma Gandhi National Rural Employment Guarantee Scheme by utilizing State and Central Government funds.
- Repairs and restoration of defunct lift irrigation schemes and construction of new schemes.
- Development of cascades, i.e., chain of tanks to divert water from surplus basins to deficit basins.
- To take up the plantation to increase the green cover and harvest rainwater through massive afforestation and soil moisture conservation works.
- Total ayacut stabilised by desilting as well as by repair to structures is 1,80,791 Acres, Total Quantity of water increased by desilting and taking up construction and as well as repairs to structures is 10.38 TMC.

***Kudimaramathu* in Tamil Nadu** is the scheme for reviving the age-old practice of self-maintenance of tanks by the community. Realizing the importance of storage of rainwater to overcome floods and droughts, the Government of Tamil Nadu has been restoring the water bodies with the help of the user communities since 2016–2017. Kudimaramathu works are entrusted to Farmers' Organisations/Irrigation Council/ Ayacutdars/Group of Ayacutdars on a nomination basis for execution with 10% of the estimated cost borne by them in the form of labour or material or cash. Maintenance of supply channels, canals, tanks, shutters, strengthening and reconstruction of surplus weirs, sluices etc. are included in Kudimaramath works. The implementation of the scheme has resulted in the increase of the irrigation capacity of water bodies by 20–30% along with an increase of 10% in cultivable land. The participatory approach has also benefited the local people by owning the scheme. The selection of the tank is made by the beneficiaries and is also allowed to use the silt in their agricultural fields. The filling up of tanks results in an increase in soil moisture and groundwater recharge.

***Jal Jeevan Hariyali* in Bihar** commenced in 2019 and is being implemented in all 38 districts of Bihar in mission mode with 11 target interventions as enumerated below:

- Identifying and removing encroachments from public water bodies, reservoirs etc.
- Restoration and rejuvenation of public water bodies such as Ponds/Tanks/Aahars/ Payeens.
- Identification and renovation of public wells.
- Construction of soak pits/recharge pits and other water conservation structures near public wells and hand pumps.
- Construction of check dams and water conservation structures near small rivers, drains and water storage structures in hilly areas.
- Channelling/delivering water from the rivers having surplus water to the water deficit areas.
- Roof Top Rain Water Harvesting Structure.
- Developing nurseries and massive afforestation.
- Adaptation of alternative agriculture, drip irrigation bio-farming and other new techniques.

Table 3.4 Salient features of the project

Status of projects	Cost of project (Billion Rs)	Committed water to minor irrigation		Proposed no of tanks	Live capacity (TMC)	Tanks filled		Water pumped (TMC)
		MLD	TMC			Tanks	Check dam	
Implemented	29.16	865	11.06	298	8.86	227	143	12.39
To be implemented	23.78	308	4.20	603	12.64			

- Promoting the use of solar energy and savings on energy.
- Jal Jeevan Hariyali awareness campaign.

Significant progress has been achieved in the state under the *Jal Jeevan Hariyali* Abhiyan. The Abhiyan is progressing forward with the objective of making the state greener, water rich and relatively more resilient to the adverse impact of climate change.

In the **State of Karnataka, the** depletion of groundwater resources and shortage of source water for recharge had prompted the Minor Irrigation and Ground Water Department, to use the secondary treated water from Sewerage Treatment Plant (STP) of Bengaluru Water Supply and Sewerage Board (BWSSB) filling up of tanks in in Kolar, Chikkaballapura, Bengaluru Urban, Bengaluru Rural and Tumkur Districts (Table 3.4) (Source: Communications of the department in official meetings).

The preliminary impact indicated a rise in water levels and an increase in greenery around the tanks and the water quality indicated by pH, CoD, NH_4–N and PO_4–P were within National Green Tribunal (NGT) standards. However, detailed analysis needs to be taken up to ascertain the impact on groundwater quality, as it is a pilot study in India for utilizing treated water for artificial recharge using the surface spreading method.

3.5 Case Study

A case study of a successful execution of artificial recharge project from a hard rock terrain in the State of Tamil Nadu entitled "*Demonstrative Project on Artificial Recharge to groundwater in Gangavalli Block, Salem District, Tamil Nadu"—is enumerated below.*

3.5.1 Project Formulation

Gangavalli Block is an over-exploited block. It was decided to take up artificial recharge projects to augment the groundwater recharge and showcase the study as demonstrative project for replication in other areas. The project was executed during 2006–2008.

The project formulation is preceded by answering three points in respect of the area, viz., need, scope and benefit cost ratio. The declining trend and higher groundwater development in Gangavalli Block indicated the need for artificial recharge to supplement the natural recharge of groundwater system. The scope was defined by the available surplus harnessable runoff and the ability of the aquifer system to accept the recharge, if additional water was directed into the system. The runoff in the block was of the order of 43.46 M.Cu.m, while the committed water requirement in the Gangavalli Block for the existing water bodies was of the order of 22.99 M.Cu.m, thereby indicating a balance available surplus of 20.47 M.Cu.m. The deeper water level indicated the available space for saturation and the water level fluctuation in wells due to recharge or pumping indicated the ability of system to accept the recharge.

The scheme for artificial recharge was formulated with a two-pronged approach, viz., augmenting the irrigation sources and drinking water sources. A reconnaissance survey was made by the Tamil Nadu Water Supply and Drainage Board (TWAD Board) responsible for water supply in the State. The Agricultural Engineering Department (AED) responsible for assisting the farmers and Public Works Department (PWD) responsible for irrigation were asked to submit their proposals for augmenting the irrigation sources.

Initially, the existing surface water bodies were plotted on the map and were subsequently visited. Appropriate structures were selected for identified sites and a comprehensive scheme proposal was made for improving the irrigation and drinking sources. In Gangavalli Block, a total of 41 artificial recharge structures were proposed and executed during 2006–2008 with a budget outlay of INR 22.22 Million. Check dams and percolation ponds were constructed to recharge the shallow phreatic aquifer while recharge wells were used to recharge the deeper fractures. In addition, desilting of existing tanks was also taken up to augment the storage capacity of tanks and incremental recharge to groundwater. A perusal of the Table 3.4 shows that 8,93,938 Cu.m could be harvested in a year and 1,78,78,751 Cu.m could be harvested during the life of these structures. The quantum of water that could be recharged in a year would be of the order of 4,89,618 Cu.m and during the life of these structures would be 97,92,367 Cu.m. The average unit cost of recharge was of the order of INR 5.14/Cu.m. The unit cost varied from INR 7.97/Cu.m for Check Dam to INR 1.41/Cu.m for desilting of existing pond. The addition of recharge borewell to supplement the recharge structures had resulted in the reduction of unit cost by INR 0.79 for percolation pond and for check dam by INR4.09. The study had shown that the check dam can be an effective structure for arresting soil erosion and regulating the flow in the stream but it is not a cost-effective artificial recharge structure.

The impacts of any artificial recharge structure/scheme (either direct or indirect) are listed below.

Direct Impacts

- **Rise in water level**—Arresting of declining water level (rate/absolute Value)
- **Increase in cropped Area**—Increase in the area cultivated with water-intensive crops or with the existing cropping pattern
- **Increase in yield of wells**—Wells sustain pumping during lean months and pumping hours are reduced for the same command area
- **Increase in pumping hours**—Longer duration of pumping or increase in number of pumping days.
- **Maintenance of water level/yield**–Even during years of deficit rainfall
- **Improvement in Quality**—Improvement in groundwater quality due to dilution effect (changes in concentrations of dissolved ions, individually or collectively).

Indirect Impacts

- Increase in vegetation cover in surrounding areas
- Increase in non-seasonal flow in the streams/rivers
- Less frequent development of cracks in soil due to increased soil moisture.

The impact assessment should commence with the initiation of construction of artificial recharge structures and should ideally continue for at least two water years after construction. The study should be inclusive of direct and indirect parametres indicating the impacts. In general, the area of influence of a single structure could be taken as 1 km^2 on the downstream side, and depending on the flow direction, the impact will be pronounced.

In this project impact on water level, cropped area, and pumping hours were analysed and given below.

3.5.2 *Water Level*

The depth to water level in wells adjacent to the artificial recharge structures have been compared for different months in the subsequent years to study the impact of these structures on water levels and are presented in Fig. 3.1. In most of the wells, the water level has risen in comparison to the previous year during the same period indicating the positive impact of these artificial recharge structures. A summary of impact on the water levels has been summarized in Table 3.5.

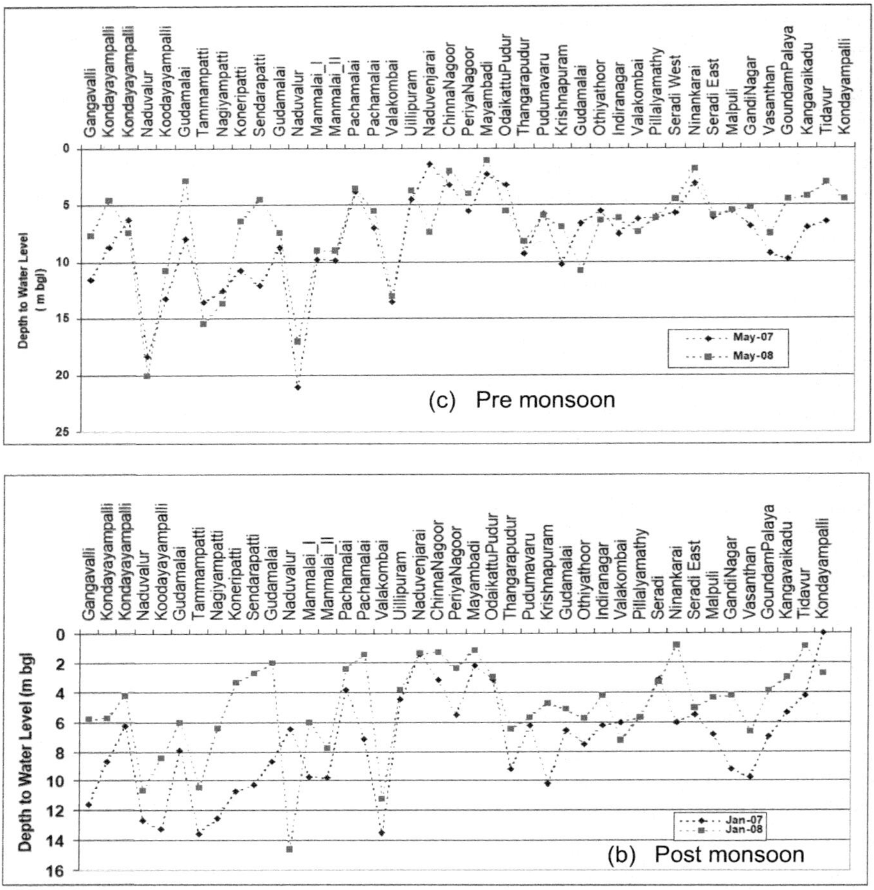

Fig. 3.1 Temporal variation in groundwater level

Table 3.5 Summary of impact on water level

Period	Range (m bgl)	Average depth to water level (m bgl)
Jul–2006	01.22–21.90	7.65
Jul–2007	01.40–21.50	7.86
Jan–2007	0.00–13.55	7.33
Jan–2008	0.85–14.60	4.93
Feb–2007	0.00–14.93	6.05
Feb–2008	1.05–15.65	6.74
May–2007	01.40–21.00	8.11
May–2008	01.00–20.00	6.95

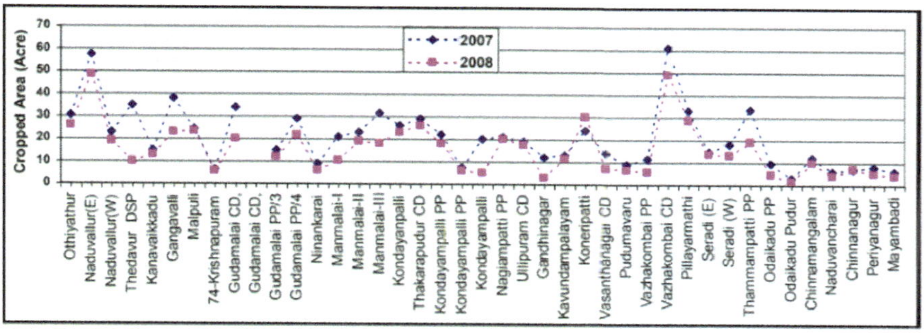

Fig. 3.2 Changes in area under cultivation (Acre)

3.5.3 *Cropped Area*

The impact can be felt either as a change in cropped area or changes in cropping pattern. The data has been collected from the farmers in the downstream areas of the structure with respect to changes in cropped area and cropping pattern during Kuruvai Season (Nov–Jan) from 2006–2007 and 2007–2008. A perusal of the data shows that there is a marginal increase in the cropped area but an appreciable increase in the conversion from Non Paddy to paddy cultivation. The changes in cropped area and area under cultivation of paddy and non-paddy crops are provided as Fig. 3.2

3.5.4 *Pumping Hours*

The impact will result in the increase in pumping hours. The data on the pumping hours were also collected from the same wells, where data on command area was collected. A perusal of the data shows that there is an increase in pumping hours in all the sites (Fig. 3.3).

3.5.5 *Regional Impact on Groundwater Regime*

In order to study the regional impact on water level in the block, depth to water level during May 2007 and May 2008 was considered and the difference was studied. The positive impact denotes the rise in water level, while the negative denotes the fall in water level. The rise in water level could be attributed to the impact of the artificial recharge structures and there is a rise in water level of the order of 1.06 m. In addition, it has also been reported that the yield of the domestic borewells has increased, thereby their pumping hours have been reduced for filing the overhead

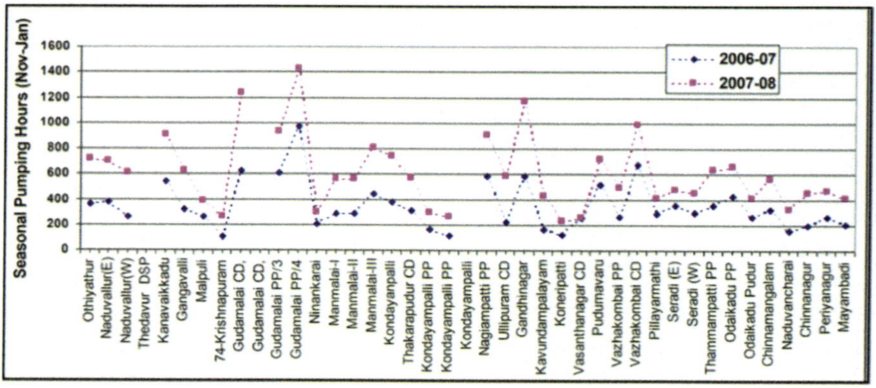

Fig. 3.3 Changes in pumping hours

tanks. The streams in the Pachamalai hills have started having flow during non-rainy season, which indirectly indicates the overall improvement in the groundwater situation in the area. The time lag between rainfall and the rise in water level is short after the completion of the project thereby indicating a quick response of the aquifer due to the intervention of the recharge wells in augmentation processes.

The data on the analysis of cropped area and number of hours of pumping has also shown a marked rise. The average change in the area between 2006 and 2007 and 2007–2008 for Kuruvai Crop (Nov–Jan) for the whole project area has been presented in Fig. 3.4 and the change in the average pumping hours for the same period has been furnished as Fig. 3.5.

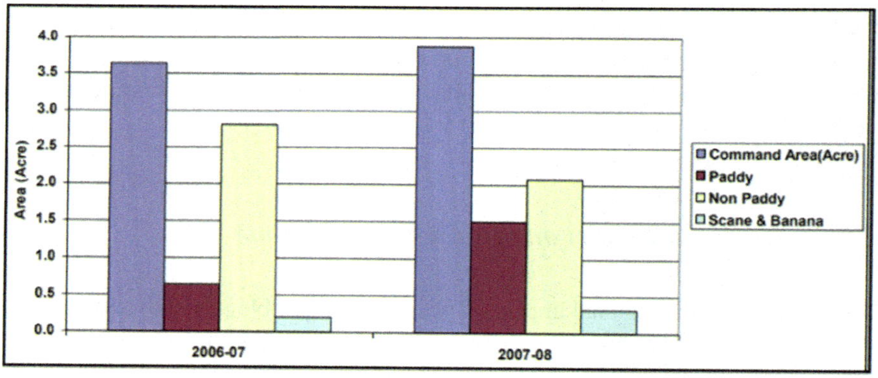

Fig. 3.4 Regional impact in terms of average area under irrigation (Acre)

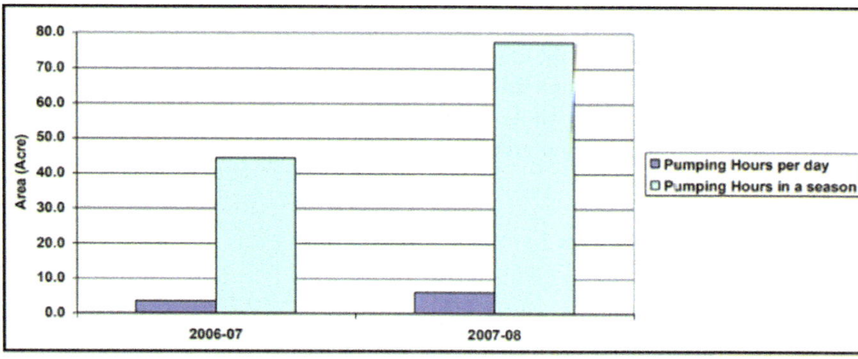

Fig. 3.5 Regional impact in terms of average pumping hours

3.6 Future Road Map

Hitherto, the Master Plan for rainwater harvesting and artificial recharge in the country prepared by CGWB has been a macro conceptual plan, providing a broad framework of feasible types of artificial recharge structures based on the terrain and hydrogeologic conditions and the numbers of such structures are limited by the availability of source water. A policy decision has been made that the next revision of the master plan will be made by respective State Agencies on the scale of 1:10,000 on GIS platform, providing the location of the existing structures and the proposed structures. The revised master plan will be an implementable plan on a larger scale catering to the siting of different structures on ground.

3.7 Conclusion

There has been a spurt in the growth of the tube wells/borewells in the country during the past 5 decades and presently there are nearly 21.7 million wells through which groundwater extraction is taking place for irrigation. Unplanned and unsustainable extraction in some parts of the country has led to a severe decline in water levels in some parts of the country. In this backdrop, rainwater harvesting and artificial recharge have assumed significant importance in the country and there has been a concerted effort on the part of the Central and State Governments in making water conservation and recharge a successful movement through people's participation. In India, the various subjects of governance are vested with either the Union or the State Government as per the constitution of India. Water is primarily a state subject and only in case of disputes between State Governments, the Union Government intervenes to resolve the issues. However, initiatives on rain water harvesting and artificial recharge (RWH & AR) are being taken up by both Central and State Governments through different schemes. The experiences from the implementation of various schemes has been documented as manuals, brochures, guides etc. and have been hosted in public domain. Considering the diversity in the aquifers, source water availability and importance of groundwater in the development of the country, a master plan for artificial recharge for the whole country is revised every five years by State and central agencies for making the groundwater system sustainable in India.

References

American Society of Civil Engineers (ASCE) (2001) Standard guidelines for artificial recharge of ground water, EWRI/ASCE 34–01 (ASCE Standard No. 34–01)

Central Ground Water Board (CGWB) (2007) Manual on artificial recharge of groundwater. Department of Water Resources, RD & GR, Ministry of Jalshakti, Government of India, p 198. http://cgwb.gov.in/documents/Manual-Artificial-Recharge.pdf

Central Ground Water Board (CGWB) (2011) Select case studies—rainwater harvesting and artificial recharge. Department of water Resources, RD & GR, Ministry of Jalshakti, Government of India, p 200

Central Ground Water Board (CGWB) (2020a) Master Plan for Artificial Recharge to Ground Water in India-2020. Department of water Resources, RD & GR, Ministry of Jalshakti, Government of India. 207p. http://cgwb.gov.in/Whatisnew/2021-06-30-Final-Approved%20Master%20Plan%202020-00002.pdf

Central Ground Water Board (CGWB) (2020b) National Compilation on Dynamic Ground Water Resources of India. www.cgwb.gov.in

Central Ground Water Board (CGWB) (2021) Guidelines for impact assessment of artificial recharge structures/schemes. department of water resources, RD & GR, Ministry of Jalshakti, Government of India, p 11. http://cgwb.gov.in/AR/Final-Guideline-IA-AR-Studies-submitted-website-hosting.pdf and http://cgwb.gov.in/Ar-reports.html

Suresh S, Subburaj A (2008) Impact assessment of demonstrative project on artificial recharge to ground water in Gangavalli Block, Salem District, Tamil Nadu. A Report of CGWB, Department of Water Resources, RD & GR, Ministry of Jalshakti, Government of India

Chapter 4
Groundwater Recharge, Rainwater Harvesting and Regulations for Sustainable Water Resources Development in Nigeria

Martin Obada Eduvie and Idris Musilim

Abstract The average precipitation in Nigeria is 1,150 mm with a total internally generated runoff of 287 billion cubic meters (BCM)/year and a surface water resources potential of about 333 BCM/year. The internally generated water resources potential is estimated at 287 BCM/year, while the total water resources potential with inflow from neighboring countries is estimated at 375 BCM/year. About 88 BCM/year of water enters from neighboring countries, which roughly indicates that almost 24% of surface water resources in Nigeria rely on inputs from neighboring countries and the rest is from inside Nigeria. The total water demand in Nigeria for the year 2010 was estimated at 5,933 million cubic meters per day (MCM/day) which is 2,165 BCM/year, while by the year 2030, it is estimated to rise to 16,585 MCM/day, hence the need for alternative water supply from rainwater harvesting and groundwater development. In addition to problems of water quantity, there are also problems of water quality. Pollution of water sources is posing a major problem for water users as well as impeding the maintenance of natural ecosystems. However, the management and development of the water sector face many challenges, including pressures on water resources and the increasing uncertainty being experienced with climate change. It is therefore recommended that the National Water Resources Policy and regulations should be aimed at providing a framework for addressing all challenges through the application of effective Integrated Water Resources Management (IWRM) and putting in place all that is necessary for effective and workable best international practices in the water sector. Finally, to meet the objectives of the policy and regulations, IWRM will be fully entrenched in all eight Hydrological Areas in the Country and should be operating successfully by the year 2030. Special variability of rainfall and the uncertainty imposed on rainfall by climate change is becoming an issue. To address that a plan on rainwater harvesting and recharging the

M. O. Eduvie (✉) · I. Musilim
National Water Resources Institute, Kaduna, Nigeria
e-mail: martinduvie@gmail.com

D. Saha et al. (eds.), *Managed Groundwater Recharge and Rainwater Harvesting*, Water Resources Development and Management
https://doi.org/10.1007/978-981-99-8757-3_3

aquifers artificially is to be framed for the entire Country. With increasing groundwater extraction, the space for recharge would also be available more as water levels are lowered, creating scope for recharging more water beyond the natural recharge.

Keywords Nigeria · Groundwater · Recharge · Harvesting · Regulations

4.1 Introduction

The Federal Republic of Nigeria has a total surface area of 923,768 km^2 and comprises 36 states and its Federal Capital Territory is Abuja. It shares land borders with Niger Republic to the north, the Republic of Benin to the west and Chad and Cameroon to the east, while its coast to the south lies on the Gulf of Guinea in the Atlantic Ocean. By the reckoning of the United Nations, Nigeria currently has a population of 225,597,156 per capita (2021) and it is estimated that by the year 2045, thirty years from now, the population would have risen to 515,112,413 which is more than twice the present population. As of 2014, the economy was worth more than $500 billion which made it one of the largest economies in Africa.

4.2 Overview of Nigeria Water Resources

Water is one of the most important natural resources and Nigeria is considered to be abundantly blessed with this resource. Water is a basic natural resource for socio-economic development. It is fundamental for various social and economic development activities such as drinking/sanitation, industrial production, irrigated agriculture, livestock farming, mineral processing, hydropower production, navigation, recreation and tourism among others. According to the Constitution of the Country, every Nigerian has the right to adequate water supply and sanitation, nutrition, clothing, shelter, basic education and health care, as well as physical security and the means of making a living.

However, there is a temporal and spatial variation in water availability as rainfall in Nigeria is diurnal. The northern parts have lower precipitation with only about 500 mm in the north-east while the south experiences higher precipitation levels with over 4,000 mm in the southeast. This high variability of rainfall in time and space is a significant characteristic of the tropical climatic belt, especially in the Sahelian part of the Country. The challenges of high variability in precipitation have manifested in the form of persistent drought in the northern parts of the Country in the past four decades with its attendant impact on the reduction in the extent of wetlands and the approximate complete loss of Lake Chad. In recent years, extreme weather conditions have led to serious flooding in many parts of the Country. On the other hand, the southern coastal areas have problems of saltwater intrusion and flooding as well.

Nigeria has two major river systems: The Niger entering the Country from the northwest and the Benue entering from the Northeast which together with their many tributaries drain half the land area of the Country. The northeast of the Country is mainly drained by rivers discharging into Lake Chad. It is worthy of note that many other rivers in the north are seasonal, having water in them only in the rainy season.

4.2.1 Geology, Hydrology and Hydrogeology

Geographically, in the far south, there are low-lying swamp forests, followed in a northerly direction by generally flat dense rain forests, followed by hilly shrublands in the middle belt, then by relatively flat savannah grasslands and finally by semi-arid areas in the far north. The central part of the Country is marked by crystalline rock outcroppings and gently rolling hills. About 60% of the Country is underlain by crystalline rocks, 20% by consolidated sedimentary materials and 20% by unconsolidated sedimentary materials (Fig. 4.1). Groundwater levels range between zero in parts of the coastal alluvium and 200 m amsl (above mean sea level) in some sedimentary areas. In the crystalline rock areas found in many parts of the north, well yields are unpredictable; where the sufficient depth of weathering exists the area may be suitable for the water supply schemes using shallow hand pump or borehole water (minimum yield of 10 L/min), but only at specific localities where deep weathering and underlying fractures coincide, there yields are likely to be sufficient for motorized boreholes.

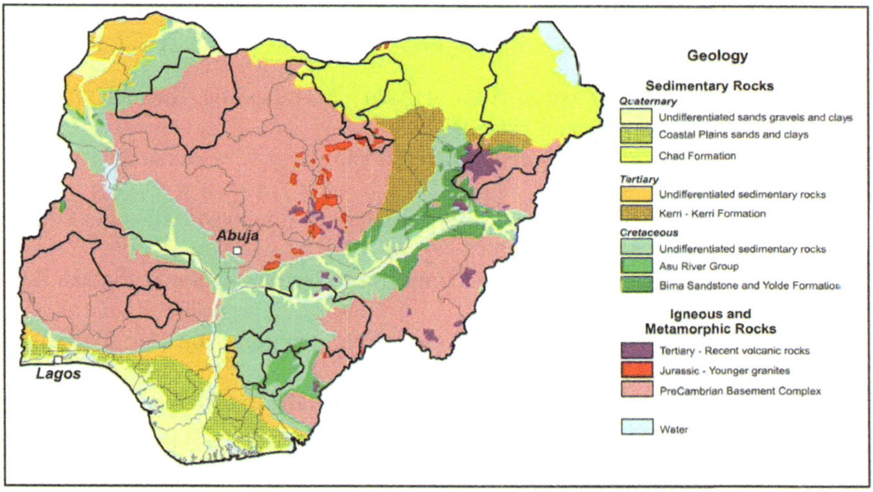

Fig. 4.1 Simplified geological map of Nigeria

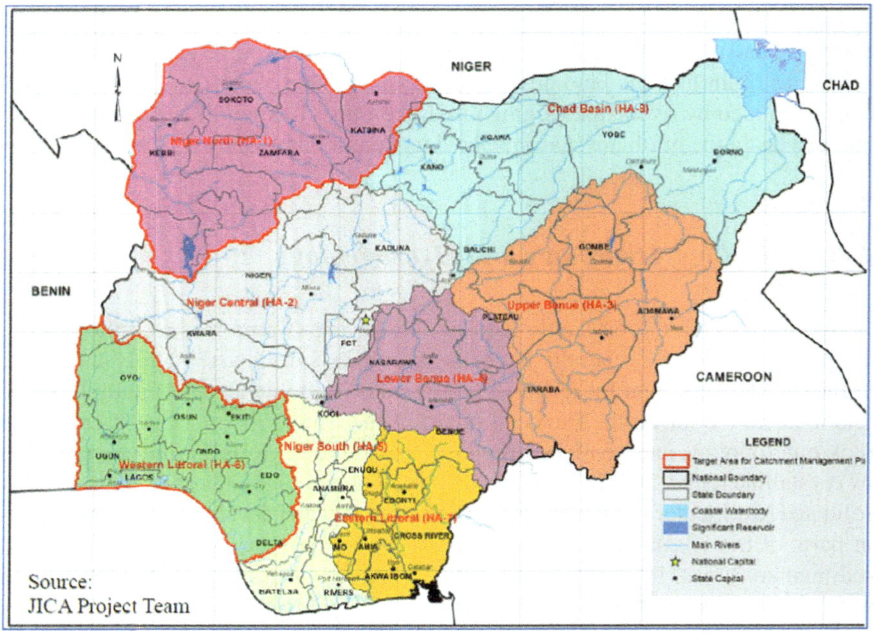

Fig. 4.2 Map of Nigeria showing the eight contiguous catchments designated into hydrological areas

For the purpose of scientific assessment and management of water resources and also as the building blocks of hydrological evaluations, Nigeria is divided into 8 contiguous catchments designated as Hydrological Areas (Fig. 4.2) as follows:

i. Niger North Hydrological Area (HA-1) encompassing Katsina, Zamfara, Sokoto, Kebbi, Niger and Kano states.
ii. Niger Central (HA-2) directly encompassing Niger, Kaduna, Plateau Kogi states and FCT.
iii. Upper Benue (HA-3) directly encompassing Adamawa, Bauchi, Gombe, Plateau, Taraba and Yobe states.
iv. Lower Benue (HA-4) directly encompassing Nasarawa, Benue, Plateau, Kogi and Rivers states.
v. Niger South (HA-5) directly encompassing Anambra, Bayelsa, Delta, Imo, Kogi and Rivers
vi. Western Littoral (HA-6) directly encompassing Edo, Ekiti, Lagos, Ogun, Ondo, Oyo and Osun States.
vii. Eastern Littoral (HA-7) directly encompassing Abia, Akwa-Ibom, Cross River, Ebonyi, Enugu and Imo.
viii. Chad Hydrological Area (HA-8) directly encompassing Bauchi, Gombe, Jigawa, Kano, Yobe and Borno

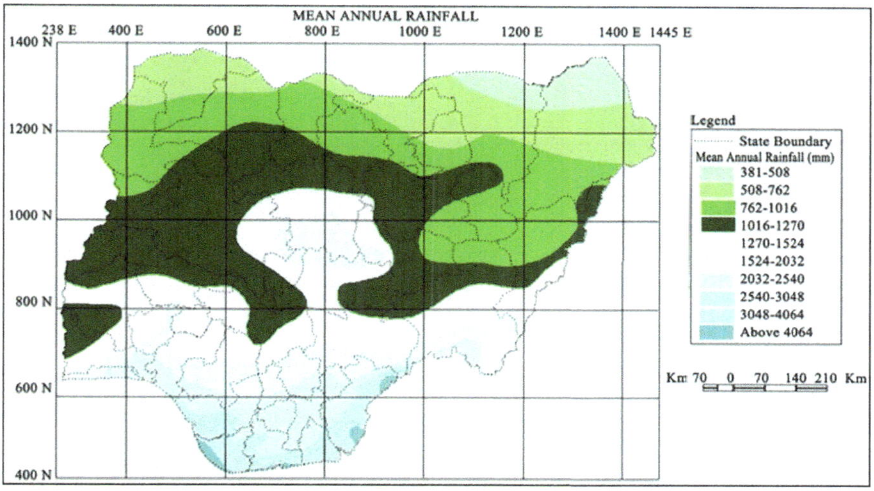

Fig. 4.3 Map of rainfall pattern in Nigeria

4.2.2 Outline of Meteorology in Nigeria

The distribution of mean annual precipitation in Nigeria is presented in Fig. 4.3. Rainwater harvesting is well suited in the central and southern parts of Nigeria where average to adequate rainfall is experienced. As per the Water Resources Master-Plan of 2013 by JICA (NWRMP 2013), the average precipitation in Nigeria is 1,150 mm with a total internally generated runoff of 287 billion cubic meters (BCM)/ year, while the total water resources potential including inflow from neighboring countries is estimated at 375 BCM/year. While 88 BCM/year of water comes from the neighboring countries, which roughly indicates that almost 24% of surface water resources in Nigeria rely on inputs from neighboring countries. The total groundwater resource potential is estimated at 156 BCM/year as a renewable source on the basis of the estimated groundwater recharge (NWRMP 2013).

4.2.3 Impact of Climate Change

In Nigeria, with more than 450 constructed dams of different capacities, there are impacts of recharges from these dams into the aquifers systems. In Goronyo Dam, for example, the area under its catchment is characterized by inconsistencies and inter-annual variability due to climate change. A conceptual diagram of the impact of climate change on groundwater is presented in Fig. 4.4. The variability of rainfall pattern is also shown in Fig. 4.5 representing major towns within the northern parts of Nigeria.

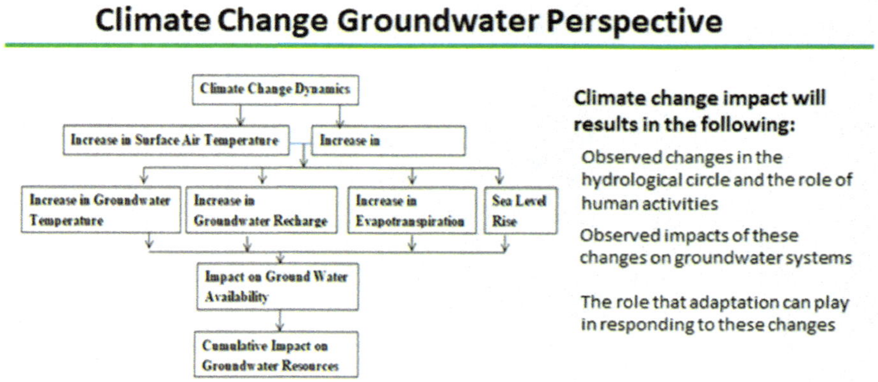

Fig. 4.4 Conceptual diagram on the impact of climate change on groundwater. Modified from Oseke et al. (2017)

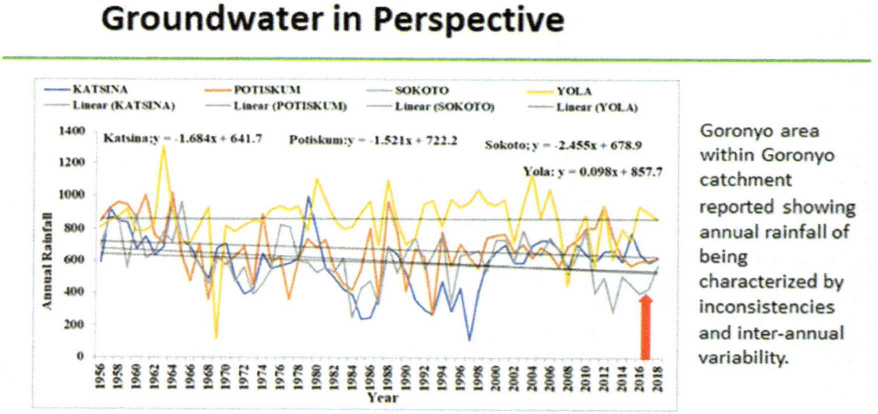

Fig. 4.5 Trend in total annual rainfall at goronyo catchment

4.3 Rainwater Harvesting in Nigeria

Collection of runoff and impounding them on the surface is called rainwater harvesting (RWH). Rainwater harvesting is a simple low-cost technique and simple to design and construct.

A typical rainwater harvesting system consists of three basic components, which include:

1. *Catchment/roof surface*: The catchment of a water harvesting system that receives rainfall directly and drains the water to the system.
2. *Delivery system*: The delivery system from the rooftop catchment consists of gutters hanging from the side of the roof sloping toward a downpipe and into the

tank. The delivery system is used to convey the rainwater from the roof to the storage tank.
3. *Storage Tank/Reservoir*: This is where the water is stored for use. It could be a simple plastic tank, metal construction or concrete tank.

Generally, rainwater harvesting has the following advantages:

I. Simple to construct, and local people can easily be trained, encourage more participation, ownership and sustainability at the community level.
II. The operation and maintenance of a household catchment system can easily be operated and maintained.
III. Water quality is generally very good and reliable.
IV. Rainwater is a renewable resource with no damage done to the environment.
V. Convenience at the point of consumption.
VI. There is no geological or topographic effect.
VII. The flexibility and adaptability of the system is quite good.

While rainwater has numerous advantages, the following are some of the drawbacks:

a. The water quality is sometimes vulnerable through air pollution, animals and bird's droppings, insects' dirt and organic matter.
b. The supply is limited in terms of the amount of rainfall, size of catchment area and storage reservoir.
c. Climate change could affect the rainfall and consequently the supply of the rainwater harvesting and
d. The application could be limited due to roofing systems and difficulty in the storage tank cleaning.

Findings on the use of rainwater in a Usugbenu community, located in Esan Central Local Government area, Edo state, in the South–south geopolitical zone of Nigeria, showed that RWH is practiced by over 80% of households, with the rooftop as the catchment area. Stored water was most commonly used for personal hygiene purposes. The majority of the 30 water samples tested had unacceptable levels (according to Nigerian potable water standards) of total coliform, while one sample had *Escherichia coli*. Therefore, health education should focus on enlightening households on the appropriate design and maintenance of RWH systems (Tobin et al. 2013).

4.4 Groundwater Recharge

Groundwater forms the basis of water supplies across much of Africa and its development is rising as demand for secure water increases. Recharge rates are a key component for assessing groundwater development potential but have not been adequately mapped across Africa, other than from global models (MacDonald et al. 2021). The quantification of long-term average (LTA) distributed groundwater recharge rates across Africa for the period 1970 to 2019 from 134 ground-based estimates have been carried out and reported and upscaled statistically (MacDonald et al. 2021; Allan et al. 2010). Natural diffuse and locally focused recharge, where this mechanism is widespread, are included but discrete leakage from large rivers, lakes, or irrigation are excluded. It was reported that measurable LTA recharge is found in most environments with average decadal recharge depths of 60 mm in arid and semi-arid areas of 200 mm.

In Africa, where considerable groundwater resources exist (MacDonald et al. 2012), groundwater south of the Sahara has not yet been widely exploited other than for low-yielding drinking-water supplies (WHO and UNICEF 2015). Only 1% of cultivated land in Africa is estimated to be irrigated by groundwater with 80% of this occurring in North Africa (Siebert et al. 2010).

Estimating groundwater recharge is challenging with no widely applicable method available that can directly and accurately quantify the volume of rainwater that reaches the water-table (Scanlon et al. 2002; Healy 2010). Instead, several different methods are relied upon to estimate recharge. The most common methods are chloride mass balance; soil physics methods; environmental and isotopic tracers; groundwater-level fluctuation methods; water balance methods (including groundwater models); and the estimation of base flow to rivers.

Recharge involves the unsaturated movement of water in the vertical direction, that is, the movement is in the direction in which the hydraulic conductivity is generally lowest (Heath 1982; Edmunds et al. 2002). Recharge occurs during and immediately following periods of precipitation and, thus, is intermittent.

In a study carried out in kaduna state, two methods were used in the assessment of recharge to the aquifers. The first method involved the analysis of borehole hydrographs as shown in Fig. 4.6 and the specific yield or storativity of the aquifers at Kwoi and Kaduna areas (Eduvie 1991). In addition, Houston (1988) remarked that the borehole hydrograph is the most direct way of quantifying recharge. The recharge is obtained from the following equation,

$$\text{Recharge (mm)} = \text{Hydrograph rise} \times \text{Specific yield} \tag{4.1}$$

The second method entails the use of the chloride mass balance method (Houston 1988). This technique relies on the fact that, except in areas of evaporite deposits, chloride is a conservative and generally non-reactive anion in natural water. Therefore, chloride in rainfall (derived from ocean water evaporation) usually passes through the soil zone to the water table without change. The ratio of chloride concentration in rainfall to that in groundwater is then a measure of recharge; through the following relationship,

$$\text{Recharge (mm)} = \text{rainfall (mm)} \times \frac{\text{MgCl}^{-1}\ \text{rainfall}}{\text{MgCl}^{-1}\ \text{groundwater}} \tag{4.2}$$

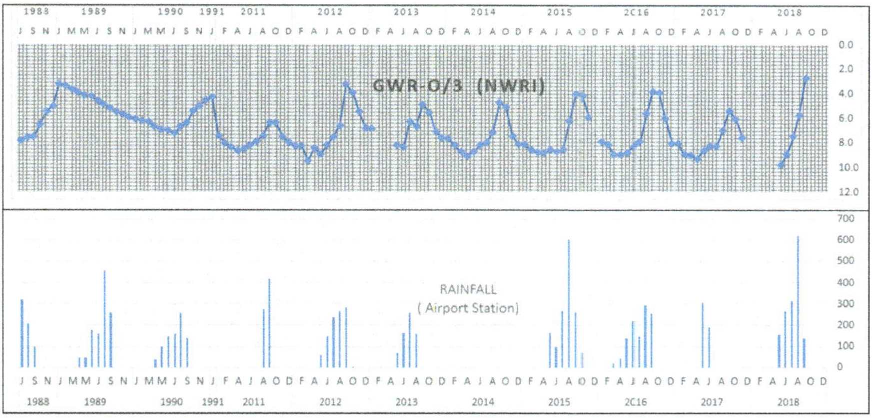

Fig. 4.6 Borehole hydrograph in the study area located at National Water Resources Institute (NWRI), Kaduna. GWR-03 is a monitoring borehole used for the construction of the borehole hydrograph

where

MgCl rainfall Concentration of chloride in rainwater.
MgCl groundwater Concentration of chloride in groundwater.

Table 4.1 is a summary of the recharge estimates according to the two methods explained above. The result of these estimates shows clearly that recharge to aquifers varies with rainfall, soil type and vegetation. The values obtained by the chloride mass balance method are generally higher compared to the well hydrograph. This could be attributed to temporal variation of rainfall, coefficient of storativity and other related parameters including soil type, vegetation and chemical variation of chloride concentration in groundwater. However, these methods could act as a guide for the estimation of recharge, the uncertainty involved in the measurement of the various parameters, notwithstanding. Therefore, there is a need to use more than one method for recharge estimation and sufficient confidence could be placed on the results for the planning purposes of water resources development.

Table 4.1 Groundwater recharge estimates in (mm) in the study area

Location	Specific Yield m^3/h	Well Hydrograph Mm	Cl mass method Mm/a	Annual Rainfall Mm/a
Kaduna	0.0015	9.0	20.1	1260
Kwoi	0.001	6.0	48.6	1685
Jura	–	–	41.10	1264
Kafanchan	0.001	14.13	22.0	1755

Data source Eduvie (2011)

Table 4.2 Comparative Water Demand Estimates by Sector for Years 2010 and 2030 (JICA 2013)

Estimated national consumption	Year 2010 (MCM/Day	Year 2030 (MCM/Day)
Water supply	3,047	8,852
Irrigation	1,926	6,245
Aqua culture	728	1,166
Livestock	233	312
Total	5,933	16,585

4.5 Policies and Regulations Review and Its Implications

These issues are discussed below.

4.5.1 Water Resources Potential of the Country

National Water Demand Projection

Water supply and sanitation, irrigation water, freshwater aquaculture, livestock water demand and mining are generally the major consumptive aspects of water resources. The other relatively non-consumptive socio-economic activities are hydropower generation, flood management and inland navigation. The total water demand in Nigeria for the year 2010 was 5,933 MCM/day, while by the year 2030, it is estimated to rise to 16,585 MCM/day as detailed in Table 4.2.

The water use rate (the ratio between the total water demand and the surface water resources potential) in 2010 was just 1.6% and is expected to grow to 4.4% by the year 2030. Thus, the total water demand is much less than the total water resource potential. However, because the water demand and water resources are unevenly distributed, the necessity of surface water and groundwater resource development should be examined through the water balance between supply and demand at local levels.

4.5.2 Surface Water Quality

Surface water bodies in and around large urban areas in Nigeria are frequently of poor quality due to pollution. In fact, in areas with high population density, there are significant levels of fecal bacteria pollution that have migrated from septic tank clusters into the groundwater. The high level of water pollution in Nigeria is mostly caused by:

i. Indiscriminate disposal of both solid and liquid household waste in storm drains that eventually discharge into these streams.

ii. Central sewerage systems with associated sewage treatment sites are rare in Nigeria resulting in the general use of septic tanks.
iii. Lack of proper sewerage dumping sites resulting in emptying septic tanks into water bodies (lagoons in Lagos) and other sites in waterways around other cities.
iv. Unregulated cottage industries and automotive repair shops resulted in industries sited around and close to urban centers discharging oils and chemicals into water bodies.
v. Poor mining control results in oil spillage and heavy ion poisoning like lead etc.
vi. Open air defecation is common in many places, especially in rural areas.

4.5.3 Groundwater Quality

On average, groundwater quality in most parts of the Country is good. Only in some areas is iron, nitrate or fluoride with concentrations above the recommended Nigerian Standard of Drinking Water Quality. There are areas where the water quality is seen to be moderately to highly corrosive. Most common sources of water for the urban are poor. The groundwater table in some areas where hand-dug wells are sited is most times as high as the bottom of the septic tank system. In addition, some of these wells are down gradient of large clusters of septic tanks in high-density residential areas.

In rural areas, the groundwater is partly polluted due to open air defecation too, but this affects surface water more seriously.

4.5.4 Operation and Maintenance

JICA 2013 Water Resources Master Plan reported that the government has become increasingly concerned by the poor level of operation and maintenance in most water infrastructures. The National Council on Water Resources during its 16th Meeting in Asaba in 2002 set up a National Committee to recommend ways and policy initiatives to address the problem to make the water resources infrastructures more sustainable. It is expedient to ensure strategic projects are embarked upon in the Water Sector. For instance, huge sums expended on the construction of Dams and Reservoirs could easily be lost, if there is a natural disaster that causes severe damage to the Dams. Also, the collateral damage that may result from this natural crisis such as loss of lives due to flooding caused by breakdown of the spillway, etc., could result in serious legal suits against the Government.

4.5.5 *Monitoring and Evaluation*

The system of Monitoring and Evaluation of water projects and infrastructure which hither to is a challenge. There are continuous efforts to enhance this system for effective project delivery and sustainability.

4.5.6 *Legal Framework*

Presently, only the following statutes are relevant in the development and management of the nation's water resources:

- Water Resources Act: CAP W2. LFN. 2004.
- Nigerian Minerals and Mining Act Cap N162, LFN, 2004,
- National Inland Waterways Act No. 47 LFN 2004.
- River Basins Development Authority Act. R9, LFN 2004.
- Nigeria Hydrological Services Agency (Establishment) Act, Cap N110A, LFN 2004,
- The National Water Resources Institute Act, Cap N83 LFN 2004,
- The Land Use Act, Cap L5 LFN 2004,
- Nigeria Environmental Standards and Regulations Enforcement Agency Act, Cap N164, LFN 2004.

4.5.7 *Funding and Financing*

The water resources sector has been underfunded with almost all financial resources coming from the Government and to some extent International Institutions and Donors. Private sector financing is limited.

4.6 Water Resource Challenges in Nigeria

Current Challenges in Water Resource Management in Nigeria

There are great differences in water availability from region to region in Nigeria, from the extremes of deserts in the far north to tropical forests and mangrove swamps in the extreme south. In addition, there is variability of supply through time as a result of both seasonal variation and inter-annual variation.

In addition to problems of water quantity, there are also problems of water quality. Pollution of water sources is posing major problems for water users as well as impeding the maintenance of natural ecosystems. Nigeria has to some extent artificially overcome natural variability by supply-side infrastructure to assure reliable supply and reduce risks, albeit at very high cost and often with negative impacts on the environment and sometimes on human health and livelihoods. The Country is now finding that "supply-side" solutions alone are inadequate to address the ever-increasing demands from demographic, economic and climatic pressures as a result, waste-water treatment and water recycling and demands management measures are being introduced to counter the challenges of inadequate supply. The traditional fragmented and top-down approach to water resources management in Nigeria is no longer viable and a more holistic approach is recommended as reported by JICA (2013). Master plan study. To counter the special variability of rainfall and the uncertainty imposed by climate change rainwater harvesting should give priority. A detailed plan is to be made for the entire Country. As the groundwater extraction would increase, the space would be created by lowering of water level, where recharge from the water harvesting structures or other artificial recharge techniques should be adopted.

During the last few decades, the desire to improve access to water resources in Nigeria has become more and more elusive due to various challenges. The challenges facing the water resources sector in Nigeria include the following:

i. Unevenly distributed water resources and demand necessitating the creation of dams and transportation of water to the areas of need.
ii. Inadequate access to usable water resources to meet the rapidly increasing domestic and industrial (economic growth) water demand. These are manifested by poor access to clean and potable water in urban, small towns and rural areas; low levels of irrigation agriculture, poor utilization of hydropower potentials and limited inland fishery.
iii. Degrading watershed and water courses as a result of widespread pollution, including the indiscriminate disposal of hazardous wastes due to poor pollution and mining control that leads to the deterioration of water quality.
iv. Fragmented and uncoordinated water resources development because of inadequate catchment management.
v. Unclear roles and responsibilities among the various levels of government, different ministries, departments and agencies at the Federal and State levels.
vi. Poor coordination (Including International Donor Coordination), mobilization and application of funds for water supply development. This often leads to duplication of efforts, wastages and inefficiency in the development and management of water infrastructure throughout the Country.
vii. Inadequate water resources data collection and management. This leads to poor planning and project designs.
viii. Limited groundwater availability in the areas underlain by crystalline rocks which cover substantial parts of the Country, while for the more productive sedimentary areas of the Country, detailed study and documentation is still premature.
ix. Poor or lack of monitoring and control of groundwater resources.

x. Escalating costs of water production and distribution for domestic and industrial water supply, irrigation, husbandry, horticulture and other uses against dwindling financial resources.
xi. Inefficient government subsidies on the provision of water services.
xii. Extreme weather conditions due to climate change result in prolonged droughts, increased flooding, widespread erosion and communal conflicts.
xiii. Vicious cycle of unreliable projects that provide services that do not meet consumer needs and for which the consumers are unwilling to pay.
xiv. Poor or inefficient management of water resources infrastructures like dams, reservoirs, waterworks with their related distribution networks, irrigation structures and navigable waterways leads to financial losses and unreliable service delivery.

4.7 Relationship of Water and Economic Development in Nigeria

Water has a significant impact on the growth and development of the Nigerian Economy. Apart from the commonly known importance of water for domestic use, human and health, there is a direct relationship between the availability of water and economic activities. There is a clear impact of water on different sectors of the Nigerian economy that include: agriculture, health, power, mining, tourism, trade, industry, education and transportation majorly.

4.8 Conclusion and Recommendations

Water is the most important natural resource in Nigeria. The relevance of water to national development has progressively increased over the last years with rapid population growth, urbanization, modernization of agriculture and industrial development. Because of its usefulness in different capacities—for direct human consumption, agricultural irrigation, fisheries, hydropower, industrial production, recreation, environmental protection and industrial effluents absorption, it is considered Nigeria's most unique and indispensable natural resource. However, the management and development of the water sector face many challenges, including pressures on water resources and from the increasing uncertainty being experienced with climate change.

It is therefore recommended that the National Water Resources Policy and regulations should be aimed at providing a framework for addressing all challenges through the application of effective Integrated Water Resources Management (IWRM) and putting in place all that is necessary for effective and workable best international practices in the water sector and necessary regulatory laws should be put in place and enforced accordingly. Harvesting rainwater and artificial recharge of the aquifers, particularly where under depletion, should be taken up in suitable areas.

Finally, to meet the objectives of the policy and regulations, IWRM will be fully entrenched in all eight Hydrological Areas in the Country and should be operating successfully by the year 2030. Since each Hydrological Area has its own peculiarities regarding water allocation priority, socio-culture of the communities, etc., an initial implementation strategy should be developed for each Hydrological Area as soon as possible which can then be refined for the various catchment levels. However, rainwater harvesting and artificial recharge of groundwater is practiced in Nigeria and no precise info is available to assess this component in Nigeria.

References

Allan RP (2010) Current changes in tropical precipitation. Environ Res Lett 5(2):025205. https://doi.org/10.1088/1748-9326/5/2/025205

Edmunds W, Fellman E, Goni IB, Prudhomme C (2002) Spatial and temporal distribution of groundwater recharge in northern Nigeria. Hydrogeol J 10:205–215

Eduvie OM (2011) Groundwater exploration and development in basement complex area: examples from southern part of Kaduna State, Nigeria. LAP Lambert Academic Publishing. Schaltungsdienst Lange O.H.G, Berlin. ISBN: 978-8433-9172-6

Goni IB, Taylor RG, Favreau G, Shamsudduha M, Nazoumou Y, NgounouNgatcha B (2021) Groundwater recharge from heavy rainfall in the southwestern Lake Chad Basin: evidence from isotopic observations. Hydrol Sci J. https://doi.org/10.1080/02626667.2021.1937630

MacDonald AM, Bonsor HC, Dochartaigh BÉÓ, Taylor RG (2012) Quantitative maps of groundwater resources in Africa. Environ Res Lett 7:024009

MacDonald AM, Lark RM, Taylor RG, Abiye T Falbs HC, Favteau D, Goni IB, Kebede S, Scanlon B, Sorenson JPR, Tijani M, Upton K; West C (2020) Mapping groundwater recharge in Africa from ground observations and implications for water security

NWRI/RWSSC (2011) training manual on alternative to water supply. Handdug well, Spring development and Rainwater harvesting

NWRMP (2013) National Water Resources Master Plan, vol 1–9. Report of Federal Ministry of Water Resources and JICA

Tobin EA, Ediagbonya TF, Ehidiamen G, Asogun DA (2013) Assessment of rain water harvesting systems in a rural community of Edo State, Nigeria. J. Public Health Epidemiol **5**(12):479–487

WHO and UNICEF (2015) Progress on sanitation and drinking water: 2015 update and MDG assessment. UNICEF, New York, p 80

Chapter 5
Farm Ponds in Semi-arid Hard Rock Terrain of India. Are They Increasing Dependency on Groundwater?

Ankita Yadav, Taufique Warsi, Eshwer Kale, Sarita Chemburkar, Marcella D'Souza, and Dipankar Saha

Abstract Farm ponds are traditional water harvesting structures to capture surface runoff and utilize this stored rainwater during the lean periods for agriculture. Benefits of the farm ponds include improvement in land productivity, supplemental irrigation to crops, reduced water logging in high rainfall events, and fish culture when sufficient water is available. Farm ponds are becoming increasingly popular in India as water storage tanks or reservoirs, designed for harvesting rainwater on agricultural land. The Maharashtra State Government as well as Central Government have been promoting Farm Pond through various schemes with the provision of subsidies. The farm ponds in Maharashtra are becoming popular after the government subsidy was made available. The present study dwells into the rapid rise in farm pond numbers and how they are impacting the precious groundwater resources in the drought-prone Ahmednagar district. of Maharashtra. The density of farm ponds in Sangamner Taluka of the district is very high, and so is groundwater extraction. Study indicates that multiple sources of water are used to fill the farm ponds, with borewells and dug wells accounting for the bulk of the supply. In order to decipher the vagaries in the Electrical Conductivity (EC) trend and eventually demarcating the zone of influence, the physio-chemical parameters were assessed to speculate on solid dilution from pumping well to surrounding non-pumping wells. Moreover, the artificial recharge zone maps prepared by Groundwater Survey and Development Agency, Govt. of Maharashtra, were utilized to understand how the farm ponds, often lined are faultily located in recharge zones, blocking the recharge. The practice of using farm ponds as storage tanks for pumped-out groundwater has imposed a negative impact on the surrounding groundwater wells that leads to inequality among the different categories of water users in villages. It is important to relook into the

---

A. Yadav · T. Warsi · E. Kale · S. Chemburkar · M. D'Souza (✉)
W-CReS (Watershed Organisation Trust (WOTR) - Centre for Resilience Studies), Pune, India
e-mail: director.wcres@wotr.org

D. Saha
MRIIRS, Faridabad, India

D. Saha et al. (eds.), *Managed Groundwater Recharge and Rainwater Harvesting*, Water Resources Development and Management,
https://doi.org/10.1007/978-981-99-8757-3_4

policy and practice of farm ponds for sustainable use of groundwater resources in this drought-prone stretch of Maharashtra.

Keywords Farm ponds · Maharashtra · Groundwater recharge · Water policy · Water equity

5.1 Introduction

As history demonstrates, civilizations flourished and declined in the Fertile Crescent by harnessing water for agricultural production, which provided economic and social stability and military power. Dry spells of long duration generally led to changes in irrigation with its negative impacts on water exploitation. In arid and semi-arid regions, climate plays a kingpin role in regulating agriculture practices and hence the economy. The vagaries of rainfall raises concern; therefore the need for water storage structure is felt by the agriculturalists. The attempt of storing surplus rainwater is an age-old practice to overcome the serious problem of water scarcity posed by nature.

Since 1970, the Government of India (GoI) has initiated different programs to assure water security in the drought-prone regions of India. Also, many programs including Drought Prone Area Program (DPAP) and the Desert Development Program (DDP) (Kerr 2002; Kumar et al. 2018) and other bilateral watershed development projects, along with many minor, medium, and major irrigation projects by Water Resource Departments, were undertaken to increase the water availability across drought-prone regions. However, these programs failed to address the water scarcity problem due to improper planning, incomplete implementation, and poor demand side management (MoRD 1994: 11). Later, in 2008 all watershed programs were brought under the umbrella of a common guideline known as Integrated Watershed Management Program (IWMP). The 'National Water Policy' that was launched in 1987 was subsequently revised in 2002 and 2012. Moreover, it is likely to be revised in the near future (Kale and Kulkarni 2022). Various water-related schemes on groundwater were launched by the State and Central Government which are being implemented in block/taluka levels. The important Central Govt Schemes are PMKSY, Atal Solar Schemes, MGNREGA whereas state-sponsored schemes are many like, PoCRA, CM-Solar scheme, Dr Babasaheb Ambedkar Krushi Yojna, Birsa Munda Yojna, and Farm Pond on demand (as translated in the local language "MagelTyalaShet Tale") are brought to cope with the current water crisis (Abhay 2016). Among all the schemes, Farm Pond on demand was depicted as a miracle strategy by the State and Central authority, which was later on popularized by the media (Kale 2017). As mentioned before, the farm ponds are expected to provide protective irrigation during lean periods, particularly in *Kharif* and *Rabi* seasons. This gives assurance to farmers to sustain their crops and have a secure income. With an increase in area under irrigation and crop yield, and support to allied livelihoods, farmers are expected to earn more as envisioned by the Central Government (doubling farmers' income). However, the system did not notice how the construction of the farm pond is being implemented on the ground. A farm pond needs to be a dug-out structure with a definite size and shape which collects rainfall water, or overland flow with one inlet and outlet to arrest the surface runoff from the farm area. The feasible sites are in a low-lying area. Farm ponds can be classified into four types depending on its location and water source: (1) Excavated or Dug out ponds, (2) Surface ponds, (3) Spring or creek-fed ponds, and (4) Off-stream storage ponds (Reddy et al. 2012).

About 70% of the geographical area of the central and southern parts of India is underlain by low-potential hard rock aquifers (Saha and Roy 2018; Saha et al. 2019). In many parts underlain by hard rock aquifers, farmers follow the practice of pumping groundwater to fill up the farm ponds which are kept lined with impervious plastic material; they then irrigate the farms. Thus farm ponds are converted into groundwater storage tanks (Bendapudi et al. 2020). The dimensions of the ponds have also started increasing depending on the farmer's affordability. This is leading to a possibility of precious groundwater losses due to surface evaporation. Besides pumping of massive amounts to fill up farm ponds causes reduced groundwater availability for the adjacent landowners. This issue is raising its head in the drought-prone areas of Maharashtra. The objective of the study is to understand the impact of farm ponds spatially and vertically on how the farm ponds are impacting the surrounding groundwater users. The groundwater vulnerability of identified groundwater users around the farm ponds is assessed through social surveys and deciphering the hydrological characteristics.

5.2 The Genesis of Farm Ponds in Maharashtra

In the last few decades, many parts of Maharashtra started facing recurrent droughts, erratic rainfall, and groundwater depletion (Udmale 2014). The Government of Maharashtra in 2010 issued two Government Resolutions (GRs) to promote the farm ponds in the state under the Maharashtra Rural Employment Guarantee Scheme and Horticulture Mission (Shah et al. 2021). Soon farmers could see the benefits of assured water stored in the pond for protecting their crops and growing cash crops, getting more income. Seeing the popularity of farm ponds, the state in 2016 came up with a special and popular program on farm ponds. *MagelTyalaShet-tale* (which translates as 'farm ponds on demand'). The scheme received huge momentum in the state by providing subsidies to farmers for constructing farm ponds. Apart from this scheme, farm ponds are also being promoted through various schemes, namely the Jalyukt Shivar and PoCRA projects, and through the support of many CSRs and NGOs. Construction of farm ponds is also promoted through Central Government programs like Pradhan Mantri Krishi Sinchai Yojana (PMKSY) and Rashtriya Krishi Vikas Yojana (RKVY). Each year the state government sets an ambitious target of building ponds. For example, for 2016–2017 state's target of farm ponds was 1,11,111 ponds and received around 3.5 lakh applications, and the target set for 2020–2021 is constructing 51,369 farm ponds. Even recently, in the state budget speech of 2022, the finance minister of the state announced an increase in the subsidy amount from earlier Rs. 50,000 to now Rs. 75,000 in Maharashtra (Kale and Kulkarni 2022).

5.3 Background and Context of Study

The groundwater stored in the farm pond is not able to percolate into the ground due to the lining and a significant part is lost to evaporation. Groundwater is extracted mercilessly, particularly in drought-prone hard rock areas by a few well-to-do farmers to sustain their cash crops and horticulture to fill up the farm ponds. But in the process other users tend to lose their share of the resource, hence the groundwater resource is unofficially getting privatized. A significant change in cropping pattern, specifically the shift to cash crops and horticulture from traditional low-water-intensive crops can be seen. The small and marginal farmers do not have the scope to change cropping patterns due to a reduction in availability and access to water resources that leads to an increase in inequity. The groundwater regime is adversely affected by water level decline generating hydrological disequilibrium and depriving small and marginal farmers with shallow wells. The efforts taken to establish the direct linkage between the delineation of hydrological zones of influence around farm ponds and socio-hydrological aspects of various users within the zone would help to identify the scale of groundwater depletion at the local level and the extent of vulnerability of the communities.

5.4 Study Area

The study villages are located in Sangamner block of Ahmednagar district which lies between longitudes 74° 02′ 30″ and 74° 16′ 30″ and between latitudes 19° 15′ 30″ and 19° 26′ 0″ (Fig. 5.1). Characteristically, the block is placed in 'rain shadow zone' with an annual average rainfall of 484. The climate of the region is characterized by hot summers and generally dry weather conditions throughout the year. These villages are selected based on their groundwater vulnerability that was assessed under hydrological investigations. Geologically, the area is mainly occupied with Deccan basalt flow with important water-bearing formation in terms of Zeolitic Vesicular—Amygdaloidal Basalt, Fractured and Jointed basalt. The major river basins are Godavari and Pravara sub-basins and the dominant soil type is black cotton soil.

The six villages taken under study include Gunjalwadi, Jawalebaleshwar, Bhojdari, Wankute, Dolasane, and Karjule Pathar (Fig. 5.2). Topographically, Gunjalwadi, Karjule Pathar, and Dolasane villages are located on plateau land whereas Jawalebaleshwar, Wankute, and Bhojdari are situated on hilly terrain. These villages have a total of 68 functional farm ponds that are converted into water storage tanks. There are three types of water sources to fill up the farm ponds: groundwater, surface water,

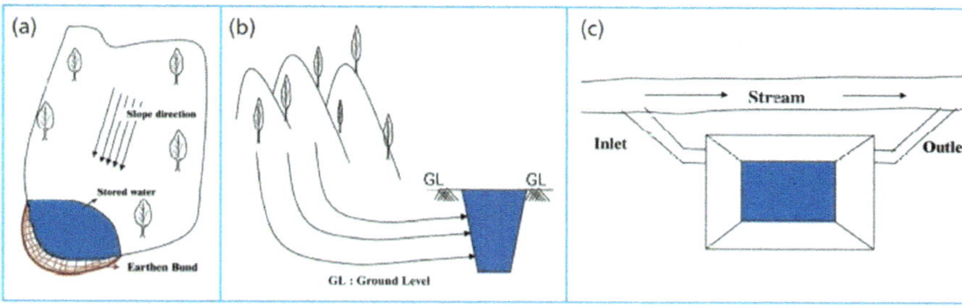

Fig. 5.1 Different types of farm ponds: **a** Surface ponds, **b** Spring or creek-fed ponds, **c** Off-stream storage ponds and their locations in the catchment. *Source* Reddy (2012)

and a combination of both. Farm ponds are mostly filled with either borewells or dug wells; however, some ponds are filled by multiple types of wells. In some cases, the farmers filled their farm ponds with nearby stream water. Increasing demands for irrigation water have triggered the construction of new borewells and the farm pond concept has accelerated the exploration of groundwater. Farm Ponds in study villages are filled with water from dug wells and boreholes, of which the numbers have recently increased. The Sangamner block falls in 'semi-critical condition' in terms of groundwater resources.

5.5 Methodology

The studies are taken up in two layers: (i) structured interviews of farm pond owners and (ii) technical and scientific investigation of sub-surface dynamics of groundwater.

5.5.1 Household Survey

A household-level survey was conducted in 2017 to identify the farm pond owners in six study villages to capture details of all farm ponds present that year (Bendapudi et al. 2020). There were 188 farm pond-owning households having 208 farm ponds (few farmers owned more than 1 farm pond). Further, the functionality of the farm ponds was studied. Functionality is defined by the users as ponds being lined with impervious plastic sheets and then filled with groundwater, surface water, or a combination of both. Finally, 61 households owning 68 farm ponds were identified as functional farm ponds from groundwater and were taken under the current study during 2019–2020.

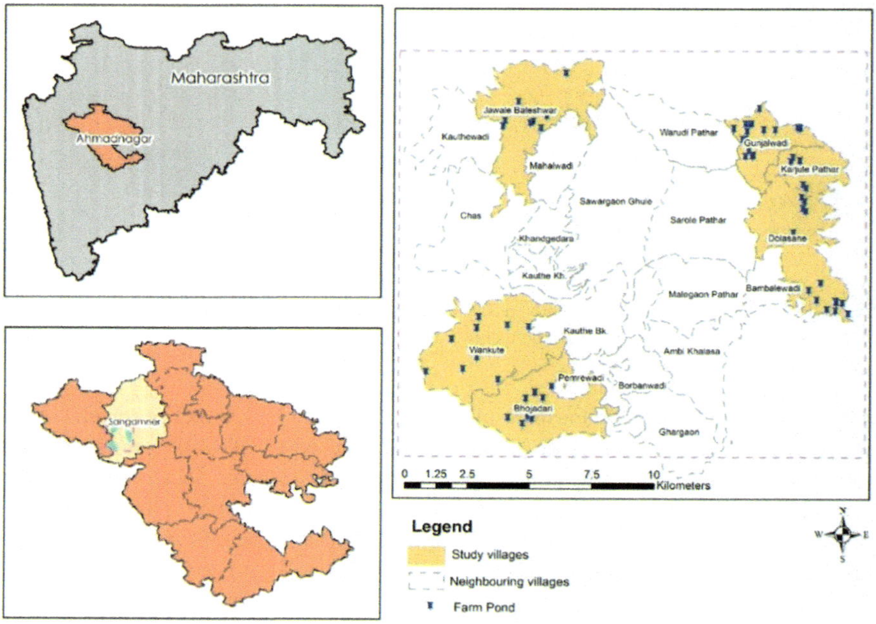

Fig. 5.2 Location map of study villages, 6 villages colored (right) are taken under study and the pinned locations represent the farm pond

5.5.2 In Situ Parameter Testing

An in situ parameter is a potential tool to understand the local hydrological boundaries on the basis of EC, Ph, TDS, and Salinity. In the study area, a total of 138 wells (borewells/dug wells) were selected downstream and around the farm pond feeding wells. The situation, however, was not ideal for each farm pond due to the limited availability of wells, the greater distance between observation and feeding wells, and the disruption of electricity for borewells. In order to determine the hydrological boundaries, the distance of downstream wells was measured from the feeding well. Subsequently, measured quality parameters were plotted against the borewell distance to understand the zone of influence with the help of a breakthrough curve.

5.6 Data Analysis and Results

5.6.1 *Farm Pond Sizes and Landownership Classification*

The 68 farm ponds were broadly categorized into three groups on the basis of land holding capacity and the size of farm ponds (Table 5.1).

The farm pond sizes are based on the categorization proposed by Mushtaq et al. (2000). According to this categorization, farm ponds having more than 10,000 m^3 storage capacity and are treated as large farm ponds, whereas with a capacity between 1000 and 10,000 m^3 as medium and less than 1000 m^3 as small (Mushtaq et al. 2000).

In order to fill these farm ponds, different water sources are being used, namely surface water, groundwater (39 Dug well and 40 Borewell), and a combination of both. Table 5.2 shows that borewells and dug wells are the most common groundwater sources. Despite this, there are farm ponds that are filled with surface water from streams.

Table 5.1 Classification of farm pond according to size and the land holding of farm pond-owning farmers

Household category	Large land holders	Medium land holders	Small land holders	Total no of farm ponds
	(4.1–15.4 ha)	(2.1–4 ha)	(0.4–2 ha)	
Large farm pond >10,000 m^3	10	7	6	23
Medium farm pond 1000–10,000 m^3	7	9	18	34
Small farm pond 64–1000 m^3	0	3	8	11
Total	17	19	32	68

Table 5.2 Different water sources used to fill the farm ponds

Farm pond category	Farm ponds (No.)	Total volume (m^3)	Dug wells	Borewells	Streams
Large farm ponds: (10,000–47,572 m^3)	23	4,47,935	13	16	3
Medium farm ponds (1000–10,000 m^3)	34	1,35,494	14	15	4
Small farm ponds (64–1000 m^3)	11	6,148	12	9	1

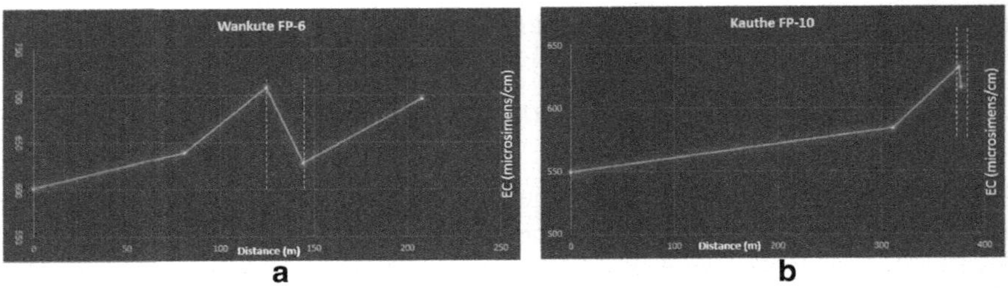

Fig. 5.3 Breakthrough curve of EC vs. lateral distance from the source well. **a** The feeding dug well of Wankute FP-6 shows the zone of influence ranges from 124 to 144 m. **b** Feeding dug well of FP-10 shows the zone of influence ranges from 374 to 377 m

5.6.2 Analysis of in Situ Water Quality Parameters and Results

Several observation wells were geotagged in the vicinity of the Pumping Well (PW). Heavy pumping of a well will cause a widening cone of depression, which will cause the water around the PW to rush toward it. This in turn will be associated with the massive mixing and dilution that will occur, thereby reducing the EC and TDS of the groundwater in the near vicinity of the well. As we move away from the PW, the dilution effect of solids will be minimized, and finally, the zone will arrive where the effect of pumping will be negligible and this point will be marked by a drastic change in the Electrical Conductivity (EC) value, thereby establishing the Zone of Influence. The well that falls immediately to the last well of the ZoI will be marked with a drastic change in the EC and TDS, hence it will define the boundary between the affected and not affected wells due to pumping.

The above outcome was taken as a showcase scenario under a zone of influence from two villages (Wankute and Kauthe) (Fig. 5.3). The zone of influence in Wankute Farm pond-6 can be observed in a range of 124–144 m. However, in the case of Kauthe Farm pond-10, it varies from 374 to 377 m, the reason being the greater depth of the dug well (i.e. 33 m).

5.6.3 Land Use Land Covers Distribution in the Study Villages

In addition to providing the information necessary for thematic maps, land cover identification establishes a baseline for monitoring (change detection). Sentinel-2 data from USGS Earth Explorer (10 m scale) was considered to generate the land use land cover (LULC) maps. Out of six study villages, Dolasane has the maximum area (434 Ha) under cultivation, followed by Wankute and the least area (191 Ha) in Karjule Pathar. Grassland is the major land cover, which occupies most of the area

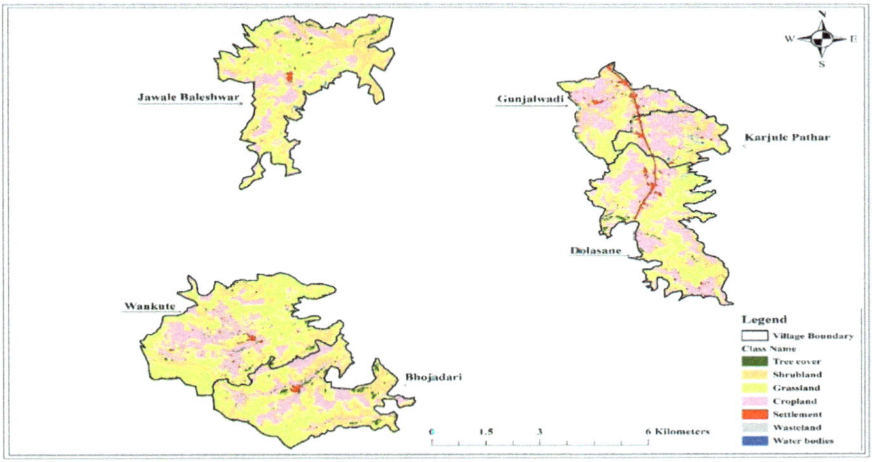

Fig. 5.4 Land use land cover map of the study villages. *Source* http://esa-worldcover.org/en/data-access

and as per the statistics, Wankute accommodates 765 Ha as grassland from the total area of 1513. However, Karjule Pathar has the minimum grassland area, occupying 124 Ha out of a total 446 Ha. Out of the six study villages, Wankute has the maximum geographical area (1513 Ha), and Karjule Pathar is the smallest one with an area of 446 Ha (Fig. 5.4).

5.6.4 Agriculture Practice of the 61 Farm Pond-Owning Households

As intended, farm ponds are supposed to provide protective irrigation during the *Kharif* and the *Rabi* seasons. In order to understand the social impact of the farm ponds under consideration, households in the sample were interviewed regarding the agricultural practices. This information includes crops cultivated, input costs and materials used in the area under cultivation, irrigated area, pattern of water use, etc. Such data were analyzed to understand the change in the cultivated area before and after the farm pond construction. The most significant change taken place is in the area under crops.

The post-farm pond crop data is for 2016–2017 while the pre-farm pond crop data has been adjusted to the calendar year 2015–2016 since farm ponds were constructed in different years between 2009 and 2015 (Fig. 5.5). From the graph, it could be interpreted that for summer and perennial crops; farm ponds are proven to be helpful interventions to increase the area under cultivation with assured irrigation facilities. Also for *Kharif* and *Rabi* seasons, farm ponds have not only secured crops but also enhanced the area under farming. This analysis points out that the farm ponds were

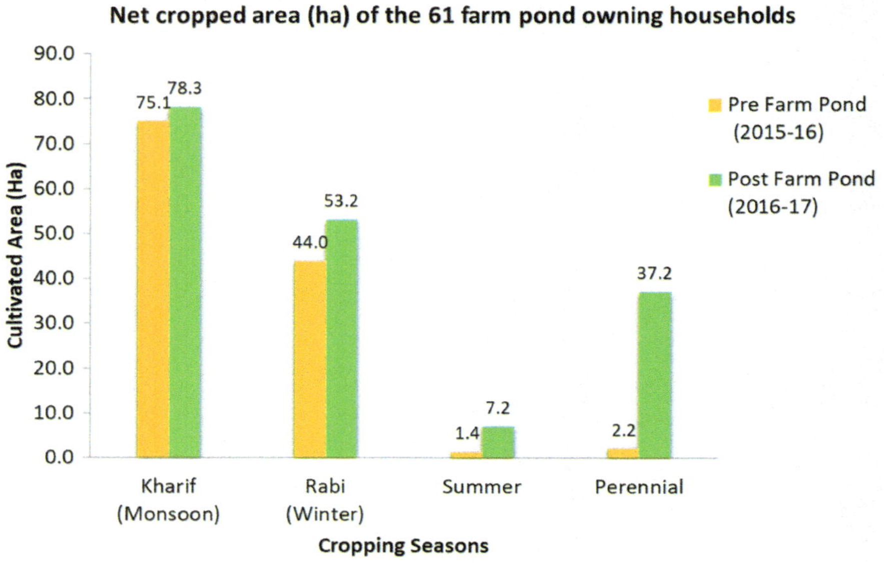

Fig. 5.5 Net cropped areas of 61 farm pond-owning households (in ha)

primarily intended to provide supplementary or protective irrigation during *Kharif* and *Rabi* seasons. However, they are mainly used in summer for additional irrigation for cash crops and horticulture.

5.7 Farm Ponds in Different Recharge Priority Zones

The water that falls on the surface does not filtrate uniformly into the soil; the rate of infiltration is controlled by rock properties and soil characteristics. In order to understand the suitability of recharge, several methods and techniques are used, which generate the map to prepare groundwater potential recharge zones (Kumar et al. 2008; Kadam et al. 2012; Ammar et al. 2016). Groundwater Survey and Development Agency (GSDA) has generated the groundwater recharge priority maps at the village level for identifying the artificial recharge zones on the basis of priorities and categorized them into four sections, viz., high priority, moderate priority, low priority, and limited scope zone (GSDA 2018). High priority zone is defined as the zone where the maximum volume of water can infiltrate into the unsaturated zone, while in the low-priority and limited-scope zones, the potential for infiltration is low.

In the study village 2 (Fig. 5.2), out of 34 farm ponds, 16 fall on a High Priority Recharge zone. On one hand, the farm ponds, being lined, prevent seepage of water. On the other hand, in the same High Priority recharge zone, there are 26 dug wells and borewells which extract groundwater to fill the farm ponds. It has been observed that

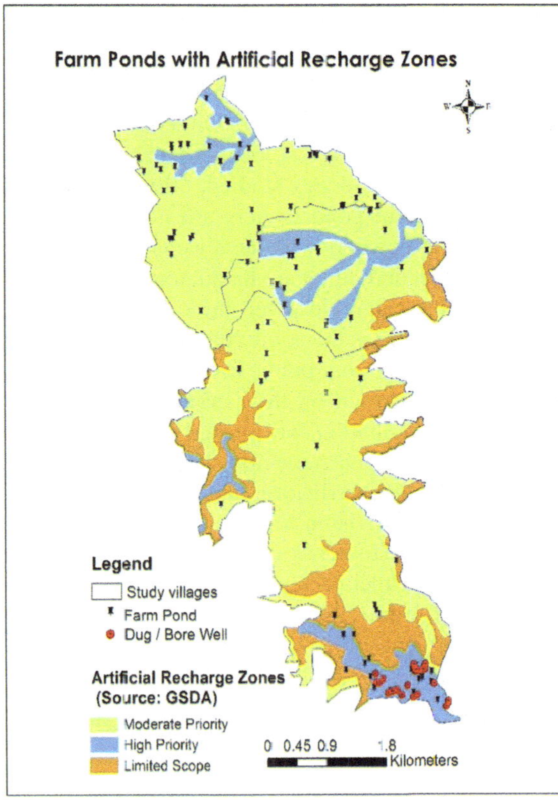

Fig. 5.6 Groundwater recharge priority map developed by GSDA. *Source* GSDA

the 13.54 Ha area is covered by impervious lined farm ponds in the high and moderate priority recharge zones. So the construction of farm ponds without understanding the above-mentioned zones will end up with consequences, which in the long run will adversely impact the recharge rate, which threatens the sustainability of aquifers (Fig. 5.6).

5.8 Policy–Practice Mismatch

Linking our analysis, we comment here on how the current practices of farm ponds by farmers are misaligned with the policy measures and objectives.

5.8.1 Clear Deviation from the Objective

Our research findings clearly show the contradiction between the objectives set behind the scheme and the actual practices of farm pond-owning farmers. Although rainwater harvesting is one of the main objectives behind the farm pond strategy, out of a total of 68 ponds studied, 60 were being filled by lifting groundwater. As a result, the farm pond owners are negatively affecting the water availability of non-farm pond owners who have wells in the zone of influence. For farm ponds as designed, inlet and outlet are mandatory for receiving the rainwater and discharging the additional water. However, in no farm ponds, we observed the inlet and outlet. The other concern is the equity of groundwater resources distribution, as the overall investment comes from public investment. The GRs issued in 2010 are very specific and equity-concerned on this. Initially, the farm ponds were promoted to ensure the protective irrigation for rain-fed crops. It was clearly mentioned in GRs on the criteria for selecting beneficiarie,s that the 'farmer should not have a well in the farm, or a well or another irrigation facility should not be sanctioned to the same farmer.' This was a fair provision for benefiting rain-fed farmers and capturing the runoff in the farm ponds. However, the practice is that the farm pond owners are also well-owners; they extract groundwater for filling farm ponds, and there is no regulation mechanism for this.

5.8.2 Privatization of Groundwater

The right to water has been protected as a fundamental human right by the Indian Supreme Court as part of the Right to Life, guaranteed under Article 21 of the Indian constitution (Narain 2009). Even the Maharashtra Groundwater Act 2009 and its central policies treat groundwater as a public trust. However, the farm pond strategy resulted in the privatization of groundwater by a handful of resource-rich farmers who can invest in digging deeper, extracting, and storing groundwater in the farm ponds. This problem has become severe, as the state has no control over the size and depth of farm ponds. Few rich farmers have built farm ponds of Olympic-sized swimming pools. This has severe equity consequences, as these farm pond owners are not only extracting groundwater below their land but also from the surrounding areas. The current study finds surrounding farmers get negatively impacted as their wells lie in the zone of influence of the well from where the farm pond owner extracts groundwater. This has also a severe consequence on water sharing with the downstream villages. Other studies also show that in drought years, the carrying capacity of water resources falls significantly, so even a low level of extraction may result in a tragedy of the commons. As the number of farm ponds is mushrooming in the villages beyond the carrying capacity of the village catchment, downstream villages are deprived of water availability.

5.8.3 Absence of Regulation on Size and the Number of Farm Ponds

Although the GR (dated 22 February 2016) on farm ponds allow a maximum dimension of 30 × 30 × 3 m, there is an absence of monitoring to assess whether sanctioned dimensions are followed in the actual construction. Although in GRs issued in 2010 clearly mentions that 'the beneficiary farmers have the responsibility to follow the stated dimension/size of the farm pond' in the program, the GR issued in 2016 is vague on this provision, allowing farmers to invest an extra amount from their pockets and increase the dimensions of farm pond at any level (even of few acres), and there is no control over this. The same is true with the number of farm ponds constructed in a watershed or in a village. While sanctioning farm ponds in a village or watershed, integrated and sustainable planning of overall water resources, the carrying capacity of that area to supply the amount of water for different uses and structures need to be seriously considered. There are many villages in the state where farm ponds are in enormous numbers. With the lessons learned on violation of farm pond size and numbers, the state department has had discussions and initiated the process to set a methodology and guideline to decide the number of farm ponds for a given geography (village or watershed). This process should soon reach a logical conclusion before it gets too late.

5.8.4 Evaporation of Groundwater from Tanks

The precious groundwater resource in tank is exposed to evaporation. Local evaporation data indicate around 20–25% of water stored in farm ponds gets evaporated annually. Moreover, in many pockets of arid and semi-arid Maharashtra, the maximum temperature in summer reaches up to 40–45 °C; in such weather conditions, the evaporation rate further increases. In the background of the drastically increasing number of farm ponds, a huge volume of groundwater gets extracted to store water in the farm ponds. In the background of a changing climate (rise in temperature), the evaporation of water from farm ponds becomes a more severe and alarming concern.

5.8.5 Water Stored in Farm Ponds is a Free Gift

The MWRRA has the mandate and is responsible for fixing the bulk water tariff for different uses in the state. The Authority revises the water tariffs every three years for drinking, irrigation, and industrial use. Past experiences show that the Authority has focused mainly on irrigation projects and canal irrigation through water user associations (WUAs). However, the changing dynamics of farm ponds have not yet been considered by the Authority. The owners in the command area fill their huge farm ponds by pumping water from canals, minor structures and by extracting huge amount of groundwater resource (which is a shared common resource). Till now, there is no provision or thought by MWRRA to price the water in farm ponds, and we propose that unless the stored water in farm ponds is not priced, there will be no deterrent for misusing the groundwater.

5.9 Conclusion and Discussion

Harvesting rainwater and improving groundwater recharge are the main purposes of farm ponds. Most farmers use farm ponds as storage structures with plastic lining at bottom and sides to prevent water loss through percolation. A farm pond site can be chosen considering local soil conditions, topography, drainage, infiltration capacity, and rainfall pattern. Farm ponds should be constructed within the specifications provided, which include an inlet for surface water flow during rains and an outlet for overflow water. But no science based guidance is found in choosing the sites and its design, which has led to inefficient performances of most of the farm ponds. Due to a lack of technical support, farm ponds are poorly constructed as farmers do not take into account these factors and build them on fallow land or along slopes where overland flow can easily reach the farmlands. The usual practice is to draw water from a borewell or a dug well to fill up the farm pond, which causes a significant amount of groundwater to evaporate. As most of the farm ponds are constructed without an inlet or outlet they cannot arrest excess run off, therefore cannot be used as rainwater harvesting structures. Furthermore farmers line up the farm pond with plastic, restricting seepage and recharge, while converting the ponds into storage structures. For lining, ISI mark plastic lining is recommended for farm ponds. Villagers prefer to purchase low quality cheaper materials e.g.- plastic lining, to save investment cost as the materials recommended by government standards are costlier. Due to poor quality, these materials fail to sustain over a longer period which results in recurring cost and high maintenance efforts. The cost of plastic lining is much higher than digging a farm pond, therefore there are many non-functional documented farm ponds, as the owners cannot bear the high cost of linings. Such farm ponds have adverse effects on the water table and accelerate water loss; this intensifies competition for extraction of groundwater from the aquifer, which is a common pool resource.

5.9.1 Recommendation

The farm pond scheme has a huge potential in addressing the drought and increasing income for farmers. The state's efforts and intention to promote it are also appreciable. However, the study clearly point out an urgent need to make a course corrections to the scheme. The farm pond activities must be truly regulated, ensuring that farmers follow the specified standards and dimensions while constructing farm ponds. The methodology for deciding the number of farm ponds in a village/watershed gets prepared and followed. More importantly, there is a need for a strict monitoring system to control the practice of groundwater extraction to fill the farm ponds, particularly in drought-prone and overexploited areas. These changes are crucial for the long-term sustainability of groundwater resources and also preserve its unique nature that it is the common pool resource. Recommendation made here to ascertain that farm ponds are used as well-adapted drought mitigation measures.

1. The carrying capacity of the local watershed needs to be determined by considering average local rainfall, runoff and groundwater recharge, terrain etc. the total number of farm ponds that may be constructed with their size and depth
2. In the areas where the current number of Farm Ponds has exceeded the carrying capacity of the respective watershed, the construction of additional farm ponds should be restricted.
3. Based on the local hydro-geological settings and soil quality, farm ponds that harvest surface runoff and without linings are to be promoted largely.
4. The inlet and outlet for farm ponds in assured rainfall areas should be mandatory. As the present design of the farm pond goes above the surface (2–3 m above the surface land) and has a lining of plastic. Hence, ways have to be explored to make appropriate changes in the design of plastic-lined ponds that allow runoff water from the land surface through the inlet.
5. Initiate a pilot/action research (a) water governance standard for incentivizing community with farm ponds to adapt sustainable practices (b) for monitoring of groundwater withdrawal and electricity use to assess aquifer performance with a view to developing sustainable governance mechanisms.

5.10 Annexure

See Table 5.3.

Table 5.3 Details of the source wells from where samples were collected for chemical analysis and the chemical results

Farm pond	Village	Type of well	Latitude	Longitude
FP1	Jawalebaleshwar	DW	19°26′ 43.44″	74°07′ 33.20″
FP2	Jawalebaleshwar	DW	19°25′ 45.09″	74°06′ 45.90″
FP3	Jawalebaleshwar	DW	19°25′ 39.86″	74°07′ 22.51″
FP4	Jawalebaleshwar	DW	19°25′ 20.35″	74°07′ 17.09″
FP5	Jawalebaleshwar	DW	19°25′ 20.35″	74°07′ 17.09″
FP6	Wankute	DW	19°20′ 02.22″	74°05′ 39.21″
FP7	Wankute	DW	19°20′ 54.38″	74°05′ 50.97″
FP8	Wankute	DW	19°20′ 57.07″	74°05′ 54.97″
FP9	Wankute	DW	19°20′ 53.14″	74°05′ 48.70″
FP10	Wankute	DW	19°20′ 15.54″	74°06′ 26.57″
FP11	Wankute	DW	19°19′ 50.91″	74°06′ 10.55″
FP12	Wankute	DW	19°20′ 21.94″	74°05′ 53.25″
FP13	Wankute	DW	19°19′ 59.56″	74°05′ 36.13″
FP14	Wankute	DW	19°20′ 36.10″	74°05′ 18.45″
FP15	Wankute	DW	19°21′ 04.69″	74°07′ 17.84″
FP16	Bhojdari	BW	19°19′ 23.10″	74°06′ 59.13″
FP17	Bhojdari	DW	19°19′ 33.76″	74°07′ 20.82″
FP18	Bhojdari	BW	19°19′ 22.76″	74°07′ 11.26″
FP19	Bhojdari	DW	19°19′ 08.89″	74°07′ 01.62″
FP20	Bhojdari	BW	19°18′ 46.81″	74°06′ 52.28″
FP21	Bhojdari	BW	19°18′ 46.17″	74°06′ 47.43″
FP22	Bhojdari	BW	19°18′ 58.47″	74°06′ 54.13″
FP23	Bhojdari	DW	19°19′ 10.58″	74°07′ 00.90″
FP24	Bhojdari	DW	19°19′ 27.84″	74°07′ 08.53″
FP25	Dolasane (Dhumalwadi)	BW	19°21′ 29.55″	74°13′ 38.01″
FP26	Dolasane (Dhumalwadi)	BW	19°21′ 29.76″	74°13′ 36.05″
FP27	Dolasane	BW	19°21′ 31.31″	74°13′ 35.30″
FP28	Dolasane	DW	19°21′ 17.70″	74°13′ 34.84″
FP29	Dolasane	BW	19°21′ 18.80″	74°13′ 29.12″
FP30	Dolasane	DW	19°21′ 20.32″	74°13′ 24.44″
FP31	Dolasane	DW	19°21′ 19.70″	74°13′ 22.49″
FP32	Dolasane	BW	19°21′ 23.86″	74°13′ 44.94″
FP33	Dolasane	BW	19°21′ 15.09″	74°13′ 51.82″
FP34	Dolasane	DW	19° 23′ 27.50″	74° 12′ 32.49″
FP35	Dolasane	BW	19° 23′ 35.49″	74° 12′ 40.83″
FP36	Dolasane	DW	19° 23′ 03.66″	74° 12′ 35.84″

(continued)

Table 5.3 (continued)

Farm pond	Village	Type of well	Latitude	Longitude
FP37	Dolasane	DW	19° 23′ 33.01″	74° 12′ 47.59″
FP38	Dolasane	BW	19° 23′ 31.86″	74° 12′ 55.40″
FP39	Dolasane (Dhumalwadi)	DW	19° 21′ 41.57″	74° 13′ 02.24″
FP40	Dolasane (Dhumalwadi)	DW	19° 21′ 31.21″	74° 13′ 11.94″
FP41	Dolasane (Dhumalwadi)	DW	19° 21′ 40.65″	74° 13′ 03.92″
FP42	Karjule Pathar	DW	19° 24′ 07.68″	74° 12′ 57.29″
FP43	Karjule Pathar	BW	19° 24′ 45.68″	74° 12′ 38.06″
FP44	Karjule Pathar	BW	19° 24′ 32.68″	74° 12′ 37.80″
FP45	Karjule Pathar	BW	19° 24′ 25.19″	74° 12′ 29.78″
FP46	Karjule Pathar	BW	19° 24′ 23.09″	74° 12′ 34.59″
FP47	Gunjalwadi	BW	19° 25′ 21.98″	74° 12′ 15.84″
FP48	Gunjalwadi	BW	19° 25′ 17.87″	74° 11′ 36.16″
FP49	Gunjalwadi	BW	19° 25′ 24.73″	74° 11′ 22.34″
FP50	Gunjalwadi	BW	19° 25′ 07.91″	74° 11′ 35.94″
FP51	Gunjalwadi	BW	19° 24′ 38.78″	74° 11′ 18.01″
FP52	Gunjalwadi	BW	19° 25′ 26.03″	74° 11′ 40.56″
FP53	Gunjalwadi	BW	19° 25′ 27.83″	74° 11′ 41.71″
FP54	Gunjalwadi	BW	19° 24′ 46.59″	74° 11′ 50.94″
FP55	Gunjalwadi	DW	19° 24′ 35.17″	74° 11′ 43.15″

References

Abhay N (2016) How to get shettale subsidy for farm pond in 2016. http://indiamicrofinance.com/shettale-subsidy-farm-pond2016.html

Ammar A, Riksen M, Ouessar M, Ritsema C (2016) Identification of suitable sites for rainwater harvesting structures in arid and semi-arid regions: a review. Int Soil Water Conserv Res 4(2):108–120

Bendapudi R, Yadav A, Chemburkar S, D'Souza M (2020) Adaptation Strategy or maladaptation–the conversion of farm ponds into surface storage tanks in semi-arid regions of Maharashtra. Adaptation at scale in semi-arid regions (ASSAR) and the collaborative adaptation research initiative in Africa and Asia (CARIAA).

Dipankar Saha AK, Agrawal, (2006) Determination of specific yield using water balance approach—a case study of TorlaOdha watershed in Decan Trap Province, Maharashtra State, India. 2006. Hydrogeol J 14:625–635

Government of India, Ministry of Jal Shakti, Department of water resources, River development and Ganga Rejuvenation, CGWB (2021)

Groundwater Surveys and Development Agency (GSDA) 2018 Ground Water Recharge Priority Map

Kadam AK, Kale SS, Pande NN, Pawar NJ, Sankhua RN (2012) Identifying potential rainwater harvesting sites of a semi-arid, basaltic region of Western India, using SCS-CN method. Water Resour Manage 26(9):2537–2554

Kale E (2017) Problematic uses and practices of farm ponds in Maharashtra. Econ Pol Wkly 52(3):20–22

Kale E, Kulkarni P (2022) Challenging today's water threats for tenable tomorrow: a review of policies and programs in the water sector of maharashtra, research report. Watershed Organisation Trust, Pune, India

Kerr J (2002) Watershed development, environmental services, and poverty alleviation in India. World Dev 30(8):1387–1400

Khambete AK (2009) The Maharashtra Groundwater (Development and Management) Act 2009 | India Water Portal

Kumar MD, Reddy VR, Narayanamoorthy A, Bassi N, James AJ (2018) Rainfed areas: poor definition and flawed solutions. Int J Water Resour Dev 34(2):278–291

Kumar MG, Agarwal AK, Bali R (2008) Delineation of potential sites for water harvesting structures using remote sensing and GIS. J Indian Soc Remote Sens 36(4):323–334. Mall, RK et al (2006) "Jun06-Water-Cc-Cgc-India-Cursci" 90(12)

MoRD (Ministry of Rural Development) (1994) Report of the technical committee on drought prone areas programme and desert development programme. Government of India, New Delhi

Mushtaq S et al (2000) An assessment of the role of ponds in the adoption of water-saving irrigation practices in the Zhanghe Irrigation System, China. Agric Water Manag 83(1–2)

Narain V (2009) Water as a fundamental right: a perspective from India. Vt l Rev 34:917

Oweis T, Hachum A (2009) Water harvesting for improved rainfed agriculture in the dry environments. Rainfed Agric: Unlocking Potential 164–182

Peacock B, Williams LE, Christensen LP (1998) Water management and irrigation scheduling. Raisin Prod Manual ANR Pub 3393:127–133

Reddy KS, Manoranjan Kumar Rao KV, Maruthi V, Reddy BMK, Umesh B, Ganesh Babu R, Srinivasa Reddy K, Vijayalakshmi, Venkateswarlu B (2012) Farm ponds: a climate resilient technology for rainfed agriculture; planning, design and construction. Central Research Institute for Dryland Agriculture, Santoshnagar, Saidabad, Hyderabad, Andhra Pradesh, India, p 60

Saha D, Marwaha S, Dwivedi SN (2019) National aquifer mapping and management plan- a step towards water security in India. In: Singh A, Saha D, Tyagi SAC (eds) Water governance: challenges and prospects 2019. Springer, Berlin. ISBN 978-981-13-2699-8.

Saha D, Roy R (2018) Groundwater resources of India: potential, challenges and management. In: Sikdar P (ed) Groundwater development and management, issues and challenges in South Asia. Springer, Berlin. https://doi.org/10.1007/978-3-319-75115-3_2

Shah SH, Harris LM, Johnson MS, Wittman H (2021) A 'drought-free'Maharashtra? Politicising water conservation for rain-dependent agriculture. Water Alternatives 14(2):573–596

Udmale P, Ichikawa Y, Kiem A, Panda S (2014) Drought impacts and adaptation strategies for agriculture and rural livelihood in the Maharashtra State of India. Open Agric J 8(1)

Chapter 6
Rainfall Recharge Wells for Groundwater Sustainability in Qatar

Mohamed Shamrukh and Abdulaziz A. Al-Muraikhi

Abstract Groundwater sustainability is a vital issue for people on planet earth. Qatar is known for its scarcity of renewable water resources including groundwater. In the last decades, there has been an escalating trend in groundwater abstraction to provide resources for social and economic development in the country. Currently, almost all water demand for municipal and industrial demand is provided by desalination while agriculture demand is almost entirely met from groundwater with a total abstraction of about 218 Mm^3/year (90% of total pumping). In 2009, a comprehensive assessment was carried out to study natural and artificial recharge of groundwater and define scenarios and designs for increasing artificial recharge. In order to design and develop an optimal artificial recharge scheme for future investments, an array of measurements and tests were carried out, including an assessment of the performance of pre-existing recharge wells and how these and new recharge wells could be optimised. Field and laboratory measurements were done for relevant parameters, and water balance modelling using the relatively simple SWAT2005 model was carried out, including rainfall-runoff and recharge analysis. Mapping the level and quality of groundwater indicated its extensive degradation. To augment the natural recharge, a total of 313 pre-existing passive artificial rainfall recharge wells facilitating the infiltration of ponding water in natural landscape depressions were identified across Qatar. However, data indicated that only 166 of these sites are located in favourable depressions and the remaining wells are not. The modelling results indicate that artificial recharge from the existing recharge wells and from natural rainfall recharge is 10.6 and 89.4 Mm^3/year, respectively. For optimal design, modelling results indicate that a total of 1,502 recharge wells (1,388 as new wells) are required in favourable depressions to infiltrate the storms, typically around 20 mm within a 5-days period, which

M. Shamrukh (✉) · A. A. Al-Muraikhi
Ministry of Environment and Climate Change, P.O. Box 7634, Doha, Qatar
e-mail: mshamrukh@gmail.com; msmahmoud@mu.edu.eg

A. A. Al-Muraikhi
e-mail: AAAl-Muraikhi@mecc.gov.qa

M. Shamrukh
Department of Civil Engineering, Faculty of Engineering, Minia University, Minya 61111, Egypt

D. Saha et al. (eds.), *Managed Groundwater Recharge and Rainwater Harvesting*, Water Resources Development and Management,
https://doi.org/10.1007/978-981-99-8757-3_5

will augment the groundwater by 33.5 Mm^3/year. Our recent estimation of groundwater resources in Qatar indicates that an abstraction volume of about 150 Mm^3/year leads to an equilibrium status under recent conditions of recharge. Comparing to actual abstraction volume (227 Mm^3/year), this confirms the need for further increasing water supply and reducing demand. This work needs to be complemented by reviewing the design of existing recharge wells and using historic long-term data from monitoring networks complemented with advanced national 3D groundwater modelling for more accurate estimation. The goal is to manage natural and artificial recharge to support the achievement of groundwater sustainability in Qatar. Other alternatives sources such as treated sewage effluent and desalination to be considered as well.

Keywords Qatar · Groundwater · Artificial recharge · Rainfall · Recharge Wells · Water balance

6.1 Introduction

Freshwater is the source of life and so vital to present and future generations. In arid regions, it sometimes becomes a sacred resource. Sustainable access to freshwater is becoming one of humanity's greatest challenges, and a potential source of conflict in many areas of the planet. The United Nations (UN) considers safe and sustainable access to water as one of the 17 Sustainable Development Goals (SDGs) of the UN 2030 Agenda, Goal 6. It also figures prominently, but indirectly, in many other SDGs.

The Gulf countries are located in one of the world's driest regions, with limited freshwater resources, high evaporation rates, and increasing demand for water due to population growth and economic development. To address the growing water scarcity, the Gulf countries have practised augmentation of freshwater sources by different methods including desalination, artificial recharge (also called managed aquifer recharge or aquifer storage and recovery), recycled water and circular economy, and even by cloud seeding (Mohieldeen et al. 2021; Kadhem and Zubari 2020; Delene and Shamrukh 2016). The geographic focus of this chapter is the State of Qatar, one of the Gulf countries. It occupies a peninsula, which projects into the Arabian Gulf and borders with Saudi Arabia. The capital is Doha. The presence of the sea influences the climate of Qatar by increasing the air humidity that is relatively high (relative humidity values are around 70%). However, as the rest of the Gulf countries in this region, rainfall is scanty. There are no natural water bodies with freshwater such as rivers and lakes and the only historical freshwater source is groundwater (Amer 2008; Shamrukh et al. 2012). This groundwater, which is brackish in places, is pumped from aquifers, especially in central and northern areas (Elsaid et al. 2017). The single largest discharge of groundwater is that from abstractions, in particular agricultural abstractions. The other discharges are discharges to the sea and internal hydraulic movement of groundwater between aquifers.

Water for agricultural irrigation is almost entirely derived from groundwater. In the 1960s, municipal use of groundwater accounted for around 25% of the abstraction but this percentage has been declining over the years (MoE 2009; Amer 2008). The total number of active/producing farms has increased from 259 in 1976 to 915 in 2015 (SoE 2021). In 2015, the volume of groundwater pumped for agricultural use was estimated to be 218 Mm^3. This abstraction is anticipated to impact the level and quality of groundwater resources (MoE 2009). More than 91% of the groundwater abstracted in Qatar is currently used for agricultural irrigation (MoE 2009). Groundwater abstraction has increased from 160 Mm^3 in 1993 to 250 Mm^3 in 2010 and continues to increase. The total number of production wells for all purposes including agriculture, domestic, and industrial is about 4,205 according to 2009 records (MoE 2009).

The main recharge source of aquifers in Qatar is the infiltration from rainfall. However, abstraction from aquifers is much higher than the freshwater recharge (SoE 2021). This unbalanced situation has resulted in a decline in average groundwater level by 0.28 m over the period 1971 to 1992 (except urban areas), together with an associated notable deterioration of water quality (MoE 2009). Irrigation, together with saline intrusion from the coast and underlying aquifers, has been the main cause of deteriorating groundwater quality. Estimates of irrigation return flows ranged between 20 and 25% (SoE 2021; Eccleston and Harhash 1982; Entec 1994). These return flows are more concentrated than irrigation water in terms of total dissolved solids (TDS) due to evaporation and it includes elevated concentrations of sodium, chloride, nitrate, and possibly pesticides (Harhash and Yousif 1985). The deterioration of water quality has led to the abandonment of farms, particularly in the coastal areas, and in other areas with salinized soils.

In the last years, Qatar's population has grown significantly, from 744,000 in 2004 to 2,774,000 in 2021 (PSA 2021; SoE 2021). This has been associated with a rapid rise in water demand for domestic, industrial, and agricultural activities. Until 1953, the population of Qatar was entirely reliant on groundwater for its potable and agricultural water. In 1953, the first desalination plant was commissioned, and the country's desalination capacity has been increasing since. Since 2007, almost all water demand for municipal and industrial use has been satisfied by desalination (Amer 2008; Abdel-Wahab et al. 2008Dillon et al. 2019). There are three major desalination plants in operation in the country, using different desalination technologies.

Nowadays, the water resources for all demands are desalination, groundwater, and treated sewage effluent (TSE). Drinking water is supplied by desalination plants installed on the Arabian Gulf. According to 2017–2018 data, the total water supplied from desalination, groundwater, and TSE are 606, 250, and 258 Mm^3/year, respectively (SoE 2021). However, there is still a potential for TSE, which is not used according to 2018 data (SoE 2021). From this data, we can estimate the groundwater abstraction from shallow/deep aquifers to be about 23% of the total annual water supply. However, exploration of very deep aquifers, like the Aruma Formation, has been carried out and results presented in the literature (Shamrukh et al. 2012).

Numerous studies were carried out to investigate the aquifers, hydrogeology, water balance, and artificial recharge in Qatar. The first comprehensive study of the hydrogeology of northern Qatar was carried out by the Qatar Petroleum Company in 1957–1959 and included drilling and geophysical investigations. Studies were jointly carried out by the Qatari Government and the United Nations Food and Agricultural Organisation (FAO) during the 1970s (Pike 1978). The main productive aquifer formations in Qatar are the Eocene-age Rus Formation and the underlying Paleocene-age Umm ErRadhuma (UER) Formation (MoE 2009; Lloyd et al. 1981, 1987). Alsharhan et al. (2001), Kimrey (1985), Entec (1994) studied the recharge of groundwater and concluded that the focused recharge is much higher than the diffuse recharge. Pike et al. (1975) produced an early water balance model for Qatar for the 4-year period 1971–1975. The water balance assumes an irrigation return rate of 35% of irrigation water. Another groundwater balance was developed for the period between 1971 and 1983 by Harhash and Yousif (1985), and a more recent water balance was presented in SoE (2021). It is worth mentioning a more recent work done by Mohieldeen et al. (2021) assessing the artificial recharge potential for enhancing the strategic water reserve in Qatar. They used GIS methods to delineate the natural areas for aquifer recharge using desalinated or highly treated sewage waters to form a strategic water reserve of about 182.8 Mm^3 to be used for water supply in emergencies lasting up to three months.

In 2009, a significant and comprehensive project was carried out by the Ministry of Environment of Qatar (MoE, today called the Ministry of Environment and Climate Change, MECC) to study and develop artificial recharge of aquifers, especially by rainfall recharge (MoE 2009). The main objective of this chapter is to present the main work, assessment, analyses, and findings of MoE (2009) study and to present an updated estimation of the groundwater balance for the state of Qatar, indicating safe well abstraction. Recommendations for future work are also provided.

6.2 Geographic and Environmental Setting of Qatar

Qatar occupies a peninsula, which projects into the Arabian Gulf and covers an area of 11,627 km^2, approximately 180 km in length north to south and 85 km wide east to west at its widest point. It is mostly surrounded by the Arabian Gulf waters except the 60 km of land border with the Kingdom of Saudi Arabia to the south (Fig. 6.1). The marine area (35,000 km^2) of the country is almost three times the land area. The salinity of seawater can be as high as 60–70 mg/l (SoE 2021) in some places and seasons. The land surface of the peninsula is generally of low relief with a maximum height of 103 m above sea level (masl). In the south and west, erosional escarpments form the highest ground in Qatar, while sand dunes and ridges in the

Fig. 6.1 Location of Qatar and its marine area

southeast contrast to the normally flat stony desert surface. Numerous sabkhas[1] are present along the coastline (MoE 2009).

[1] A sabkha is a coastal, supratidal mudflat or sandflat in which evaporite-saline minerals accumulate as the result of semiarid to arid climate.

6.2.1 Hydrology and Hydrogeology of Qatar

6.2.1.1 Climate and Hydrology

The climate of Qatar can be described as subtropical dry, hot desert climate with low annual rainfall, high temperatures in summer and large differences between maximum and minimum temperatures. In the inland areas, monthly temperatures range from 17 °C in January to 48 °C in August. Heavy localised rainfall on one day may account for more than the long-term average annual rainfall. The average annual rainfall is approximately 75 mm with rainfall confined to the months of October to May. There is a gradient in rainfall from north to south, with more rainfall in the north. Rainfall is very irregular and hence extremely unpredictable, and temporal variability of annual rainfall is large (Fig. 6.2) (SoE 2021).

Given the arid setting, surface water bodies and channels are very short-lived, and permanent inland fresh surface water features are rare. The drainage system in Qatar may be described as endoreic. The drainage network is such that the majority of surface water runoff is channelled, via wadis and overland flow, to inland surface water accumulations in terminal depressions. The largest surface water bodies are associated with the brackish or saline sabkhas, and these are restricted to the coastal areas of Qatar. Sabkhas receive surface water inflows from the surrounding higher topography as well as influx provided by marine water incursion and lateral and upwelling groundwater seepages. Historically, the watershed of Qatar has been divided into 11 larger catchments with a total area of 6,980 km^2 (Sect. 4.2.1 and Fig. 4.10).

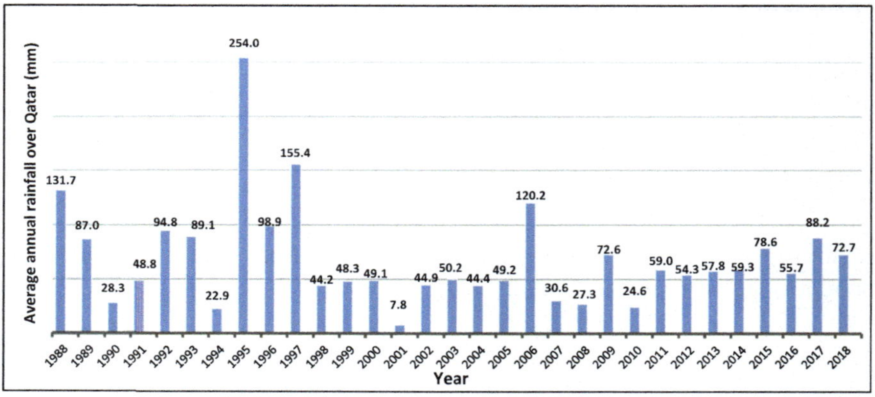

Fig. 6.2 Annual rainfall in Qatar (1988–2018) (SoE 2021)

6.2.1.2 Geology

The formations outcropping in Qatar are entirely of Palaeogene to Quaternary age. Older Mesozoic-age rocks have been found in boreholes and include the Upper Jurassic-age carbonate sequence of the Arab Formation, which forms the oil reservoir of the Dukhan anticline, and which in turn is overlain by an 800 m thick Cretaceous-age sequence dominated by calcarenite, limestone, and sandstone. The manifestation of the deeper geology on the surface is mainly through karst processes. Karst features are present throughout most of the peninsula and include depressions, sinkholes, caves, and solution hollows. An analysis of aerial photographs identified a total of 9,736 depressions, caves, and sinkholes (Sadiq and Nasir 2002). These karst features are mostly associated with dissolution of gypsum and anhydrite deposits of the Rus Formation and, to a lesser extent, the dissolution of the carbonate sequences of the Dammam Formation (Table 6.1). The general directional trend of these features coincides with the regional joint and fracture systems.

The karst depressions play a key role in the project discussed in this chapter, as the recharge wells are or will be located in these areas. Depressions are former sinkholes which have been partly or completely infilled with autochthonous material. Simple depressions have a single centre while compound depressions are the product of several depressions joining one another. They are normally circular in shape and range from 100 s to 3,000 m in diameter and reach depths of up to 25 m relative to the surrounding plateau. Most depressions and sinkholes are concentrated south of Northing 384,000 Qatar National Datum, QND (latitude 25° 20′), which is the approximate northerly limit of the subsurface deep gypsum and anhydrite layers interbedded within the Rus Formation (Sadiq and Nasir 2002). Analysis of Digital Elevation Model (DEM) data for the whole of Qatar has identified 21,786 topographic depressions covering a total area of 1,799 km^2. It is considered that the vast majority of these topographic depressions are likely to have formed by dissolution-induced subsidence, processes occurring over millions of years. It is apparent that most of the largest depressions are located in the southern part of Qatar.

6.2.1.3 Hydrogeology

Table 6.2 shows the hydrogeological and aquifer units in Qatar. There are two formations and two members within another formation, which form the four main aquifer bodies in Qatar exploited for groundwater. These are: the Rus Formation, the Umm Er Radhuma Formation (UER), and the Simsima and Abarug-Allat members of the Dammam Formation. The Rus and UER Formations are the most important aquifers in terms of storage capacity and extent and underlie the whole of the Qatar peninsula. The Simsim a member of the Dammam Formation outcrops across 80% of the surface area of Qatar, yet its elevation and thickness make it a relatively small aquifer as it is largely unsaturated. The Abarug-Allat member of the Dammam Formation outcrops mainly in the southwestern part of Qatar and forms a small aquifer body in this area.

Table 6.1 Generalised geologic successions in Qatar

Age	Epoch	Formation	Member	Lithology
Quaternary	Holocene pleistocene			Sand dunes, beach sediments, sabkhas
Tertiary	Pliocene Miocene	Upper (Hofuf)	–	Residual gravel
		Lower (Dam)	–	Sandstone, clay, and gypsum layers (up to 30 m thick)
	Eocene (Middle) (Lower)	Upper Dammam	Abarug	Clayey dolomitic marl
			Allat	Chalky marl and dolomite
			Simsima	Dolomite and limestone layers (~30 m)
		Lower Dammam	Dukhan Alveolina	Chalky Limestone
			Midra Saila	Shale with thin argillaceous limestone
			FhaihilVelates	Limestone
		Rus (Upper and Lower)	–Khor limestone –Carbonate facies –Sulphate facies	Chalky limestone with intercalations (commonly up to 4m thick) of marl and gypsum with minor anhydrite. Outcrops in the core of the central arch and the gypsum are typically absent in the Northern Province of Qatar
	Paleocene Maastrichtian	Umm ErRadhuma		Dolomite with intercalations of chert, marl, shale, and anhydrite over calcareous shale at the base (~300m)
		ArumaGroup	Simsima Ruilat Halul Lafan	Dolomite Dolomite, limestone Grainstone, packstone Green blue shale

Table 6.2 Hydrogeological and aquifer units in Qatar

From surface	Stratigraphic unit	Description
1	Quaternary deposits	Limited outcrops
2	Dammam Formation	Limestone and dolomite (30–50 m thickness)
3	Midra Shale	Aquitard, extent limited to southern part of Qatar (4–10 m)
4	Upper Rus Formation (carbonate)	Fractured/karstic limestone (20–120 m thickness)
5	Lower Rus Formation (gypsum)	Aquitard, mainly very low permeability gypsum, extent limited to southern part of Qatar
6	Umm Er Radhuma Formation (Upper and lower)	Deep strata, fractured/karstic carbonate aquifer, most probably unconnected to the shallow groundwater system in the south. Limit of interest is defined at the upper surface of the chert layer that is found in this formation (270–370 m thickness)

In northern Qatar, where the Rus Formation is thinner and dominated by the carbonate facies, there is vertical hydraulic continuity between the Rus and UER Formations to such a degree that they effectively form one aquifer (Lloyd et al. 1987, 1981; Eccleston and Harhash 1982). Where the thicker gypsiferous, argillaceous facies at the bottom of the Rus Formation occur over most of Qatar (except for the Dukhan structure) hindering this connectivity, connectivity is restricted to areas around depressions where dissolution of the underlying evaporite sequences have resulted in collapse and fracturing of the aquifer materials. However, Streetly and Kotoub (1998) suggest that the level of hydraulic continuity between these two formations in the northern part of Qatar is dictated predominantly by the presence or absence of a thin interbedded layer of clay situated at the top of the UER Formation and, to a lesser extent, a chert layer. This clay layer is interpreted from geophysical logs, which show a significant gamma spike at the top of the UER Formation (MoE 2009). A key characteristic of the near surface aquifers of Qatar is karst as already discussed in the geology section above. Hydrogeological cross sections are presented in Fig. 6.3.

From the above review, there are two main aquifer units that have been used to provide fresh groundwater. The Rus aquifer, comprising the Rus and Dammam formations (30–50 m thickness) and the UER aquifer, which is the major and regional aquifer throughout the Gulf region.

6.2.1.4 Groundwater Provinces

The early division of Qatar's area into groundwater provinces highlights natural differences between the south and the north, as well as differences between urban areas and the remainder of the country (SoE 2021). The aerial extent of the aquifers

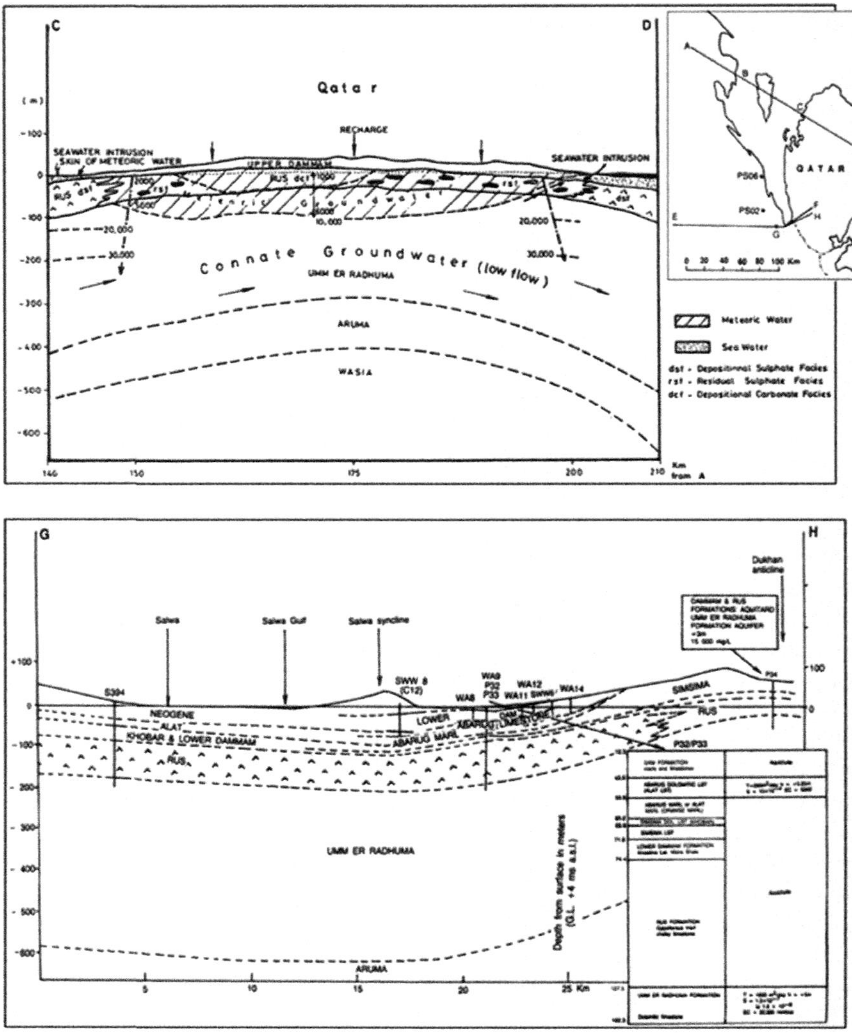

Fig. 6.3 Hydrogeological cross sections of Qatar (Eccleston and Harhash 1982)

of Qatar has been subdivided into two main hydrologic provinces: Northern and Southern Basins (Fig. 6.4). The boundary between the two provinces coincides closely with the southern limit of the Lower Dammam Midra Shale member in the north and the northern limit of the Lower Rus gypsum member in the south. In the southern province, the UER is confined at its top by massive gypsum of the lower Rus Formation. In the Northern Province, the lower Rus Formation gypsum unit is generally absent. These two provinces form the V-shape boundary between the depositional carbonate facies in the north and the depositional sulphate facies in the south. The two basins are different in the hydrochemistry of the groundwater, with a less saline groundwater being present in the northern zone, which largely reflects the higher incidence of rainfall occurring in this region. Smaller groundwater provinces have subsequently been added to this overall conceptual framework.

Figure 6.4 shows the main groundwater provinces, which are as follows (MoE 2009):

> **Northern Basin**, where available data indicate the massive gypsum beds are generally absent, and the aquifers are less hydraulically separated. The Rus and upper UER is considered as an unconfined aquifer, and it is considered the most important sources of groundwater in Qatar.
>
> **Southern Basin**, where the shallow Rus and deep UER aquifers are generally hydraulically separated by a thick sequence of gypsum-confining bed with low

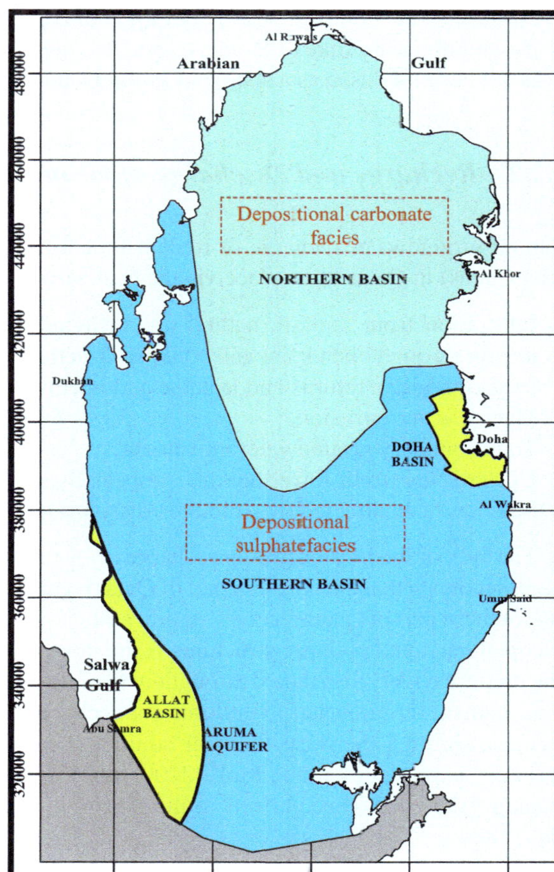

Fig. 6.4 Main groundwater provinces in Qatar, Qatar National Grid 20 km (after MoE 2009)

permeability (UER is confined). This basin is confined by a layer of the Midra Shale member, which is part of the overlying Dammam Formation.

Abu Samra Province or Southwestern Province, which is dominated by an artesian aquifer in the units of the upper Dammam Formation. Referred to as the Allat Aquifer, the groundwater body sits above the Abarug Marl and is generally brackish in quality.

Doha Province, which is defined to reflect the enhanced recharge due to anthropogenic impacts in the Doha city area. The water, which occurs as a lens above the impermeable Midra Shale member, is generally encountered during the excavation of building foundations in the city and requires dewatering. Significant amounts of shallow groundwater are collected in a pipeline system and continuously discharged to the sea from this zone.

Aruma Groundwater Basin, which underlies the whole of Qatar, is a separate deep regional basin, which can be considered another groundwater province. It constitutes the confined Aruma Group with recharge interpreted as occurring in the neighbouring state of Saudi Arabia. It is understood that this confined aquifer is not currently used as a source of groundwater in Qatar (Shamrukh et al. 2012).

6.2.2 Recharge and Discharge of Groundwater

Natural groundwater recharge or replenishment in the ambient environment (non-urban areas) in Qatar takes place via the following methods:

Infiltration from rainfall, natural and artificial depressions, including lagoons, mostly through open karst and collapse structures.

Irrigation/urban returns and leakage and injection.

Marine water intrusion.

Transboundary groundwater movement.

Upward flow from lower aquifers, especially where the intactness of confining beds has been affected by structures/dissolution/subsidence.

The bulk of freshwater aquifer recharge in Qatar is derived from rainfall. Previous studies indicate that the northern part of Qatar receives the bulk of the recharge, while the southern part of Qatar receives significantly less rainfall recharge. The infiltration via open karst and collapse structures is the most important form of recharge given that they are widely distributed throughout the country, cover a large area, and provide a mechanism for rainwater to infiltrate and recharge the aquifers. The transboundary recharge through the deep regional aquifer of the Aruma Groundwater Basin is very low as presented in the MoE (2009) study. In addition to natural groundwater recharge, there have been efforts to instal rainfall recharge wells in Qatar since the late 1980s.

In urban areas, five main types of recharges can be defined:

Direct or diffuse recharge from widespread infiltration of rainwater at or near to the point where rain falls.

Localised recharge (also called indirect or focused recharge) where surface water runs off and accumulates in localised depressions with no surface water outlet.
Indirect recharge from leaking networks of potable water, TSE, and sewage systems, and on-site water lagoon disposal systems.
Artificial recharge from soakaways and injection wells.
Irrigation's return flows from irrigation of parks, landscaping along roads, house yards, and gardens.

In contrast, the main discharge outputs from aquifers are:

Groundwater abstractions
Evaporation losses from sabkha areas and other areas of shallow groundwater
Natural outflows to the sea
Aquifer transfer (mainly from the underlying UER to the overlying Rus Formation). Depending on the aquifer in question, such transfer may be a gain or a loss.

6.3 Materials and Methods

6.3.1 The Artificial Recharge Project in Qatar

As part of the national strategy for managing water resources in the State of Qatar, a comprehensive project was commissioned (2008) to investigate the natural and artificial recharge by recharge wells, entitled 'Studying and Developing the Natural and Artificial Recharge of the Groundwater Aquifer in the State of Qatar (GWAP)' (MoE 2009). Schlumberger Water Services (SWS) was contracted to carry out the overall work programme. The main goal of the project was to assess the current conditions of groundwater underlying Qatar and to evaluate methods to enhance groundwater recharge and storage. The main results of this project, along with more updated studies, are presented here, with a focus on identifying options to enhance fresh groundwater availability through artificial recharge from rainfall and runoff.

The project was composed of three components:

Component 1: **Well and Groundwater Inventory**, providing a survey of wells and groundwater abstraction and recharge.

Component 2: **Rainfall-Runoff and Recharge Model**, to study the relation between rainfall, runoff, and recharge.

Component 3: **Groundwater Information System**, to collate information relevant to managing Qatar's groundwater resources, hereafter referred to as the Qatar Groundwater Resources Information System (QaGWRIS).

In this chapter, focus is on Module 1 and 2.

6.3.2 SWAT2005 Modelling

In regions where excess rainfall causes significant amounts of runoff, the modelling of rainfall-runoff relationship can be a useful tool for water resources evaluation. A SWAT2005 (Soil and Water Assessment Tool) model (Neitsch et al. 2005) has been constructed to simulate the rainfall, runoff, and recharge processes daily for the whole country for the period 1993–2007 (14 years). The input data and information for SWAT2005 modelling for the MoE (2009) study involves significant files. Modelling with SWAT2005 starts out with delineating subbasins and identifying the course of the main surface water channels by analysing the DEM. The fundamental hydrological process in Qatar is the accumulation of surface water runoff in topographical depressions.

Inputs: Input for the model combines surface topography, spatially distributed soil hydraulic properties, land use, and meteorology data, including observed daily rainfall to quantify the potential for generating overland flow. The Pothhole feature was selected to represent the accumulated storage capacity of surface water runoff for all actual topographic depressions in a given subbasin. Using the topographic depressions in the numerical model, an original 21,786 depressions defined by DEM and ArcSWAT were filtered to 5,899 depressions in the final numerical model applying the criteria of depression volume for favourable depressions >10,000 m^2. The model area is separated into hydrologic response units (HRU). An HRU is delineated from the spatially distributed soil properties and land surface and slope. It is the smallest modelled unit of land for which daily water balance calculations are carried out and the number of HRUs is equal to 3,941 for the 65 subbasins, according to DEM analysis. Time series of daily rainfall for each rainfall gauging station (48 stations) as well as daily values of meteorological variables that are required to calculate potential evapotranspiration have been incorporated into the model. Recharge well infiltration capacity has been determined from 30 field-based injection tests across the country. This is a critical variable, which has large variability and uncertainty.

Using the input data, SWAT2005 simulates and calculates the generation of overland flow using the Soil Conservation Service (SCS) Curve Number method and daily rainfall amounts. The model generates water routing through open channels defining the velocity and rate of flow using Manning's equation. Water is routed through the channel network using the variable storage routing method. Finally, the model generates the groundwater recharge and flow using a simplified representation of the groundwater system. The total groundwater recharge is simulated by the model from water-filled depressions (focussed recharge) and from direct rainfall (diffuse recharge.

Calibration and Sensitivity Analysis: The model was calibrated by comparing the model outputs with observations of available/limited flood/ponding events. The comparative data consists of (a) the timing and duration of flooding/ponding events, and (b) estimated flood/ponding volumes for a limited number of storm events. Both calibration datasets have been deduced from the measured data of hydrograph responses and water level time series in 59 recharge wells that are equipped with automatic groundwater level recorders. However, some of these data have gaps limiting the comparison. Other observed data such as soil moisture content and water level in recharge wells were used to evaluate the model performance. Model inputs were adjusted from the base model according to the results of the calibration process to have an acceptable matching between model results and observed data. Coupled with a spreadsheet-based model to process the model output, estimates of surface water ponding, evaporation from ponded water, and the natural and artificial recharge to the groundwater beneath depressions were made. In addition, a sensitivity analysis was carried out to assess the value and impact of varying input parameters on model results.

6.3.3 Water Balance

The water balance of Qatar was calculated using the SWAT2005 model and a spreadsheet-based post-processing tool. The post-processing workflow of this modelling work provided the estimates of the maximum number of wells necessary to route surface water runoff within 5 days, selecting only those optimal depressions that have physical properties and dimensions that are favourable for receiving and storing runoff volume. A selection process was developed that enables the user to delineate the most favourable locations for future recharge well installation. This selection is based on eight criteria: thickness of the unsaturated zone, volume of depression, area ratio, shape factor of depression, flushing factor, long-term average annual storm rainfall, probability of rainfall events >10 mm/day, and the total dissolved solids (TDS) of groundwater (area contained within the 1,000 or 3,500 mg/l TDS contour). In this project, the selection process is based on only three of those eight criteria, which are the thickness of unsaturated zone (>10 m), volume of depression (>1,000 m^3), and minimum ponding depth >0.25 m. However, these criteria may be modified to optimise a phased recharge well instillation programme or cost–benefit analysis.

The water balance calculations were carried out for four scenarios: (1) without any artificial recharge wells, i.e., natural recharge; (2) with the current number of recharge wells; (3) with the estimated maximum number of recharge wells; and (4) with the optimum number of recharge wells. However, the total volumes of artificial recharge through the existing and the optimal number of recharge wells for the whole country for different hydrological years were obtained and are presented in this chapter.

Table 6.3 Work completed for the well inventory

Well inventories	DGPS[a] surveys	Pumping tests	Wells sampled	Injection tests	EC[b] profiling	Abstraction measurements
8,509	2,504	13	80	30	93	377

[a]DGPS: Differential Global Positioning System, [b]EC profiling: Electrical Conductivity profiling

Table 6.4 Count of wells by well status

Well status	Pumping (operational)	Pumping (non-operational)	Dry	Cancelled	Recharge	Deep	Total wells
No. of wells	4,205	2,625	534	808	313	24	8,509
%	49.4	30.8	6.3	9.5	3.7	0.3	100.0

6.4 Results and Discussion

6.4.1 Wells and Groundwater Status

6.4.1.1 Well Inventory

From the comprehensive MoE (2009) study, a complete survey and mapping of the existing wells in Qatar were carried out for the first time. Table 6.3 gives details of the work completed under this component.

Results show that there are only 4,205 operational pumping wells (close to 50%) out of the total 8,509 wells (Table 6.4). There are many wells that are either dry or cancelled. Table 6.4 summarises the count of wells according to well status. The distribution of operational pumping wells is shown in Fig. 6.5.

Table 6.5 gives the number of wells according to groundwater use. It is clear that the majority (74%) of wells are for farm and agricultural purposes (MoE 2009).

6.4.1.2 Groundwater Abstraction and Aquifer Characteristics

Abstraction measurements were done from 1975 mainly concerning the farm wells. Before the well survey of MoE (2009), measurements were done with rudimentary inaccurate tools. From 2009, an ultrasonic flowmeter was used to determine in an accurate way the pumping rates. In total, 377 pumping wells with different uses were measured in order to have the most representative samples covering the state. The abstraction values were used in the simulation model. Table 6.6 presents the total estimated abstraction by well use. It is anticipated that these values are valid till 2021.

Figure 6.6 shows the total estimated abstraction and total well count versus time from 1975 to 2009, including the share of farms. Both the estimated abstraction and well count are increasing with time (MoE 2009).

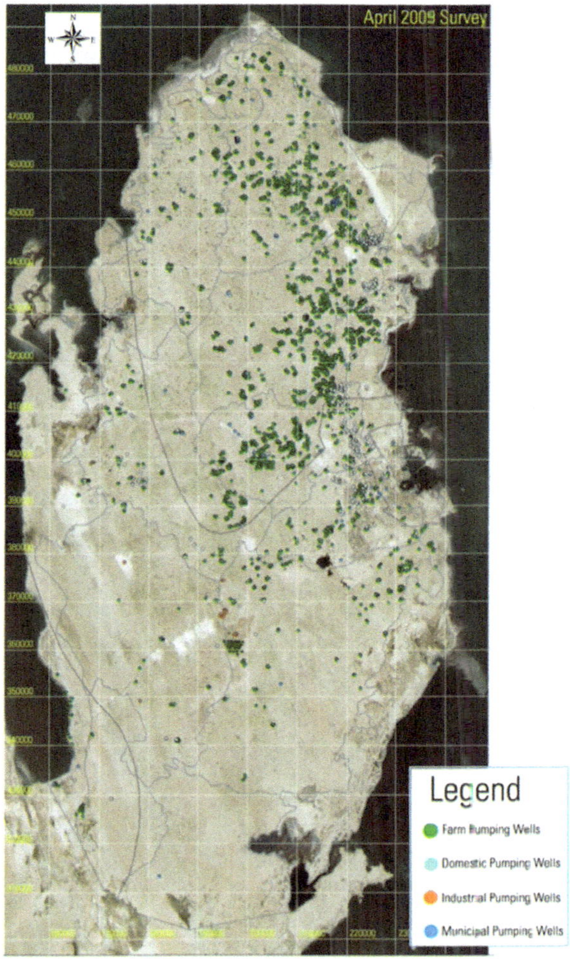

Fig. 6.5 Distribution of operational pumping wells in Qatar (MoE 2009)

Table 6.5 Count of production wells by well use

Type of well	Municipal well	Farm well	Domestic well	Industrial well	Total wells
No. of wells	1,265	6,299	665	280	8,509
%	14.9	74.0	7.8	8.3	100

Table 6.6 Total estimated abstraction by well use

Well use	Abstraction rate		Percentage of total (%)
	M^3/day	Mm^3/year	
Farm	866.4	226.0	90.9
Domestic	38.1	9.9	4.0
Municipal	35.7	9.3	3.7
Industrial	13.1	3.4	1.4
Total abstraction	953.3	248.6	100

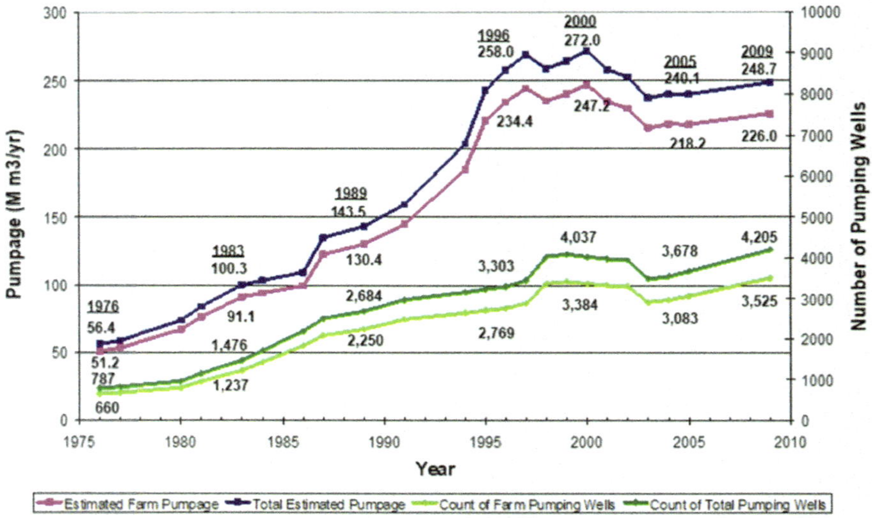

Fig. 6.6 Total estimated abstraction and well count versus time (1975–2009) in Qatar (MoE 2009)

Based on the analysis of old data and new measurements from 13 pumping tests done in the project, the aquifer hydrogeological characteristics were calculated (Table 6.7).

Table 6.7 Main hydraulic characteristics of aquifer basins in Qatar (Combined Rus and UER Aquifers where relevant)

Groundwater basin	Transmissivity (m^2/day)			Storativity (–)		
	Minimum	Maximum	Mean	Minimum	Maximum	Meann
Northern basin	19.7	1,320	305	0.000326	0.314	0.0423
Southern basin	0.265	26.2	1.44	0.000172	0.00357	0.00187
Southwestern basin	–	–	70.7	–	–	0.00102

6.4.1.3 Groundwater Level and Quality

Groundwater over-abstraction is considered the main driving force for the continuous general decline in groundwater levels and quality in Qatar. Annual groundwater abstraction in Qatar has increased from an estimated 5 Mm^3/year in the early 1960s to around 250 Mm^3/year in 2009 (MoE 2009).

For annual groundwater monitoring, a national network of observation wells was established during 1970 and 2009 and includes about 186 wells of which 148 wells were in operation in 2009. Analysis of groundwater level data obtained from 47 of the 73 telemetered recharge and observation wells during the period 2002–2007 shows an overall declining trend of on average 0.41 m during this 5-year period. Some wells show declines of over 1.0 m. In 2009, measurements of water level and salinity, the latter expressed in TDS, were collected at 2,669 locations during the well inventory. Interpretation of these groundwater level and salinity data shows that the majority of monitored wells have a long-term decline in groundwater level and a general degradation of groundwater quality indicated by increase in TDS values. Depths to the water ranged from less than a metre below land surface at Allat Aquifer in southwestern Qatar to approximately 80 m below land surface at Dukhan anticline in western Qatar. A potentiometric surface map was created from these data (Fig. 6.7). It is clear that the groundwater head is almost the same as the sea level near the coastline and slightly increasing going inland, except for deeper depressions (pink and red in Fig. 6.7) where might concentrated pumping is taking place.

Most critically, the distribution of freshwater, brackish, and saline water was well defined throughout the country at the project time. A groundwater salinity map was created from project data (Fig. 6.8). The freshwater (<1,000 mg/L) areas are shrinking all over Qatar, especially in the Northern Basin. For most of the aquifers, TDS ranges from 800 to 9,000 mg/L (MoE 2009). The TDS values of areas near the sea are very high and almost the same salinity as the seawater, which indicates a strong intrusion of seawater, possibly exacerbated by brine disposal from desalination plants. The freshwater lens located in the north-central part of Qatar has declined in area, from 15% in 1971 (data not shown) to approximately 2% of the country in 2009 (Fig. 6.8). The freshwater lens in 2009 is approximately 11% of the size it was in 1971.

Figure 6.9 is a time trend plot showing the areas underlain by TDS concentrations of <1,000, <2,000, and <3,000 mg/L between 1971 and 2009. This plot shows that the areas of freshwater (TDS <1,000mg/L) and brackish water suitable for irrigation (TDS <2,000–3,000 mg/L) are declining. Water with TDS concentration between 1,500 and 3,000 mg/L is considered very high in terms of irrigation hazard, and water with TDS higher than 3,500 mg/L is generally considered unsuitable for irrigation.

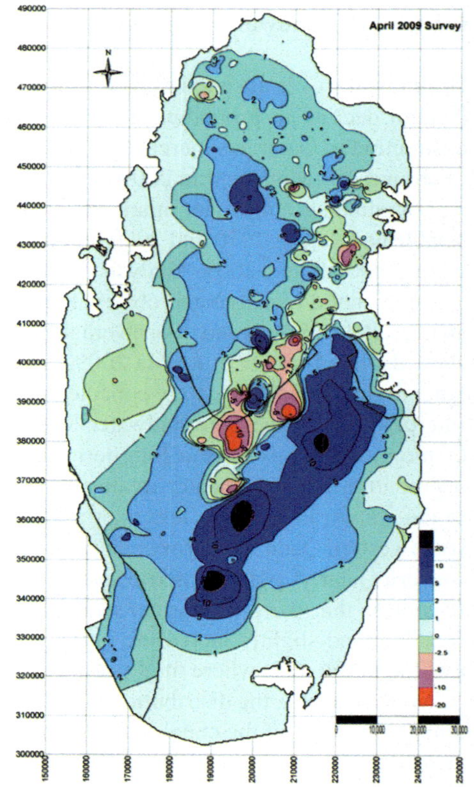

Fig. 6.7 Potentiometric surface map of Qatar in 2009 (mamsl) (MoE 2009)

6.4.2 Rainfall-Runoff and Recharge Modelling and Assessment

6.4.2.1 Natural Groundwater Recharge from Rainfall (Scenario 1)

For the purposes of this study, a drainage network of Qatar was generated using the watershed delineation component of the ArcSWAT2005 software package. DEM grid data, with an average spacing of 30 m, were used to delineate the drainage network resulting in 11 main catchments and about 65 sub-catchments. The watershed delineation is based solely on topographic elevation and although streams may not be present on the ground surface or discernible using satellite imagery, the watershed delineation by ArcSWAT2005 is considered to be a reliable indicator of the direction of surface water flow. The watersheds delineated show good correlation with the water courses that are discernible from satellite imagery. The total abstraction estimated for all wells is distributed in the model on the 65 subbasins following the distribution of wells done in phase one. The percentage of water abstracted from

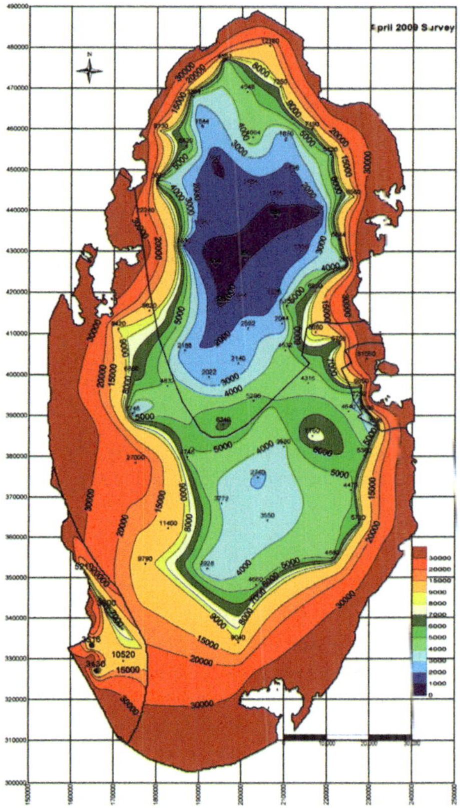

Fig. 6.8 Total dissolved solids (TDS) iso-concentration map of Qatar in 2009 (mg/L) (MoE 2009)

each basin is proportional to the number of existing pumping wells within that basin (MoE 2009).

For rainfall recharge, previous estimates have shown that the contribution from focused recharge appears to be the most important recharge mechanism in Qatar being, on average, 5 times greater than diffuse recharge (Kimrey 1985; Entec 1994). This is considered to be due not only to the concentration of surface runoff in the depressions, but also due to the elevated storage capacity and permeability of the formations underlying the depressions. Modelling carried out as part of this study indicated that focused recharge is locally between 4 and 9 times that of diffuse recharge (MoE 2009).

Natural groundwater recharge from rainwater was estimated every year according to rainfall quantity and infiltration rate into the subsurface (SoE 2021). Before the project, rainfall quantity over Qatar was estimated according to 18 rainfall stations installed across the country. For this purpose, Qatar was divided into 10 main inland hydrological catchment areas based on the surface water runoff (A to J in Fig. 6.10).

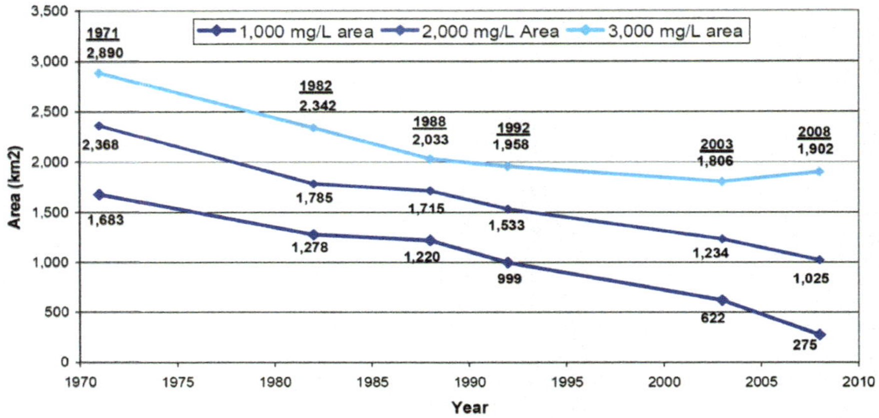

Fig. 6.9 Areas of freshwater and brackish water in North Qatar (1971–2009). Numbers indicate areas with TDS concentration below the given cut-off level (1,000, 2,000, and 3,000 mg/L) (MoE 2009)

It is worth noting that large coastal areas are Sabkhas, which generally operate as net evaporation areas rather than net infiltration areas (catchment K, no. 11). The average annual natural groundwater recharge from rainfall (in non-urban areas) for the period 1990–2018 was estimated at 68.5 Mm^3/year (SoE 2021).

6.4.2.2 Infiltrometer Experiments

A total of 81 double-rings oil infiltrometer tests were carried out at 46 locations throughout Qatar. The purpose of the infiltrometer testing was to validate the method of deriving saturated permeability (or constant infiltration capacity, Ksat) of soil based on particle size distribution data (Rosetta method) that were obtained previously for 586 points throughout Qatar (MoE 2009). The values of infiltration rate ranged from 0.012 m/day to 0.112 m/day, with an average of 0.039 m/day. The Ksat values ranged from 0.05 to 3.0 m/day. There is a moderate positive correlation of infiltrometer test results (correlation coefficient of 0.5) with the predicted average Ksat values derived using the Rosetta software. The large spread in the scatter plot of Ksat from the infiltrometer tests versus the Rosetta estimation could be caused by spatial separation of the test sites for the two methods and hence differences in soil structure and texture as well as conceptual differences in the experimental and derivation approaches. The Ksat derived from the infiltrometer tests was used for estimating recharge.

Results indicate that there is a localised, large spatial variability in constant infiltration capacity (up to an order of magnitude over short distances, e.g., 50 m). The range of constant infiltration capacities obtained from the testing indicates that no mapping unit or soil type has a distinguishable or unique range in infiltration rate.

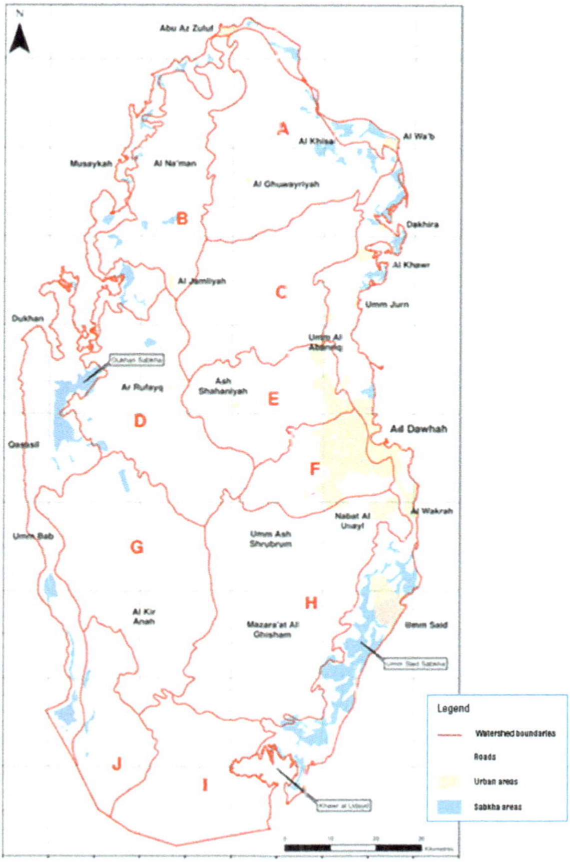

Fig. 6.10 Eleven main hydrological catchments of Qatar (MoE 2009) Catchment K (not indicated) includes the coastal *sabkha* zones as one catchment

Hence, there is no discernible trend suggesting that constant infiltration capacity is higher in e.g., the south compared to the north of Qatar. However, the infiltrometer test results show that the central parts of depressions generally have a lower maximum and constant infiltration rate than the peripheral parts of the depressions, presumably because of in fill of finer-grained material. This also indicates that recharge wells in depressions will significantly increase recharge especially in these locations.

6.4.2.3 Rainfall-Runoff and Recharge Analysis

The estimation of surface runoff is considered of great importance as it contributes significantly to indirect recharge, which is a main contributor to aquifer recharge in Qatar. Surface runoff estimates have been reported in previous studies (MoE 2009), though no direct observations of surface runoff have been made and all runoff estimations are understood to be based on mass balance calculations. Figure 6.11

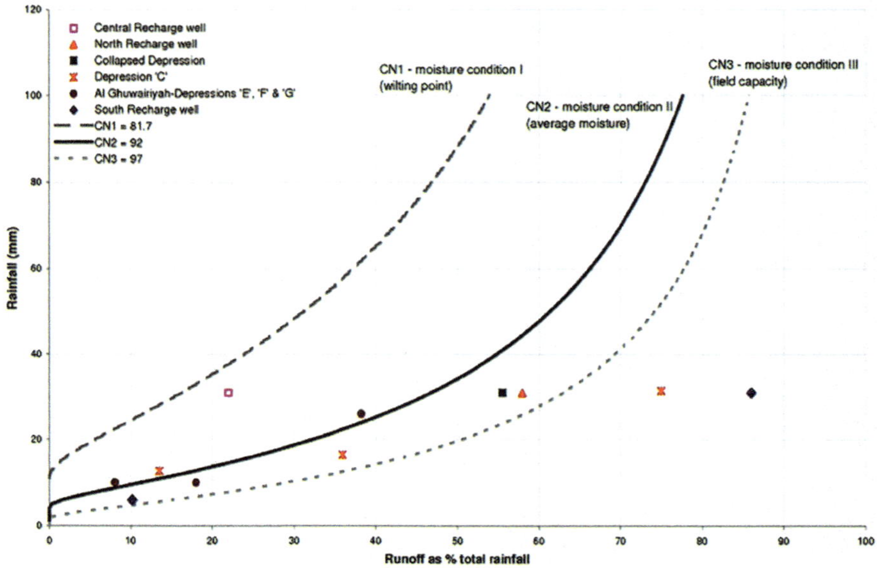

Fig. 6.11 Rainfall-runoff relationships for soils within depressions in Qatar (MoE 2009)

presents rainfall-runoff curves for soil with depressions and different rainfall events at several locations in Qatar (data after Gemmell 1976).

The review shows that the generation of surface runoff is determined by many factors including soil physical properties, antecedental soil moisture, vegetation, rainfall distribution and intensity characteristics, and macro/micro-topography. Gemmell (1976) suggests that runoff is likely to occur when a rainfall event exceeds 10 mm. Gemmell (1976), Eccleston et al. (1981) suggest that the distribution of rainfall over the catchment appears to be of greater significance than the total amount of rainfall. Two storm events were observed in a catchment in the northern central part of Qatar. Both storms produced a mean rainfall of 19 mm over 24 h hours over the catchment and the runoff amounts generated were 18 and 8%, respectively. Surface runoff can be expected to be between 15 and 25% of rainfall events of more than 10 mm per day (Pike et al. 1975).

An estimate of surface runoff has been made using data obtained from the national groundwater monitoring network and field monitoring of recharge wells. The automated pressure transducers show rises in water levels within the recharge wells when surface runoff enters the wells. Data indicate the depth of ponded water above the recharge wells following rainfall events and show that some runoff occurs after a total rainfall of only 5.9 mm within 24 h without any preceding rainfall events having occurred in the area. This contrasts with the findings of previous studies where at least 8–10 mm are required before runoff is initiated. It was found that the most intense rainfall occurred during a three-minute interval, during which a total of 1.7 mm rainfall was recorded, which equates to a rainfall intensity of 0.57 mm/min (34.2

mm/hr). When compared to the soil permeability of 0.31 mm/min, the rainfall intensity during this short period exceeded the soil permeability. Hence, surface runoff is likely to have been generated.

Other analyses show that although the rainfall intensities appear to be below the soil infiltration rate, surface runoff still occurs. This agrees with the findings of previous studies where runoff is understood to occur after less than 10 mm of rain has fallen in any 24 h period. However, the cumulative amount of rainfall should theoretically not affect the timing of surface runoff initiation. If the rainfall intensity is less than the saturated hydraulic conductivity then runoff should not occur. To summarise, from the analysis of data, it is considered in our analysis that surface runoff is likely to be initiated when:

- the rainfall rate or intensity exceeds the infiltration rate of the soils; and/or
- the cumulative amount of rainfall exceeds around 10mm in any 24-h period.

As a result, long-term average annual natural recharge of the aquifers is estimated in this study (based on the statistical average of the whole simulated period of 14 years) at around 89.4 Mm^3 derived from data on rainfall and natural infiltration over the whole of Qatar and in topographic depressions. This is somewhat higher than the previous estimate by SoE (2021) at 68.5 Mm^3/year.

6.4.2.4 Artificial Recharge Through Recharge Wells

Prior to the MoE (2009) survey, a total of 313 passive (gravity driven) recharge wells were installed across Qatar. Recharge wells are 'passive' gravity recharge wells in which no active injection is undertaken. Rainfall water that enters the recharge well infiltrates into the aquifer under the force of gravity. The recharge is facilitated and enhanced by the recharge well due to the penetration and short circuiting of less permeable layers by the well. Recharge wells are normally located in the lowest topographical point of depressions and along trajectories of shallow wadi systems. About 188 of these pre-existing recharge wells are located in depressions identified in this study and the remaining 125 are located outside. It is noted that 179 of the 188 recharge wells are located in depressions larger than 10,000 m^2. The majority of recharge wells are concentrated in the northern part of Qatar. The estimated long-term average recharge through the recharge wells based on modelling is 10.6 Mm^3/year (MoE 2009).

(a) **Water balance of existing recharge wells**

Modelling this scenario with the array of 188 existing recharge wells located in depressions was carried out. In order to do this, components of the groundwater balance have been integrated, as appropriate, within the numerical model in order to assess natural and artificial recharge. This modelling estimates the additional recharge (in addition to the natural recharge estimated above) resulting from existing recharge wells to 10.6 Mm^3/year as a long-term average across the country. The estimation of urban recharge was 28.8 Mm^3 as an annual average. The annual volumes of artificial recharge amounts are given as following.

Wet year (1995–1996): 33.5 Mm^3
Dry year (1999–2000): 0.73 Mm^3
Long-term average (1993–2007): 10.6 Mm^3/year.

(b) **Water balance of optimal recharge wells**

The model was applied to estimate the maximum number of recharge wells required to infiltrate all surface water accumulation within 5-days after the most extreme rainfall event in the average hydrological year (1996–1997). Only those depressions that were retained after applying selection criteria related to the unsaturated zone and physical dimensions of depressions have been selected to determine the most favourable depressions. The model calculates that 1,502 recharge wells would be required to infiltrate all ponding water in an average hydrological year in the most favourable depressions within a 5-day period. It is observed that there are already 114 recharge wells located in favourable depressions. Hence, 1,388 additional new recharge wells are required. The model indicated that 47 existing recharge wells are not required in the optimal total number of recharge wells, but we can keep them. Figure 6.12 shows the predicted number of recharge wells and their respective recharge volumes from the water balance model for the whole of Qatar in the optimal design of artificial recharge.

The new wells would be installed in the 1,146 selected favourable depressions distributed over Qatar. The model simulates that the 1,549 recharge wells would contribute an additional volume of 33.5 Mm^3 of water to the groundwater as an annual

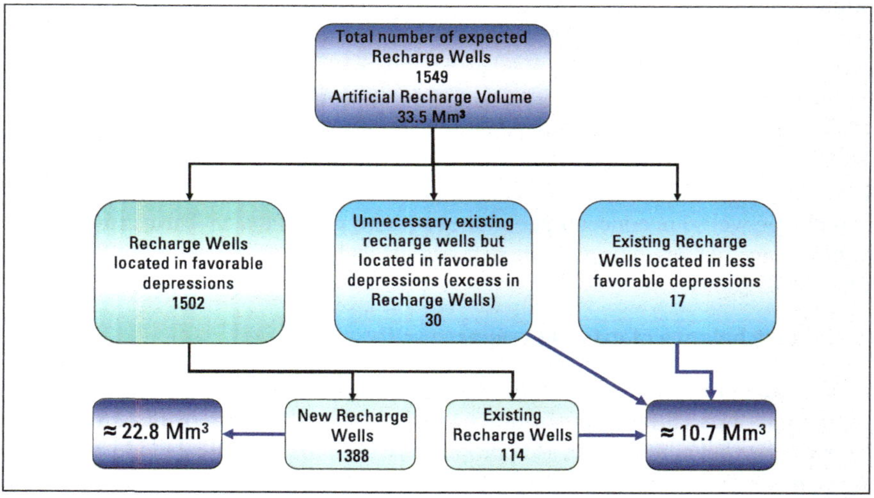

Fig. 6.12 Predicted number of optimal recharge wells and their respective recharge volumes from the water balance model (MoE 2009)

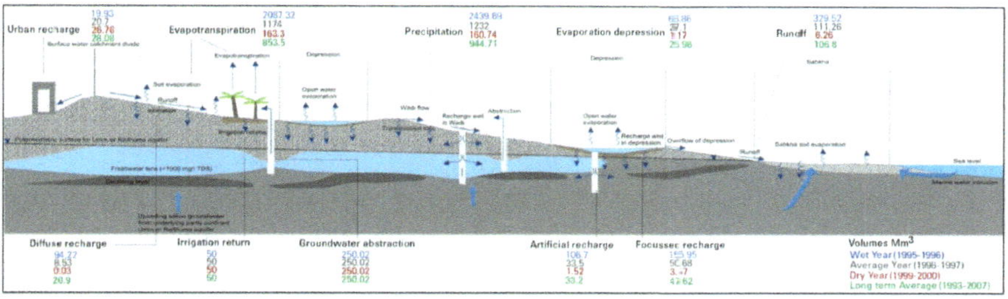

Fig. 6.13 Modelling results with the optimum number of recharge wells (MoE 2009)

average. The annual volumes of artificial recharge are calculated for characteristic hydrological years and given in Fig. 6.13.

Wet year (1995–1996): 106.7 Mm^3
Dry year (1999–2000): 1.52 Mm^3
Long-term average (1993–2007): 33.5 Mm^3/year.

A sensitivity analysis revealed that model output (runoff volumes, flood frequency, and percolation from the soil) is sensitive to the runoff curve number value, the permeability of the subsurface, the allowed time of ponding of water in depressions, the selection of favourable depressions, and the storm rainfall amount used in the design. It is important to note that the model results show that the southern part of Qatar has equal potential as the northern part of Qatar for artificial recharge although this is dependent on the temporal and spatial rainfall distribution (MoE 2009).

Summarised results of the model four scenarios calculating the recharge values are given in Table 6.8.

In conclusion, the current digital rainfall-runoff-recharge model for Qatar is an exploratory tool for identifying regions of particular interest, which can then become the focus of further detailed field and modelling studies. The consistent collection of high-quality runoff volumes on a number of key locations in Qatar is crucial for constraining model parameters and evaluating the performance of the model in the future. Finally, including uncertainty assessments of the modelling results, based on the sensitivity analysis, would be another important future exercise.

Table 6.8 Estimated long-term average recharge under different modelling scenarios

Scenario	Recharge from	Number of recharge wells	Added recharge (Mm^3/year)
1	Natural rainfall	0	89.4
2	Existing wells	188	10.6
3[a]	Maximum wells	4399	52.4
4	Optimal wells	1549	33.5

[a]Results of Scenario 3 are not described in this paper

6.5 Well Functionality

6.5.1 Design of Recharge Wells

Recharge wells are normally around 14 inches in diameter and have a 5–6 m length of perforated steel casing in the uppermost 3–4 m of the subsurface with the remaining length of casing above the ground surface and surrounded by a very coarse gravel pack (average diameter of gravel pack particles is around 10 cm) composed of locally derived limestone. The recharge wells are open boreholes to their maximum depth which, when they were drilled, reached the water table. The construction details of a typical recharge well are presented in Figs. 6.14 and 6.15. Figure 6.16 shows the spatial distribution of the existing 313 recharge wells. Most recharge wells are concentrated in the northern part of Qatar (reaching the UER aquifer) while the southern region has comparatively few (reaching the Rus aquifer). This is understood to be due to the apparent higher incidence of rainfall and higher water demand in the northern part of Qatar.

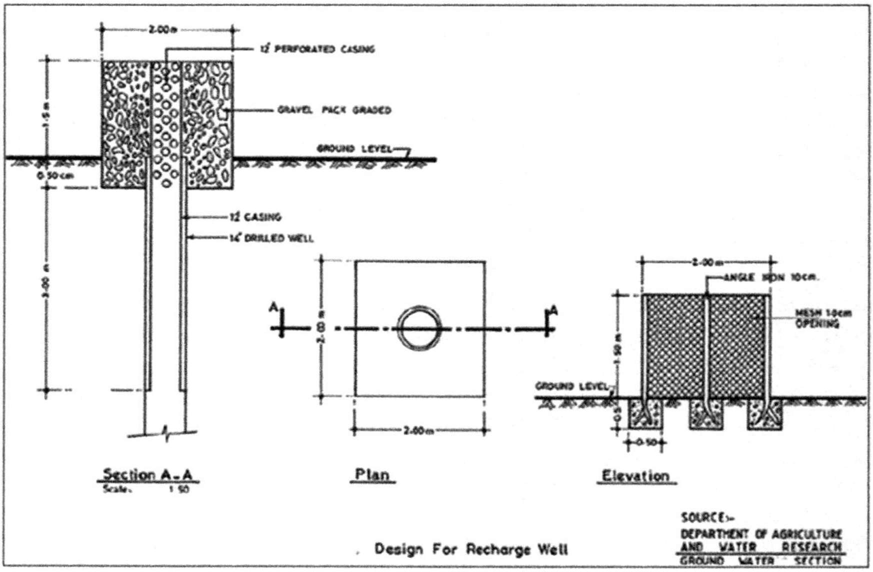

Fig. 6.14 Typical design of an open passive artificial rainfall recharge well (MoE 2009)

Fig. 6.15 Photo of a typical rainfall recharge well in Qatar (MoE 2009). *Note* Qatar National Grid. Horizontal and vertical grid lines are 10 km apart

6.5.2 Recharge Well Behaviour

When surface water runoff reaches the well, water percolates through the previous well casing and falls into the open well chamber and gradually fills it up. With a high enough flow rate of water entering the well, and the water level in the well often reach the ground surface, depending on the volume of the well chamber, the amount of surface water runoff generated. The flooded well chamber then temporarily stores the water until the water percolates into the unsaturated and saturated zones below (Fig. 6.17).

The head of water in the well will most typically be higher than the head in the surrounding aquifer and will result in water being driven into the aquifer until the head in the well equilibrates with the head in the aquifer. As this water dissipates into the surrounding aquifer, the groundwater level in the aquifer will increase. This groundwater level increase will be localised around the recharge well creating a mounding of groundwater surrounding the recharge well which will dissipate by migrating radially outwards away from the recharge well and eventually equilibrate with, and raise, the regional groundwater level.

Datalogger records from the 59 recharge wells, which have been equipped with automatic groundwater level measuring instruments, and covering the period 2001–2008, have been reviewed to determine the behaviour of recharge wells during rainfall events. Hydrographs have been plotted together with rainfall data from local manual measurements or from the closest available rainfall station. Many of the hydrographs show some response to rainfall events either as large head increases over short periods

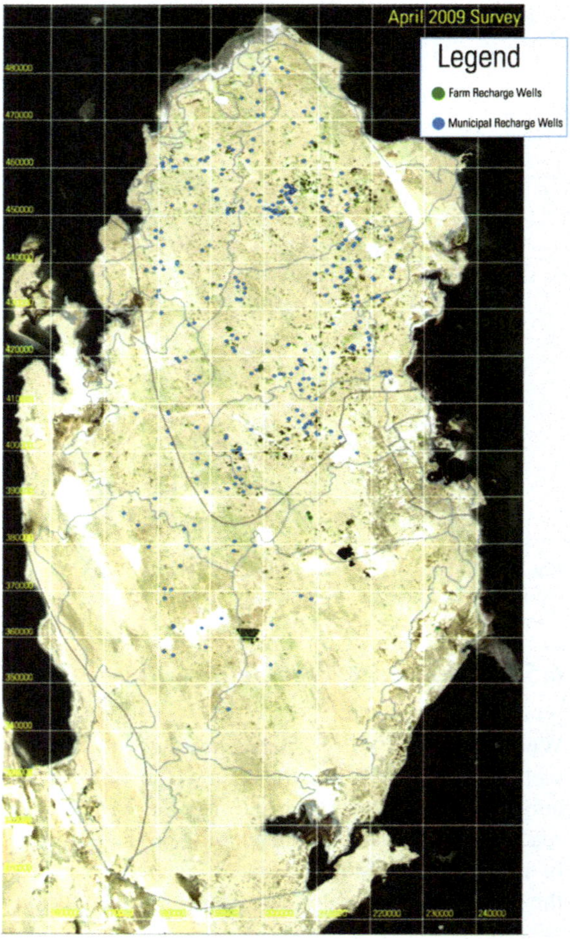

Fig. 6.16 Spatial distribution of installed recharge wells in Qatar (MoE 2009)

of time (e.g., 20 m head difference within 1 h of rainfall onset) or as a gradual, prolonged increase in water level. A number of different typical responses have been identified during the analysis of the hydrographs and these typical responses have been assigned to 'Type' responses. Eight types of responses were recorded from this work (MoE 2009).

6.5.3 *Recharge Well Efficiency*

There is no standardised approach to assessing the efficiency of passive artificial recharge wells of this nature and in this hydrogeological setting. For the purposes of

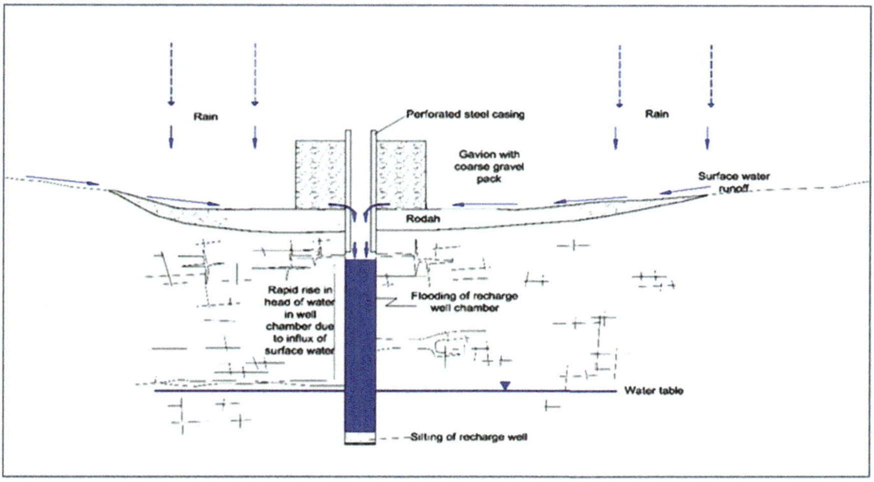

Fig. 6.17 Conceptual figure of artificial recharge through a recharge well (onset of ponding and flooding of well chamber) (MoE 2009)

this study, a methodology to assess the efficiency of recharge wells, based on natural recharge event data collected by pressure transducers, has been developed. The rate of decline in water level, or dissipation rate, in the recharge well is considered to be representative of the efficiency of the recharge well. One of the main findings from the analysis of well responses to recharge events is the reduction in the well dissipation rate which is observed to occur within days. However, the well dissipation rate is seen to recover during periods where no rainfall or recharge events are observed. The period of time required for a well to return to its original state of efficiency is a complex function of many parameters including the period of time the well has been flooded, the specific yield of the aquifer material in the unsaturated zone, matrix pore size (i.e., drainability under gravity), fracture morphology (density, connectivity, filling, opening, etc.) and the amount of diffuse recharge which may occur prior to, during, or in between well flooding events (the latter will influence the moisture content in the unsaturated zone).

Analysis of groundwater level data from 44 recharge wells in Qatar for the period 2001–2007 and injection tests from 27 recharge wells show that the capacity of a recharge well to infiltrate surface water varies widely. Recharge well infiltration capacity, referred to as efficiency in this study, has been determined from 30 injection tests in Qatar and is found to vary between 32 m^3/day and 2616 m^3/day with an average of 1,056 m^3/day. The lower injection rates are likely to be due to the nature of the aquifer material which may have a low permeability or low density and connectivity of fissures or karst features. By targeting areas of the subsurface known to have elevated permeability and by attaining inflow rates equal to or above the average of 1,056 m^3/day anywhere in Qatar is likely. Of the 73 recharge wells equipped with telemetered automatic logging systems, 44 showed responses to rainfall events. The efficiency of the recharge well is observed to decline with increasing surface water inflow. During prolonged periods of no inflow and no rainfall the well efficiency is observed to recover fully. This phenomenon is understood to be attributable to the decrease of suction forces in the unsaturated zone during recharge events and then a return to original suction forces during dry periods.

Future work improving the current design of recharge wells should focus on enhancing the potential for artificial recharge reducing the number of additional recharge wells required. Therefore, it is recommended that the design of any proposed future recharge well is reviewed. In addition, the selection of favourable recharge well location is important. Factors such as accessibility, land ownership, spatial variability of groundwater demand, and cost–benefit analysis must be considered when determining the number of and favourable locations for future recharge wells.

6.6 Current Status of Groundwater and Balanced Pumping

6.6.1 Groundwater Recharge and Sustainability

Since the 2009 project, there have been recent efforts to instal the recommended optimal 1,388 new recharge wells. Assessment of environmental impacts is necessary, especially the impact on green areas in these depressions due to the accelerated draining of the areas. This claimed negative environmental impact could be mitigated by making the lower part of the recharge well impermeable. This allows a portion of ponding water within the depression to be accumulated for some height and enhances the development of green area, though at the expense of recharge, as more water is used for evapotranspiration.

It is worth mentioning that even with installing the optimal recharge wells, the added amount of recharge water is relatively small, at 33.5 Mm^3 per year, and while this will be significant in a low and rare rainfall country such as Qatar, this will not overcome the deficit of the annual groundwater balance due to intense exploitation. Other interventions are necessary for sustainable groundwater management in Qatar, including both abstraction reduction and augmenting the groundwater recharge through other means (SoE 2021).

6.6.2 Current Groundwater Balance

Numerous groundwater balances have been produced for Qatar (MoE 2009). A recent groundwater balance was presented in SoE (2021) to estimate the groundwater abstraction level that achieves equilibrium status of the aquifers. To achieve quasi-equilibrium status, the discharge (both natural and human-driven) must equal the recharge over an extended time. In this case, the intrusion from seawater and

Table 6.9 Long-term groundwater recharge-discharge balance, 1990–2018 (average)

No	Balance component	Replenishment (Mm^3/year)	Discharge (Mm^3/year)
1	Recharge from rainfall (plus urban areas)	68.5 (plus 28.8)	
2	Recharge from existing artificial rainfall recharge wells	10.6	
3	Recharge from groundwater irrigation return flows (25% of irrigation volume)	46.0	
4	Recharge from TSE irrigation (fodder farms + landscape)	19.7	
5	Recharge from lateral transboundary aquifers	2.2	
6	Discharge by pumping of irrigation wells		227.0
7	Discharge by pumping of industrial, household, and municipal wells		9.4
8	Discharge by lateral, vertical and horizontal seeping		18.0
9	Total (average 1990–2018)	175.8	254.4

lower aquifers is almost nil, groundwater storage change is minimal, and groundwater salinization is minimised.

Using the data published in MoE (2009), SoE (2021), an updated assessment of the recharge-discharge groundwater balance in Qatar was done using data from 1990 to 2018 (Table 6.9). There is a current deficit of 78.6 Mn^3/year. This entails that the groundwater pumping should be reduced from 227.0 Mm^3/year to 227.0–78.6 Mm^3/year, or 148.4 Mm^3/year (Table 6.9).

6.7 Conclusions and Recommendations

A comprehensive project to study and design artificial recharge was carried out in Qatar. Groundwater abstraction over the last decades greatly exceeds the recharge and this continues to result in declining groundwater levels and degradation of groundwater quality through saline water intrusion. Indirect recharge is generated from surface runoff when the intensity of rainfall exceeds the rate of soil infiltration or the cumulative amount of rainfall exceeds 10 mm in any 24 h period. Modelling results show that the average annual artificial recharge from existing 188 recharge wells is 10.6 Mm^3/year. In the optimal scenario, modelling calculates that 1,388 new recharge wells would be required in addition to the existing recharge wells in depressions in order to infiltrate ponding water that is generated during an average hydrological year and within a 5-day period. The resulting additional artificial recharge volume would then amount to 33.5 Mm^3. Even with the optimal scenario, there is a deficit of 78.6 Mn^3/year on average in groundwater resources. An updated groundwater balance was calculated, in which the discharge equals the recharge, and the balanced groundwater pumping is about 150 Mm^3 for equilibrium status. More detailed work is needed to update the rainfall-runoff-recharge model to improve the recharge estimations. Methods such as geo-referenced radar-based data on rainfall coupled with automatic and manual rain gauge data would provide a more accurate estimation of generated recharge. The rainfall and groundwater monitoring network and equipment installed should be reviewed so that the optimum amount of data is obtained for precise recharge calculations. 3D groundwater modelling is very useful for groundwater balance calculations. Improvements for the design of the future recharge wells are important to avoid environmental impacts of green areas of recharge depressions.

Acknowledgements The authors are grateful to the rest of the staff of the MoE (2009) project for their inputs and contributions and to the editors of this book for their generous invitation to contribute a chapter to this publication. The view expressed in this paper is the authors' own and may not reflect that of their institutions and affiliations.

References

Abdel-Wahab A, Amer K, Shamrukh M (2008) An advanced technology for groundwater treatment in the northern aquifer of the State of Qatar. In: WSTA 8th gulf water conference. Manama, Kingdom of Bahrain

Alsharhan AS, Rizk ZA, Nairn AEM, Bakhit DW, Alhajari SA (2001) Hydrogeology of an arid region: the arabian gulf and adjoining areas. Elsevier

Amer KM (2008) From fully natural to fully artificial groundwater recharge, Qatar experience. Unpublished presentation slides, Water Research Section, Department of Agricultural and Water Research, Ministry of Municipal Affairs and Agriculture, Kingdom of Bahrain

Delene D, Shamrukh M (2016) Investing in rainfall enhancement: an innovative plan for arid regions. In: ARC'16, Qatar foundation annual research conference. Qatar

Dillon P, Stuyfzand P, Grischek T, Lluria M, Pyne RDG, Jain RC, Bear J, Schwarz J, Wang W, Fernandez E, Stefan C, Pettenati M, van der Gun J, Sprenger C, Massmann G, Scanlon BR, Xanke J, Jokela P, Zheng Y, Rossetto R, Shamrukh M, Pavelic P, Murray E, Ross A, Bonilla Valverde JP, Palma Nava A, Ansems N, Posavec K, Ha K, Martin R, Sapiano M (2019) Sixty years of global progress in managed aquifer recharge. Hydrogeol J Springer 27(1):1–30

Eccleston BL, Harhash IE (1982) The hydrogeology of Qatar. Ministry of Industry and Agriculture, Dept. of Agricultural and Water Research

Eccleston BL, Pike JG, Harhash I (1981) The water resources of Qatar and their development. Technical Report No. 5, Water Resources and Agricultural Development Project (FAO Funds-in-Trust), Ministry of Industry and Agriculture

Elsaid K, Shamrukh M., Elkamel A, Abdel-Wahab A (2017) Assessment of groundwater quality in State of Qatar for agriculture and domestic water supply applying membrane processes. In: WSTA 12th Gulf Water Conference. Manama, Kingdom of Bahrain

Entec (1994) Artificial recharge of groundwater in Northern Qatar. Report prepared for Ministry of Municipal Affairs and Agriculture, Dept. of Agricultural and Water Research, EntecHydrotechnica

Gemmel B (1976) Rainfall/run off, Infiltration and Recharge Data Obtained from Flooded Depressions in Qatar, February and March 1976. Report produced as part of 'Project for water and soil investigations for Agricultural Development' (QAT/73/007)
Harhash IE, Yousif AM (1985) Groundwater recharge estimates for the period 1972–1983. Ministry of Industry and Agriculture, Dept. of Agricultural and Water Research
Kadhem GM, Zubari WK (2020) Identifying optimal locations for artificial groundwater recharge by rainfall in the kingdom of Bahrain. Earth Syst Environ 4:551–565
Kimrey JO (1985) Proposed artificial recharge studies in Northern Qatar. Open-File Report 85–343, United States Geological Survey, Orlando, Florida
Lloyd JW, Pike JG, Eccleston BL, Chidley TRE (1987) The Hydrogeology of Complex lens Conditions in Qatar. J Hydrol 89(3–4):239–258
Lloyd JW, Eccleston BL, Pike JG (1981) The hydrochemistry of the groundwaters of Qatar. Technical Note No. 14, Ministry of Industry and Agriculture, Water Resources and Agricultural Development Project, State of Qatar, Food and Agriculture Organization of the United Nations
Ministry of Environment (MoE) (2009) Studying and developing the natural and artificial recharge of the groundwater aquifer in the State of Qatar—Project Final Report. GWAP project, Ministry of Environment (MoE), Qatar
Mohieldeen YE, Elobaid EA, Abdalla R (2021) GIS-based Framework for artificial aquifer recharge to secure sustainable strategic water reserves in Qatar arid environment Peninsula. Scientific Reports. Nature
Neitsch SL, Arnold JG, Kiniry JR, Williams JR (2005) Soil and water assessment tool theoretical documentation version 2005: Draft-January 2005, USDepartment of Agriculture-Agricultural Research Service, Temple, Texas
PSA (2021) Annual statistics of Qatar, Planning and Statistics Authority (PSA). Doha, Qatar
Pike JG (1978) The agroclimatology of Qatar. United Nations Development Programme, Food and Agriculture Organization of the United Nations. Technical Note 1 (new series), p 71
Pike JG, Harhash I, Gemmell BAP (1975) Rainfall and groundwater recharge-Qatar. Technical Note No. 24, Ministry of Industry and Agriculture, Integrated Water and Land Use Project, State of Qatar, Food and Agriculture Organization of the United Nations, United Nations Development Programme
SOE (2021) State of the environment report, Qatar. Ministry of Municipality and Environment, Doha, Qatar
Sadiq AM, Nasir SJ (2002) Middle Pleistocene Karst evolution in the State of Qatar, Arabian Gulf. J Cave Karst Stud 64(2):132–139
Shamrukh M, Al-Muraikhi AA, Al-Hamar, YI (2012) Exploring of deep groundwater in the Southwest Aquifer of Qatar. In: WSTA 10th Gulf Water Conference. Qatar
Streetly MJ, Kotoub S (1998) Determination of aquifer properties in northern Qatar for application to artificial recharge. Q J Eng Geol Hydrogeol 31:199–209

Chapter 7
Managed Aquifer Recharge in a Semi-arid Basin: A Case Study from the Souss Aquifer, Morocco

Soumia Gouahi, Mohammed Hssaisoune, Mohamed Qurtobi, Mohamed Nehmadou, Brahim Bouaakaz, Hicham Boudhair, and Lhoussaine Bouchaou

Abstract In the Souss River Basin (SRB), Morocco, with a semi-arid climate, the long-term droughts combined with groundwater overexploitation, cause serious concern about groundwater depletion and quality. Limitations in water supplies and increasing demand for water have caused greater interest in artificial groundwater recharge as a means of underground storage of seasonally variable surface runoff. The managed aquifer recharge (MAR) program was one of the main programs to ensure the sustainability of water resources, especially during drought periods. In the SRB, the MAR consists of direct recharge techniques through retention sills in the Souss River and its tributaries, in addition to rainwater harvesting. For the first time, MAR was carried out by the Hydraulic Agency of Souss-Massa Basin, from the Aoulouz

S. Gouahi (✉) · M. Hssaisoune · H. Boudhair · L. Bouchaou
Applied Geology and Geo-Environment Laboratory, Faculty of Sciences, Ibn Zohr University, Agadir, Morocco
e-mail: soumiagouahi@gmail.com

M. Hssaisoune
e-mail: m.hssaisoune@uiz.ac.ma

L. Bouchaou
e-mail: l.bouchaou@uiz.ac.ma

M. Hssaisoune
Faculty of Applied Sciences, Ibn Zohr University, Ait Melloul, Morocco

M. Qurtobi
Centre National de L'Energie, Des Sciences Et Des Techniques Nucléaires, Rabat, Morocco

M. Nehmadou · B. Bouaakaz
Hydraulic Agency of Souss-Massa Basin, Agadir, Morocco
e-mail: hichamboudhair@gmail.com

L. Bouchaou
International Water Research Institute (IWRI), Mohammed VI Polytechnic University (UM6P), Ben Guerir, Morocco

D. Saha et al. (eds.), *Managed Groundwater Recharge and Rainwater Harvesting*, Water Resources Development and Management,
https://doi.org/10.1007/978-981-99-8757-3_6

(in 1990) and Imi El Kheng (in 1993) dams, by regulating water resources through scheduled releases with an average volume of 83.7 Mm^3/year since its implementation. The maximum infiltration is highlighted between the middle and the upstream part of the basin where the hydraulic conductivity is very high. The recharge occurs mainly along the bed of the Souss River and its streams which is confirmed by the water level variability in some piezometers. The isotopic tracers are used to investigate the signal of natural recharge. The results show that the groundwater had experienced evaporation through the unsaturated zone before infiltration and that the recharge altitude ranged from 400 to 2350 according to a gradient of −0.18%/100 m.a.s.l. In addition, rainwater harvesting in hillside lakes and small reservoirs have also an important role in relieving the pressure on water resources, especially in the isolated areas in the upstream part of Souss. Aquifer enhanced recharge remains still insufficient to meet the water needs, there for other strategies and alternative solutions have been suggested in this study.

Keywords Managed aquifer recharge · Souss river basin · Groundwater recharge · Stable isotopes

7.1 Introduction

In arid and semi-arid areas, the groundwater level is continuously decreasing and is of increasing concern, especially in the Middle East and North Africa (MENA) region where several countries are already suffering from variable levels of water stress (Droogers et al. 2012; Schilling et al. 2012). Increasing water demand due to population growth has intensified the water shortage problems and put further stress, especially on groundwater (GW) resources which constitute the strategic resources in many arid areas (Hssaisoune et al. 2020a, b). This situation requires proper and innovative management such as rainwater harvesting, reuse of wastewater, and seawater desalination (Ghaffour et al. 2013; Waha et al. 2017).

As a geostrategic part of the MENA region, Morocco is characterized by an arid to semi-arid climate, with limited water resources and increasing needs (Hssaisoune et al. 2020a, b). The mobilization of water resources is among the main objectives of the country to fight against drought and to meet the water needs of all sectors of society (Ait Kadi and Ziyad 2018). This study concerns the Souss River Basin (SRB), which has the most important alluvial aquifer in southern Morocco and is strategically important for the country. The main problem of water resources management in this area is the adequacy of supply and demand. Indeed, the population growth and the agricultural development of the region induce water demand increases. This situation is aggravated by climate change impact, showing an increasingly deficient global balance (Hssaisoune et al. 2016). Moreover, the current trend of water resources in the Souss plain is alarming. According to Bouragba et al. (2011), the aquifer of Souss recorded an average annual deficit of 155 and 260 Mm^3 between 1968 and 1986 and between 1986 and 1994, respectively. The groundwater depletion is continuously decreasing to reach a deficit of 283 Mm^3 in 2007 (AEHSM 2007), and 292 Mm^3 in 2018. This decrease in groundwater quantity has affected the quality of water resources which became vulnerable to human pollution (Bouchaou et al. 2008; Tagma et al. 2009) and salinity increase (seawater intrusion, rock-water-interaction, etc.). Thus, artificial recharge and the use of non-conventional waters have accordingly become a necessity.

Managed Aquifer Recharge (MAR) with different sources of water, (e.g., reclaimed treated wastewater, rainwater harvesting, water desalination, etc.) is now being widely practiced to improve water scarcity in various parts of the world (Han et al. 2019; Dillon et al. 2009), especially in the arid and semi-arid regions (Rahman et al. 2013). According to Lerner et al. (1990), 3 types of MAR mechanisms are recognized: (i) direct recharge by vertical percolation into the vadose zone of water supply to the groundwater over soil moisture, (ii) indirect recharge by infiltration to the groundwater via the beds of the surface water channel and (iii) localized recharge by an intermediate process of groundwater recharge resulting from the near-surface water concentration in the absence of well-defined channels (de Vries and Simmers 2002). Tuinhof et al. (2002) had established a common glossary for different types of MAR to avoid the continual redefinition of terms.

This chapter aims to summarize different methods of MAR used in the SRB and evaluate the amount of recharge from the headwater in the aquifer. In our study, we will assess the effect of MAR on the water table, estimate the natural recharge altitude using isotopic data, and suggest other management techniques for mitigating the effect of droughts, groundwater overexploitation, and seawater intrusion within the SRB.

7.2 Presentation of the Study Area

7.2.1 Geographical Setting

The SRB is located in the central-southwest of Morocco and covers a surface of 16200 km^2 (Fig. 7.1). It is limited to the north by the High Atlas Mountains, by the Anti-Atlas to the south, by the Siroua to the east which is a mountain link between the two previous chains (High and Anti-Atlas), and by the Atlantic Ocean to the West (Fig. 7.1). The plain and mountain area represents 25 and 75% of the total basin area, respectively.

7.2.2 Climate Setting

The Souss-Massa area has a typical semi-arid to arid climate. The spatiotemporal distribution of rainfall is very important showing a clear decrease from the mountains

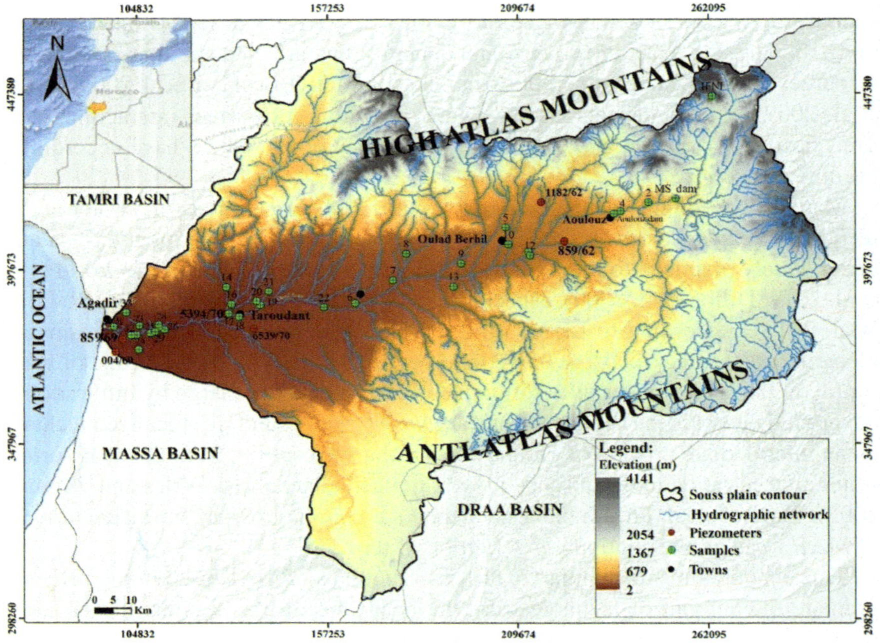

Fig. 7.1 Geographical localization of the Souss-Massa river basin

to the plain (Fig. 7.2). The average rainfall is 250–300 mm/year in the plain and around 500–600 mm/year in the mountain (Abahous et al. 2018, 2020). It is relatively scarce and varies considerably with location due to the variable topographic features and climatic conditions (Bouchaou et al. 2011). The annual mean temperature ranges between 14 and 18 °C in the High Atlas and Anti-Atlas. The average temperature in the plain is higher and can reach up to 20 °C (Bouragba et al. 2011).

The rainfall period extends from November to March, and the dry season can cover up seven months (April to October). Although the potential evaporation is very high in the plain and can reach up to 2,000 mm/year (Bouchaou et al. 2011), the rainfall exceeds the evaporation during winter in the mountain areas that become the main recharge areas. Figure 7.2 shows the seasonal and spatial variations of rainfall in the Souss area.

7.2.3 *Water Resources in the SRB*

7.2.3.1 Surface Water Resources

The principal river in the region is the Souss which drains a watershed of 65% of the total area of the Souss-Massa region (16 200 km^2). According to the Hydraulic Basin

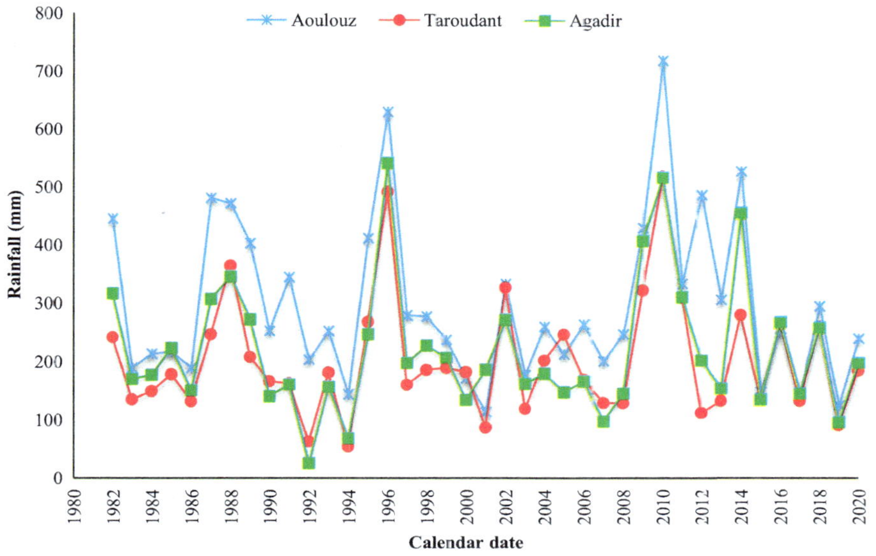

Fig. 7.2 Annual variation of precipitation in three main meteorological stations (Agadir, Taroudant, and Aoulouz) in the Souss Basin (From 1982 to 2020)

Agency of Souss-Massa, the mean contributions drained by the various tributaries of the SRBin a normal hydrological year, represent 429 Mm^3/year distributed as follows:

Oued Souss at Aoulouz: 182 Mm^3/year.
Right bank tributaries of the Souss: 189 Mm^3/year.
Left bank tributaries of the Souss: 58 Mm^3/year.

These resources are very sensitive to climate variability; indeed the rivers are dry during the year except for the violent and brutal floods that occur during the rainy days characterized by torrential rains very spaced in time. Nevertheless, these rivers can have a permanent regime in the high altitudes of the High Atlas and Siroua as a result of the snowmelt and the springs.

7.2.3.2 Groundwater Resources

The structure of the aquifer is a synclinal faulted oriented East–West, which is filled with deposits extremely heterogeneous from Neogene to Quaternary (Hssaisoune et al. 2012). The general flow is from the East (Aoulouz) to the West (Ocean) and the average hydraulic gradient decreases from upstream to downstream. The overall recharge of the Souss aquifer is estimated at an average of nearly 268 Mm^3 in a year of normal rainfall. Indeed, there is an excellent exchange interaction between surface water and groundwater through the bed of the Souss River (Bouchaou et al. 2008; Bouragba et al. 2011). The recharge of the Souss plain aquifer is mainly based on runoff, especially from the floods of the dense hydrographic network and locally from the rainwater infiltration. The outputs of the aquifer's groundwater are the flow toward the ocean, evaporation, the resurgence of the Souss River, and the extraction by pumping.

7.3 Data and Methods

Our first step in this study was the collection and updating of existing datasets. Academic and professional works of literature and field observations spanned a period of more than 20 years within the SRB. We focus especially on PDAIRE 2007 (ABHSM 2007), and other artificial recharge unpublished reports.

Dataset used in this work is provided by the Hydraulic Agency of Souss-Massa Basin, it corresponds to the annual precipitation from 1980 to 2020, the historic evolution of the water table from different piezometers within the Souss plain, in addition to the amount of flood of Souss river. These data will be used to assess the reaction of the piezometric level to the different water inflows in the basin (e.g., dam water release, flood, etc.).Three piezometers have been selected to represent respectively the upstream, the Middle, and the downstream part of the SRB; 859/62, 5394/70, and 859/6; their total depths are respectively 108.8, 72.23, and 56.70 m.

To better characterize and track the recharge signal, we use the isotopic tracers as indicators of the recharge area, the origin of water, and processes carried out during the percolation. The data were collected during November, October, January, and March, respectively from 2020 to 2021, along the Souss river, from the upstream of the basin to the estuary of the Souss basin. For surface water, 3 samples were taken respectively from the Mokhtar Soussi dam, the Aoulouz dam, and the river downstream of the Aoulouz. For groundwater, 29 samples were taken at different wells located on the right and left bank of SRB, and 3 samples of rainwater were taken respectively from High Atlas Mountains (Ifni Lake), Middle Souss (Taroudant), and the downstream of the SRB (Agadir). The analyses of the stable isotopes ($\delta^{18}O - H_2O$ and $\delta^2H - H_2O$) were carried out using the Cavity Ring-Down-Spectroscopy laser spectrometer (Picarro L2120-i) in the laboratory of Applied Geology and Geo-Environment (LAGAGE) at Ibn Zohr University.

The analyses of isotopes were performed using standard samples from the International Atomic Energy Agency (IAEA). The analytical accuracy of the laser spectrophotometer (Picarro) is on the order of $\pm 0.1\%$ for oxygen 18 and $\pm 1.5\%$ for Deuterium. The values of the isotopic results are presented in standard notation per thousand (%) compared to V-SMOW (Vienna-Standard Mean Ocean Water).

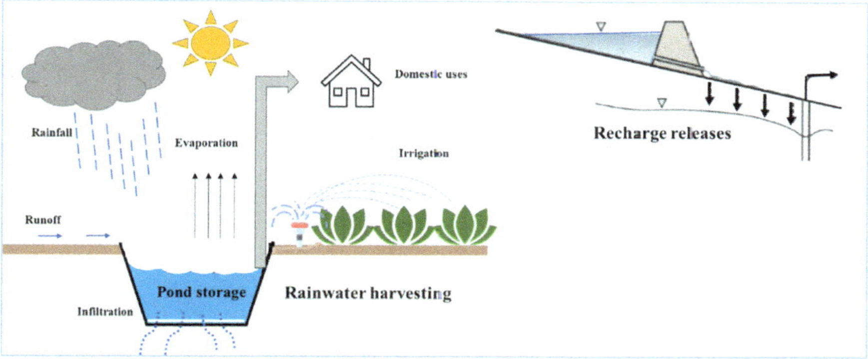

Fig. 7.3 Types of MAR used in the Souss river basin (After Dillon 2005, modified)

7.4 Results and Discussion

7.4.1 The Approach of MAR in SRB

The MAR of the Souss aquifer began in September 1991. It includes Aoulouz and Mokhtar Soussi regulation dams, built on the Oued Souss; the Imi El Kheng dam constructed on the Oued Talekljount (a tributary of Souss River), and more than sixteen infiltration sills on Souss River and its tributaries (Fig. 7.4). The principal goal of this operation is to introduce a certain volume of water into the aquifer through releases.

The objectives assigned to the MAR of the Souss aquifer are as follows:

Stopping the decrease of the water table through the infiltration of water stored in the Aoulouz dam;
Reducing losses of water upstream of Aoulouz dam and mobilizing them in the useful zone of the recharge (center of the plain);
Reducing or even canceling the losses into the sea by storing water during flood periods and releasing it appropriately.

The decision to release water is taken by the ABHSM which sets the volume and the flow rate. Two types of techniques have been identified and used by the Hydraulic Basin Agency of Souss-Massa to recharge the aquifer (Fig. 7.3).

7.4.2 Recharge Releases from Dams

This scenario aims to reduce the deficit recorded on the water balance of the aquifer, and at raising the piezometric level in the depressed sectors, with a range of flow from 25 to 50 m^3/s. Along the Souss River and its tributaries we can distinguish sills

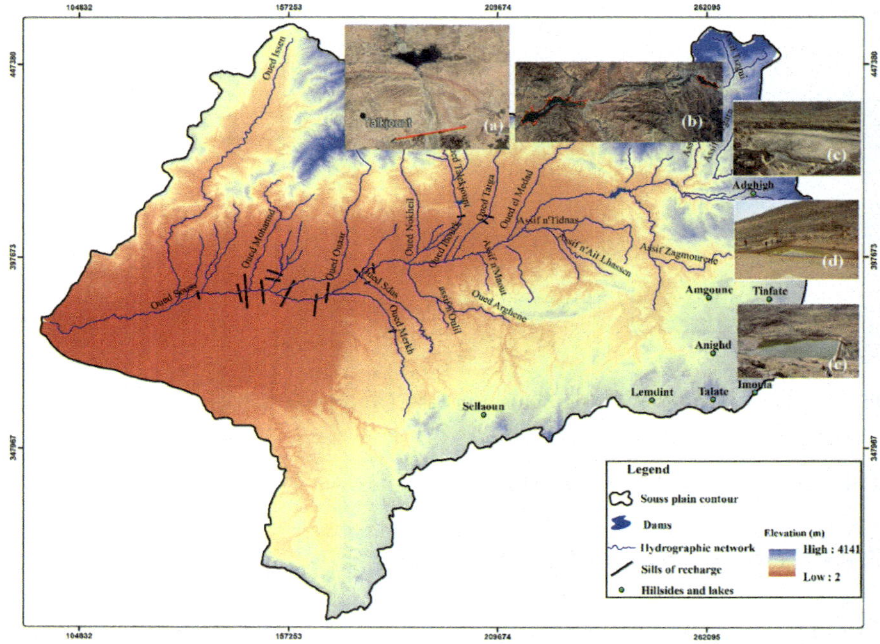

Fig. 7.4 Location of different recharge sills along the SRB and its tributaries, and the hillsides **a** Talkjount sill and Imi Lkheng dam, **b** Aoulouz and Mokhtar Soussi satellite image, **c**, **d**; and **e** represent images of Adghigh, Tinfat, and Anighd hillsides, respectively

slowing down the flow of water and increasing the wetted surfaces along the beds; and diversion of flood to drainage areas (or possibly water storage structures).

(i) **Sills delaying floods (with or without spreading)**

The following tributaries of the SRB are not developed and offer usable contributions for MAR, Oued Boussriouil, Oued Messgui, Oued Bni Mhamed, Oued Ouziki, and Oued Ouaougdi on the right side, Oued Aourga, Oued Sdas, Oued Araghan, and Oued Tangrfa on the left side of the SRB (Fig. 7.4). The water resources targeted are the floods of the said tributaries, evaluated at 16.2, and 12.3 Mm^3/year for the right side and the left side, respectively.

(ii) **Retention sills in the bed of the SRB**

To reinforce the infiltration in the Souss aquifer, 17 retention sills were built along the bed of SRB (Lalla Tamsit, Hamria, Kharouba, Bahssia, Frieja upstream, Frieja downstream, Msafer, Sidi Amara, and Souiguia). The aim is to delay the flow and allow the spreading of the water wave over a large area and thus promoting the infiltration of water. Some of these sills are equipped with Seguias for water diversion for irrigation (Sebt El Guerdane). Figure 7.4 shows the location of the infiltration sills. However, most of these sills are damaged and require rehabilitation.

7.4.3 Rainwater Harvesting

Rainwater harvesting and MAR have been a traditional solution to water scarcity problems in the arid and semi-arid regions for thousands of years, especially in the Middle East and North Africa areas (Hashemi 2015). Rainwater harvesting could provide some additional resources, although minimal in quantity, but important, given their spatial distribution in the most isolated areas in mountains. A set of structures that are part of the MAR in the SRB, also contribute to the mobilization, on a smaller scale, of water resources. As well as the small reservoirs and hill lakes, these schemes are not considered as regulation works, they are considered rather for their role in the aquifer recharge, protection against floods as well as local functions such as livestock watering. These hill dams were built in the 90s with the support of the authorities in areas that do not have alternative resources for alluvial aquifer recharge, irrigation, and watering of livestock. According to ABHSM (2007), these lakes and hillsides provide an amount of 5 Mm^3/year. These hillsides are illustrated in Table 7.1 (Alahiane 2017; Elmouden et al. 2016).

7.4.4 Groundwater Depletion

Groundwater depletion is the main issue of water management in the SRB due to climate variabilities (successive years of drought) and intense pumping, especially for agriculture purposes. The groundwater level is continuously decreasing, and the amplitude of depletion becomes more important, as illustrated in Fig. 7.5 (IRE 1182/62), due to a lack of precipitations, a sharp drop of about 2 m/year in the piezometric level, in the upstream Souss, has been recorded between 2011 and 2020. In the Middle Souss (IRE 6539/70), the piezometric level shows a reduction of the water table of about 30 m in only 10 years (2008–2018), as for the downstream part of the basin (IRE 04/69), the drop in the piezometric level is particularly pronounced after the year 2000, some small piezometric fluctuation can be noticed but does not modify the trend of this depletions.

The groundwater depletion in the Souss basin can be explained essentially by the successive years of drought and the intense pumping, especially for agriculture.

Table 7.1 Synoptic characteristics of hill dams (Alahiane 2017)

	Name: ADGHIGH Coordinates: 30°43′7.08 ″N/7°46′28.87 ″W Year of construction: 2000 Affluent: Oued Tassa Type of dam: Small earthen dam Normal capacity: 18,000 m^3. Use: Livestock watering and irrigation
	Name: IMGOUNE Year of construction: 1995 Affluent: Oued Tassa Type of dam: Earthen dyk Normal capacity: 30,000 m^3 Use: MAR, Livestock watering, and irrigation
	Name: TINFATE Year of construction: 1996 Type of dam: small concrete dam Normal capacity: 1424 m^3 Use: Recharge, watering of livestock, and flood control
	Name: ANIGHD Year of construction: 1996 Type of dam: Small concrete dam Normal capacity: 22,000 m^3 Use: Livestock watering, MAR
	Name: IMOULA Year of construction: 1999 Type of dam: Concrete dam Normal capacity: 3500 m^3 Use: MAR

(continued)

Table 7.1 (continued)

	Name: TALATE Year of construction: 1999 Type of dam: Concrete dam Normal capacity: 1680 m^3 Use: Irrigation, recharge
	Name: LEMDINTE Year of construction: 1999 Type of dam: Concrete dam Normal capacity: 4083 m^3 Use: Irrigation, recharge

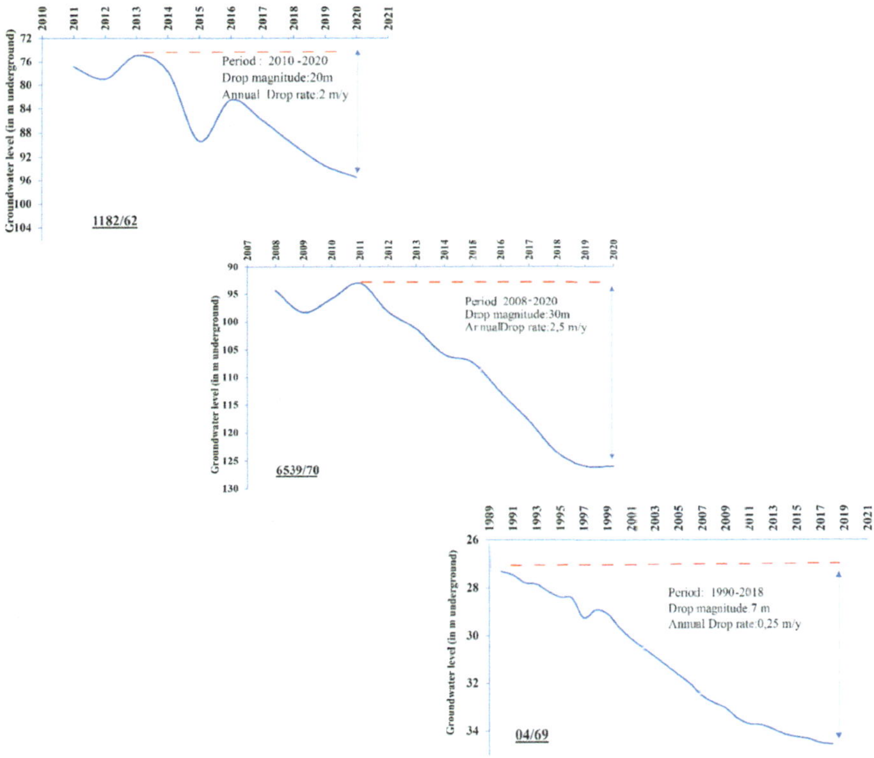

Fig. 7.5 Evolution of groundwater level in the upstream part (1182/62), in the middle part (6539/70), and the downstream part (04/69) of the Souss River Basin

7.4.5 *Assessment of Groundwater Recharge from Aoulouz Dam*

7.4.5.1 History of Water Releases

Figure 7.6 shows both the volumes released from the Aoulouz dam since its implementation and the amounts of floods in three main hydrological stations from SRB.

In this figure, we also notice the effect of the dam on the river discharge at upstream (Aoulouz station) compared to Ait Melloul, where the river can receive the contribution of other tributaries mainly from the High Atlas part.

To identify the recharge amount, the monitoring is done by differential controlling discharge in three main hydrological stations: (i) Aoulouz in the upstream near the dam, (ii) Taroudant in the middle of Souss plain, and (iii) Ait Melloul in the downstream part (Fig. 7.6).

For optimal infiltration, the determination of the flows to be released from the Aoulouz dam (without any loss in the ocean), is controlled by several factors among others, the duration and quantity of rainfall in the plain, and the duration of the stream flow. The frequency and the volumes of the releases are dependent on the inflows to the Aoulouz dam. After floods, the tributaries lead to an oversaturated alluvial deposit, which consequently reduces the infiltration capacity of the bed. Thus, a flow of 60 m^3/s can reach the bridge of Ait Melloul, however, during dry periods; a flow of 100 m^3/s may not reach the ocean.

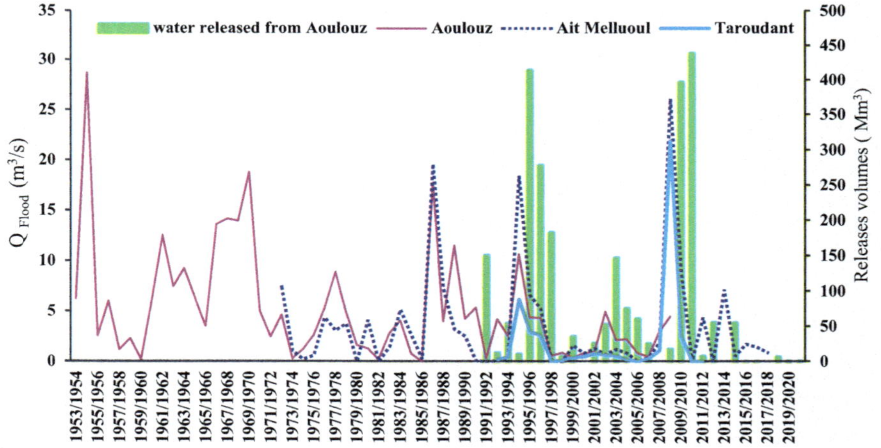

Fig. 7.6 The flow of floods in Aoulouz (upstream), Ait Melloul (downstream), and Taroudant (Middle) versus the volume of water released from the Aoulouz dam since 1991

Before the implementation of the Aoulouz dam, the frequency and intensity of the floods are noticeable, especially in the upstream part (Aoulouz). The years 1995–1996 and from 2009 to 2011 are years of excess rainfall as well as the flow contribution and the volumes of the releases are important (60, 5 Mm^3 between the period of 19–29/12/2010) (Fig. 7.6).

7.4.5.2 Groundwater Level Reaction

To evaluate the impact of these volumes released by the Aoulouz dam on the water table, the response of three representative piezometers close to the river is monitored in time and space after water releases (Fig. 7.7).

We focused particularly on the releases during periods of a lack or an absence of precipitation, and those with excess rainfall volumes to compare the reaction of the water table in different situations (Fig. 7.7).

The effect of the releases can be seen after the 1991s, which corresponds to the implementation of the dam. The positive effect of the artificial recharge persists in 1992–1993 when there was little rainfall, particularly in the upper and middle Souss, represented respectively by the piezometers 859/62 and 5394/70 (Fig. 7.7), indeed, according to the ABHSM (1991), a volume of 150 million m3 was released from the Aoulouz dam and completely seeped into the aquifer without suffering any losses to the Atlantic. After 1998, a significant fall in the water table was recorded throughout the Souss plain (Fig. 7.7), and the water table no longer responds to the artificial recharge, which remains very poor given the drought of this period. However, the years 1995–1996 and 1996–1997 received an average of about 422 mm of precipitation in the plains and even more in the mountains, which exceeds the normal average of 250 mm for the Souss plain. Therefore, the releases of the dam are also important, which was reflected in the rise in the groundwater level, which rose to 4 m in the upper and middle Souss (859/62 and 5394/70, respectively) and about 2 m in the lower Souss (859/69). The same reaction was observed in 2009–2010

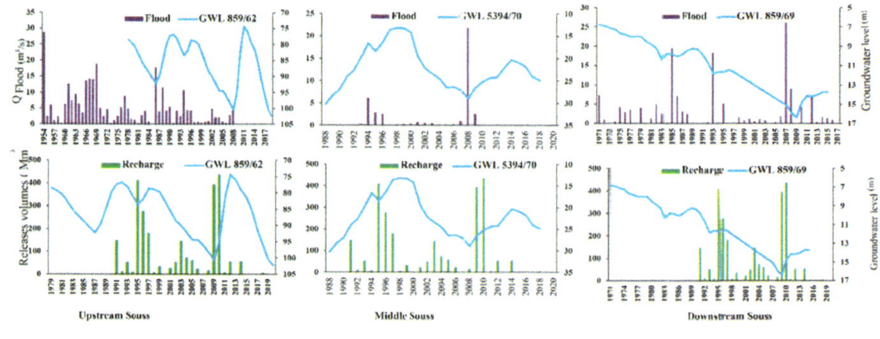

Fig. 7.7 Evolution of water table level in the Souss basin versus the flood amounts and the release volumes from Aoulouz

and 2010–2011 when the rains were very heavy and therefore the largest amount of water (480 Mm^3) was released from the dam. As the graph shows, the piezometric levels respond less significantly to the flood volume than to the dam releases. This is mainly due to the shorter residence times of the flood. It is also noted that the frequency and intensity of flooding have decreased after the construction of dams, in addition to the role played by the sills installed along the river to improve flood delay and encourage water infiltration, in particular in the upstream part of the basin from Aoulouz to Taroudant. In general, this release has contributed to the mitigation of the drop in piezometric levels and allowed farmers to irrigate by diverting water through traditional Seguias (irrigation channels).

A strong spatial variability of the recharge appears in the study area but remains difficult to quantify. According to the lithological structure and morphology of the bed of the Souss river and the differential discharge measured in the three gaging stations, we can say that the upstream part benefits the most from the natural recharge process compared to the central part and the downstream part. This is due to the granulometry of these areas as well as their location regarding the recharge zones. Therefore, it is necessary to develop alternative recharge processes that promote the replenishment of aquifers in downstream and middle Souss.

7.4.6 Infiltration Area

According to the ABHSM (2004), the data analysis of releases in each section along the Oued Souss shows that nearly 14% of the volume of releases is taken directly for irrigation by channels between Aoulouz and Taroudant. The remaining part of 86% is divided into an infiltrated part that recharges groundwater and evaporated part throughout the river. The amount of infiltration within the Souss River bed after dam releases is influenced by the lithology and the structure of the aquifer (Hssaisoune et al. 2012, 2016). The volumes infiltrated during the releases along the SRB are distributed by section as follows (Fig. 7.8):

29% Between Aoulouz and the confluence with the Oued Lamdadd.
57% Between the confluence of the Oued Lamdad and the Taroudant bridge.
7% Between the bridge of Taroudant and Oulad Taima bridge.
5% Between the Oulad Taima and Ait Melloul.
3% Between the gaging station of Ait Melloul and the ocean.

The observed variation in infiltration rate can be explained primarily by the decrease in sediment gradation along the river from upstream to downstream and the tectonic structure of the plain (Hssaisoune et al. 2012, 2016).

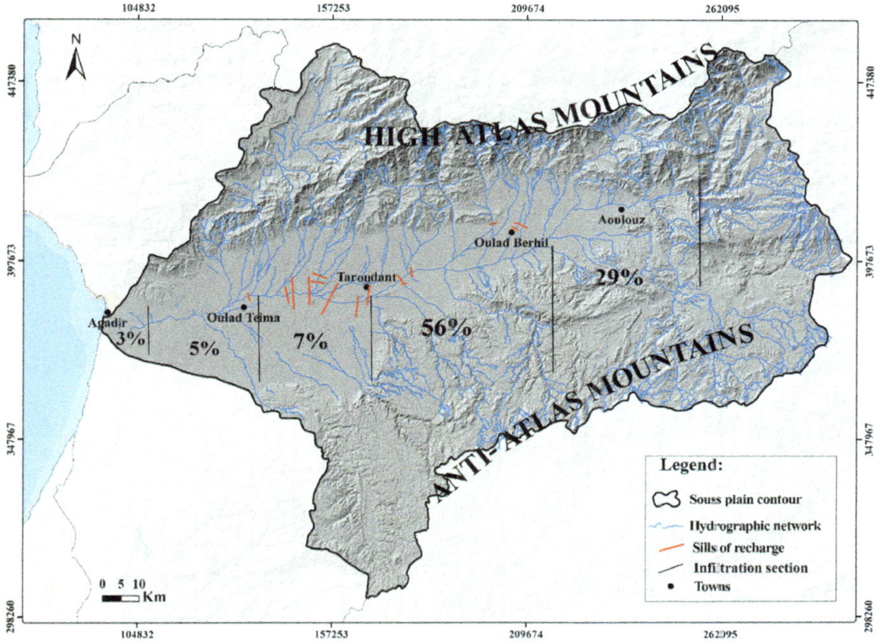

Fig. 7.8 Infiltration rate in the Souss River Basin after Aoulouz dam releases

7.4.7 Natural Recharge

7.4.7.1 Stable Isotope Contribution

Values of stable isotopes ($\delta^{18}O - H_2O$ and $\delta^2H - H_2O$) at the Souss Basin-scale range from −2.45 to –8.3% for the δ ^{18}O and from −16.1 to −53.2% for the δ^2H (Table 7.2). Enriched values were measured in the downstream part of the basin and relatively depleted values were recorded upstream and in the center. The graph of the relationship between $\delta^{18}O - H_2O$ and $\delta^2H - H_2O$ (Fig. 7.9) shows that most of the samples are close to the Global Meteoric Water Line (Craig 1961) (Eq. 7.1) and the Local Meteoric Water Line for the High Atlas Mountains (N'da et al. 2018) (Eq. 7.2), except a sample of groundwater (sample 31) from the downstream in the right bank of Souss that represents an enrichment of ^{18}O and ^2H related to evaporation effect.

$$\text{GMWL} : \delta^2H = 8 * \delta^{18}O + 10\% \tag{7.1}$$

$$\text{LMWL} : \delta^2H = 8 * \delta^{18}O + 13.5\% \tag{7.2}$$

Table 7.2 Mean isotopic composition of surface and groundwater and estimated recharge area of samples collected from the Souss Basin

Sample ID	Elevation (m)	$\delta^{18}O$ (% vs. VOW)	$\delta^{2}H$ (% vs. VOW)	Estimated recharge altitude (m)
R.Ifni	2400	−8.3	−47.9	2384
R.Taroudant	254	−7.92	V53.2	2180
R. Agadir	63	−6.47	−41.02	1374
MS dam	921	−7.08	−44.3	1711
Aoulouz dam	766	−6.29	−38.9	1272
11	468	−5.53	−33.5	850
8	326	−7.45	−50.9	1917
2	788	−5.43	−35.3	794
4	711	−6.05	−38.5	1139
5	523	−8.23	−50.4	2350
6	211	−7.17	−44.9	1761
7	273	−7.35	−45.7	1861
9	379	−7.41	−46.1	1894
10	469	−7.14	−44.1	1744
12	479	−6.95	−43.8	1639
13	361	−6.94	−43.2	1633
19	129	−7.55	−47	1972
20	114	−7.36	−44.5	1867
21	186	−7.5	−45.2	1944
22	185	−7.54	−47.7	1967
14	138	−5.33	−31.9	739
15	90	−7.78	−49.1	2100
16	92	−8.01	−50.5	2228
17	94	−6.07	−36.4	1150
18	112	−6.01	−36.4	1117
26	98	−6.81	−42.2	1561
23	60	−4.71	−29	394
24	59.5	−5.48	−32.5	822
25	21	−7.33	−46.1	1850
27	65.8	−5.7	−33.5	944
28	66.5	−5.71	−34.4	950
29	59.75	−6.7	−41.7	1500
30	70.5	−7.6	−47.9	2000
33	120.5	−5.94	−34.8	1078
32	61.24	−4.19	−24.8	106

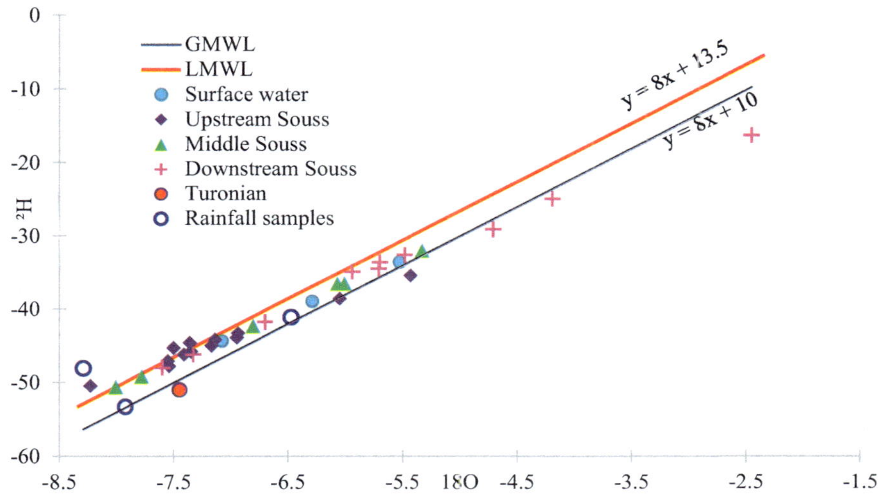

Fig. 7.9 δ^2H versus δ^{18}O values of groundwater as compared to the global meteoric water line (GMWL : $\delta^2\text{H} = 8\delta^{18}\text{O} + 10$), Local meteoric water line (LMWL : $\delta^2\text{H} = 8\delta^{18}\text{O} + 14$)

The majority of these samples are located along the meteoric lines suggesting an oceanic origin of these waters, this position implies that the water evaporates only slightly or not at all (Bouragba et al. 2011; N'da et al. 2016).

7.4.7.2 Recharge Area

According to Craig (1961), the recharge of meteoric water that originated from high altitude would be associated with relatively low values of δ^2H and δ^{18}O. To estimate the altitude of the groundwater recharge of the area, we used the isotope altitude gradient defined by Ait Brahim et al. (2016) in the High Atlas Mountains by the equation below:

$$\delta^{18}\text{O}\% = -0.0018 * \text{Alt} - 4 \tag{7.3}$$

The projection of different samples on the altitude line gives the mean altitude of different recharge zones. Figure 7.10 and Table 7.2 show the estimated altitude of recharge which varies between 400 and 2390 m. Compared to the altitude of sampling, the recharge altitude represents higher values corresponding especially to the High Atlas Mountains.

In our case, the lowest isotopic value corresponds to the Ifni station which shows the highest recharge altitude corresponding to the High Atlas Mountains.

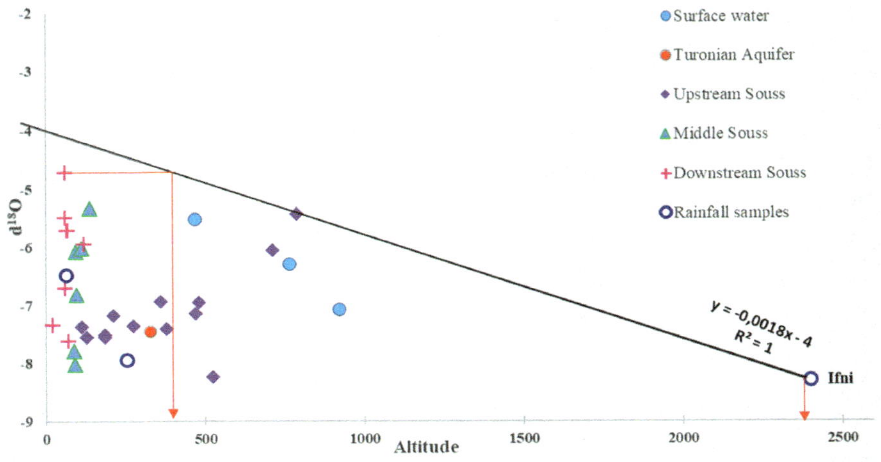

Fig. 7.10 Determination of groundwater recharge elevation for groundwater samples from wells and boreholes. The solid black line defines the relationship between $\delta^{18}O$ composition of precipitation and elevation ($^{18}O = 0.0018 * Z_m - 4$) was determined from the analysis of rainwater (Ait Brahim et al. 2016)

The stable isotope and precipitation rate suggest that the artificial recharge of the aquifer system should be scheduled in the tributaries coming from the right side of the Souss River, i.e., the High Atlas tributaries.

In summary, the isotopic tracers indicate the major contribution of the High Atlas Mountains with relatively high rainfall and snowmelt to the recharge of groundwater in the Souss aquifer throughout the infiltration along the main river and its tributaries. This is consistent with the significant rate of infiltration along the rivers in the area.

7.5 Future Perspective and Recommendations

Managed aquifer recharge appears in the future as one of the tools that can be used for enhancing the sustainability of groundwater resources under climate changes and human activities impacts with acceptable cost. In order to replenish the Souss aquifer system, we recommend some additional managed aquifer recharge strategies that could be useful for policymakers:

(i) Injecting water directly into the aquifer from the existing dams in the Souss basin, and from the existing and future wastewater treatment plants (WWTPs) that are planned to be constructed in Taroudant city and other villages;
(ii) Adopt tertiary treatment for sewage plants whose wastewater is infiltrated directly into the water table to avoid the risk of contamination, especially by nitrate and fecal coliforms;

(iii) Construction of additional hillside dams and elevate some dams for rain-water harvesting, especially in the upstream part where the permeability of the unsaturated zone is higher;
(iv) Construction of recharge trenches especially in the areas where permeable strata are available at shallow depth;
(v) In the coastal area, the aquifer should be protected from seawater intrusion by a line of injection wells where water from WWTP can be injected, in order to create a separate groundwater barrier between fresh and saline water;
(vi) Promote studies of treated wastewater reuse and seawater desalination;
(vii) Apply the groundwater contract which constitutes an innovative tool for local governance to meet sustainable management of water resources.

Before the implementation of any recharge strategy, mapping of potential managed aquifer recharge sites is needed to provide a scientific basis for the implementation of suggested strategies. Based on the analysis of existing and/or collected data during the investigation, it is necessary to (i) define the underground geology, (ii) determine the presence or absence of impermeable layers that can impede percolation, (iii) define depths to the water table and groundwater flow directions and (iv) establish the maximum rate of recharge that could be achieved at the site.

To evaluate the existing groundwater recharge strategies and those proposed in this study, systematic monitoring of groundwater quality and water table level in the concerned areas should be taken into consideration.

7.6 Conclusion

In the context of adaptation to climate change and management of water resources, Morocco has experienced a success story in managed aquifer recharge by adopting a set of techniques and strategies to mitigate the effect of recurrent droughts, groundwater overexploitation, and seawater intrusion; and ensure the sustainability of groundwater resources.

Results had shown a positive reaction of the aquifer to water released from the Aoulouz dam, especially in rainy periods, but it remains still low compared to the water demand and the deficit in the water table balance, where the necessity to improve additional strategies for managed aquifer recharge and develop the process of wastewater treatment as an important unconventional water resource, for recharge and irrigation. As found by several studies using stable isotope and altitudinal gradient, this work confirms that the recharge waters of shallow aquifers come mainly from the High Atlas Mountains, which are the potential sites for artificial recharge.

Acknowledgements The research has been carried out as part of the Coordinated Research Projects (CRP) of the International Atomic Energy Agency (IAEA). The authors would like to acknowledge and express their gratitude to Dr. Aggarwal Dr. B. Kumar and all the staff of the Isotope Hydrology Section at IAEA (Vienna, Austria) for the facilities and support provided during the research contract

F33019. Furthermore, we would like to extend our gratitude to the Hydraulic Agency of Souss-Massa Basin (Agadir, Morocco) for the access facilities to their database. Additionally, we would also like to recognize the contributions of the Agreemed and Geantech projects in directing the outcomes of this research.

References

ABHSM (2007) Plan Directeur d'Aménagement Intégré des Ressources en Eau dans les bassins de Souss et Massa. Unpublished Report, vol 13

Abahous H, Guijarro JA, Sifeddine A, Chehbouni A, Ouazar D, Bouchaou L (2020) Monthly precipitations over semi-arid basins in Northern Africa: Homogenization and trends. Int J Climatol. https://doi.org/10.1002/joc.6569

Abahous H, Ronchail J, Sifeddine A, Kenny L, Bouchaou L (2018) Trend and change-point analyses of annual precipitation in the Souss-Massa Region in Morocco during 1932–2010. Theoret Appl Climatol 134(3–4):1153–1163. https://doi.org/10.1007/s00704-017-2325-0

Ait Brahim Y, Bouchaou L, Sifeddine A, Khodri M, Reichert B, Cruz FW (2016) Elucidating the climate and topographic controls on stable isotope composition of meteoric waters in Morocco, using station-based and spatially-interpolated data. J Hydrol. https://doi.org/10.1016/j.jhydrol.2016.10.001

Ait Kadi M, Ziyad A (2018) Integrated water resources management in Morocco. In: World water council (ed) Global water security, water resources development and management, pp 143–163. https://doi.org/10.1007/978-981-10-7913-9_6

Alahiane N (2017) Méthodologie de présélection de sites potentiels des petits barrages après l'évaluation des ouvrages existants: cas du haut et anti -atlas occidentaux (Maroc). Ibn Zohr University, Faculty of sciences

Bouchaou L, Michelot JL, Vengosh A, Hsissou Y, Qurtobi M, Gaye CB, Bullen TD, Zuppi GM (2008) Application of multiple isotopic and geochemical tracers for investigation of recharge, salinization, and residence time of water in the Souss-Massa aquifer, southwest of Morocco. J Hydrol 352(3–4):267–287. https://doi.org/10.1016/j.jhydrol.2008.01.022

Bouchaou L, Tagma T, Boutaleb S, Hssaisoune M, El Morjani ZE (2011) Climate change and its impacts on groundwater resources in Morocco: the case of the Souss-Massa basin. In: Climate change effects on groundwater resources: a global synthesis of findings and recommendations. CRC Press, p 414; International Association of Hydrogeologists 27:129–144

Bouragba L, Mudry J-N, Bouchaou L, Hsissou Y, Krimissa M, Tagma T, Michelot JL (2011) Isotopes and groundwater management strategies under semi-arid area: Case of the Souss upstream basin (Morocco). Appl Radiat Isot 69(7):1084–1093. https://doi.org/10.1016/j.apradiso.2011.01.041

Craig H (1961) Isotopic variations in meteoric waters. Science 133(3465):1702–1703. https://doi.org/10.1126/science.133.3465.1702

de Vries JJ, Simmers I (2002) Groundwater recharge: An overview of process and challenges. Hydrogeol J 10(1):5–17. https://doi.org/10.1007/s10040-001-0171-7

Dillon P (2005) Future management of aquifer recharge. Hydrogeol J 13(1):313–316. https://doi.org/10.1007/s10040-004-0413-6

Dillon P, Pavelic P, Declan Page HB, JW (2009) Managed aquifer recharge. Naunyn-Schmiedebergs Archiv für Experimentelle Pathologie und Pharmakologie 186(2–5). https://doi.org/10.1007/BF01929660

Droogers P, Immerzeel WW, Terink W, Hoogeveen J, Bierkens MFP, Van Beek LPH, Debele B (2012) Water resources trends in the Middle East and North Africa towards 2050. Hydrol Earth Syst Sci 16(9):3101–3114. https://doi.org/10.5194/hess-16-3101-2012

Elmouden A, Alahiane N, El Morjani ZE, El Faskaoui M (2016) Dams siltation and soil erosion in the Souss–Massa river basin. In: The handbook of environmental chemistry (2016th ed.). Springer, Berlin. https://doi.org/10.1007/698_2016_70

Ghaffour N, Missimer TM, Amy GL (2013) Combined desalination, water reuse, and aquifer storage and recovery to meet water supply demands in the GCC/MENA region. Desalin Water Treat 51(1–3):38–43. https://doi.org/10.1080/19443994.2012.700034

Han Z, Huang S, Huang Q, Leng G, Wang H, Bai Q, Zhao J, Ma L, Wang L, Du M (2019) Propagation dynamics from meteorological to groundwater drought and their possible influence factors. J Hydrol 578:12. https://doi.org/10.1016/j.jhydrol.2019.124102

Hashemi H (2015) Climate Change and the Future of Water Management in Iran. Middle East Critique 24(3):307–323. https://doi.org/10.1080/19436149.2015.1046706

Hashemi H, Berndtsson R, Persson M (2015) Managed aquifer recharge by floodwater spreading estimated by water balances and groundwater modeling in arid Iran. Hydrol Sci J 60(2):336–350. https://doi.org/10.1080/02626667.2014.881485

Hssaisoune M (2014) Etude hydro-structurale de l'aquifère du Souss-Massa: Analyse géophysique et conséquences hydrogéologiques. Ibn Zohr University, Faculty of Sciences, p 180

Hssaisoune M, Bouchaou L, Sifeddine A, Bouimetarhan I (2020a) Moroccan Groundwater Resources and Evolution with Global Climate Changes. Geosciences (switzerland). https://doi.org/10.3390/geosciences10020081

Hssaisoune M, Bouchaou L, Sifeddine A, Bouimetarhan I, Chehbouni A (2020b) Moroccan groundwater resources and evolution with global climate changes. Geosciences (Switzerland) 10(2). https://doi.org/10.3390/geosciences10020081

Hssaisoune M, Boutaleb S, Mohammed B, Bouaakkaz B (2016) Physical geography, geology, and water resource availability of the Souss-Massa river basin physical geography, geology, and water resource availability of the Souss-Massa river basin. https //doi.org/10.1007/698

Hssaisoune M, Boutaleb S, Mohammed B, Beraaouz EH, Tagma T, M'Hamed el faskaoui, Bouchaou L (2012) Geophysical and structural analysis of the Souss-Massa aquifer: Synthesis and hydrogeological implications geophysical and structural analysis of the Souss-Massa aquifer. Geo-Eco-Trop 36:63–82

Lerner DN, Issar AS, Simmers I, Verlag Heinz H (1990) Groundwater recharge. A guide to understanding and estimating natural recharge. Int Contrib Hydrogeol 8(3):395–403. https://doi.org/10.1111/j.0021-8774.2005.00541.x

N'da AB, Bouchaou L, Reichert B, Hanich L, Ait Brahim Y, Chehbouni A, Beraaouz EH, Michelot JL (2016) Isotopic signatures for the assessment of snow water resources in the Moroccan High Atlas Mountains: contribution to surface and groundwater recharge. Environ Earth Sci 75(9). https://doi.org/10.1007/s12665-016-5566-9

N'da B, Bouchaou L, Reichert B, Hanich L, Danni SO, Ait Brahim Y, Chehbouni A, Beraaouz E, Michelot JL (2018) Chemical tracers and stable isotopes mixing models for groundwater quality and recharge study in the Moroccan high atlas mountains, pp 235–244. https://doi.org/10.1007/978-3-319-69356-9_27

Rahman MA, Rusteberg B, Uddin MS, Lutz A, Saada MA, Sauter M (2013) An integrated study of spatial multicriteria analysis and mathematical modelling for managed aquifer recharge site suitability mapping and site ranking at Northern Gaza coastal aquifer. J Environ Manage 124:25–39. https://doi.org/10.1016/j.jenvman.2013.03.023

Schilling J, Freier KP, Hertig E, Scheffran J (2012) Climate change, vulnerability and adaptation in North Africa with focus on Morocco. Agr Ecosyst Environ 156:12–26. https://doi.org/10.1016/j.agee.2012.04.021

Tagma T, Hsissou Y, Bouchaou L, Bouragba L, Boutaleb S (2009) Groundwater nitrate pollution in Souss-Massa basin(South-west Morocco). Afr J Environ Sci Technol 3:301–309

Tuinhof A, Theo O, Jan PH, Vries J de (2002) Infiltration basins and other recharge systems in Southern Africa. IAHS-AISH Publication (4)

Waha K, Krummenauer L, Adams S, Aich V, Baarsch F, Coumou D, Fader M, Hoff H, Jobbins G, Marcus R, Mengel M, Otto IM, Perrette M, Rocha M, Robinson A, Schleussner CF (2017)

Climate change impacts in the Middle East and Northern Africa (MENA) region and their implications for vulnerable population groups. Reg Environ Change 17(6):1623–1638. https://doi.org/10.1007/s10113-017-1144-2

Chapter 8
Managed Aquifer Recharge in Chile: A Promising Alternative to Enhance Water Security

Gabriella Bennison and Edmundo Claro

Abstract Chile currently faces several challenges related to water security, such as over-exploitation and contamination of freshwater supplies and high competition for water use in productive central valleys. These challenges have been further exacerbated by the ongoing drought, which has gripped central Chile since 2010. This has pushed the country to search for new ways to manage and govern its water resources. Managed Aquifer Recharge (MAR) has progressively gained interest as a potential alternative to the more traditional forms of water storage and management, particularly in Chile's northern and central zones, which have deep underlying aquifers in their central valleys between the Andes Mountains to the east and the coastal mountains to the west. While to date there have only been around 11 MAR schemes identified, the majority of which are limited to small-scale pilot schemes, there have been numerous national and local seminars on the topic, and its implementation has been included in both national and local plans to improve water security. Despite an ongoing lack of specific legislation related to MAR in Chile's Water Code, an Operational Framework for Artificial Aquifer Recharge Projects (MAR Guidelines), loosely based on the Australian approach, were published in 2021 by the National Irrigation Commission. These guidelines principally focus on the agriculture sector and encourage the development of simple and small-scale MAR schemes, with limited associated risk, which do not necessarily require lengthy consultation reviews by the national or local Water Authority. While the guidelines are not normative, it is a step closer to removing a number of barriers related to MAR implementation in Chile and instilling much-needed confidence in the user, which could be individual farmers, agricultural cooperatives or water user organizations. This paper presents an overview of the gaining interest in MAR in Chile and the various challenges currently faced from legislative, institutional, technical, and environmental and socio-economic perspectives. It also discusses the experience of preparing national guidelines to encourage the development of low-risk MAR projects and finally it presents

G. Bennison (✉) · E. Claro
Fundación CSIRO, Chile Research, Santiago, Chile
e-mail: gabriella.bennison@environment-agency.gov.uk

D. Saha et al. (eds.), *Managed Groundwater Recharge and Rainwater Harvesting*, Water Resources Development and Management,
https://doi.org/10.1007/978-981-99-8757-3_7

the authors' recommendations for further promoting economically viable, environmentally sustainable, and equitable MAR in Chile while overcoming the obstacles observed.

Keywords MAR · Water security · Water resources management · Agriculture · Irrigation · Chile

8.1 Introduction

Problems associated to water scarcity are becoming ever more frequent in Chile. The factors behind this are various, including increasing demand across multiple sectors due to growing economic activities and population growth, increasing awareness and legal requirements for environmental water (Biblioteca del Congreso Nacional de Chile 2022), contamination of water sources, excessive extraction of groundwater and the unbalanced allocation of water resources (Escenarios Hídricos 2030 Chile 2019; Donoso et al. 2020); (Correa-Parra et al. 2020). This challenge is further complicated by the "mega drought", which has affected Chile's central zone for over a decade (2010–2021) (Garreaud et al. 2019). In addition, there are expectations of more frequent and severe extreme events (Sheffield and Wood 2008; Polade et al. 2014), competing water uses, and interdependencies between multiple sectors in the water-energy-food-environment nexus. These pressures are challenging the way in which water resources have traditionally been managed and governed in the country.

For decades, the increase in demand has occurred across all water users: agriculture, mining, energy, domestic, tourism, and others. As the National Water Authority (DGA[1]) stated a decade ago, the "growing demand for water resources in different sectors has generated pressure on those resources, manifesting itself as worsening conflicts between uses and users" (Jiménez 2012a, b). On the supply side, the decadal drought has been reflected in a sharp decline in surface water availability, largely attributed to a reduction in precipitation and a corresponding decline in river flows. Irrespective of this current dry period, the World Resources Institute (2015) identified Chile as one of the most water-stressed countries in the world, placing it 24th on a global scale.

Moreover, it is very likely that climate change will further exacerbate water availability over the coming decades, particularly in central Chile. According to the Economic Commission for Latin America and the Caribbean, CEPAL,[2] the environmental, social, and economic losses that are projected to occur in Chile as a result of climate change are estimated to reach 1.1% of annual GDP by 2100 under least favorable conditions[3] (CEPAL 2012). Another, more recent study found that worst case projections estimate reductions to the national GDP by up to 29.6% by 2100

[1] Dirección General de Aguas/General Directorate of Water.

[2] Comisión Económica para América Latina y el Caribe.

[3] This would indicate a total loss of 320 billion US dollars during the period analyzed.

(Hernandez and Madeira 2022). Climate change projections indicate a warming trend across the country toward the end of this century, with increases in temperature in the coming decades of between 2 and 3.5 °C between Atacama and Los Lagos regions in central Chile (FCCyT 2012; Vázquez 2019a, b). At the same time, there are signs that rainfall and runoff will decrease for most of these central regions. Between 2010 and 2039, rainfall and runoff are forecast to decrease by 7.9 and 10.3%, respectively (compared to the period between 1976 and 2005) and then they will further decrease significantly over the second half of this century (up to 30.4% reduction in rainfall and 40.2% reduction in runoff) (Bozkurt et al. 2017).

Along with the decrease in total rainfall, it is expected that there will be an increase in long-term (over 5 years) droughts, and at the same time the intensity of the rains will increase. In other words, annual rainfall is expected to decrease and be concentrated in fewer but higher intensity events, especially during the winter months (Ministerio de Medio Ambiente 2015). The temporal persistence and spatial extent of the current mega drought are extraordinary when compared to the historical record. Indeed, an event such as this hasn't been found in the last millennium, according to climate reconstructions based on the growth of tree rings (Center for Climate and Resilience Research 2020). This could therefore be a foretaste of a new water normality facing the country.

Groundwater is an important resource in Chile, representing approximately 55% of the total volume of potable water produced in urban areas, which accounts for 80% of the country's population (SISS 2021). Groundwater also represents 22% of all permanent and consumptive water rights[4] in the country, which is equivalent to 14.3 km^3/yr. Despite, or because of, Chile's current and increasing dependence on this important water source, 47% of the 375 defined hydrogeological sectors[5] across the country are under restriction or prohibition, which means they are subject to restrictions concerning the creation of new water rights (Dirección General de Aguas 2016a). This highlights the urgent need for and importance of managing the water resources more conjunctively and sustainably.

Considering this scenario, the country faces an important challenge regarding sustainable water resource management. Moreover, there is an increasing demand for preserving water resources for ecological purposes, rather than solely productive uses, as citizens become more environmentally conscious. This is catalyzing a search for more nature-based solutions, rather than solely relying on large infrastructure projects. This more traditional form of water management is still dominant in Chile, like in many countries around the world, reflected in the prioritization and execution

[4] Water rights in Chile are constituted or recognized by an appropriate authority. They are legal rights and allow the owner to use or dispose of them. They can be permanent or temporal water rights (with a specified duration) and they can be continuous, discontinuous, or alternated between several people. Consumptive water rights allow for their total consumption, whereas non-consumptive water rights require that the used but unconsumed water return to the water source as specified by the individual Act of the water right.

[5] A Hydrogeological Sector of Common Use (*Sector Hidrogeológico de Aprovechamiento Común*) in Chile corresponds to an aquifer, or part of one, which is defined according to geological, hydraulic and administrative criteria.

of large dams with the aim of increasing water supply to different sectors. However, these projects take a long time to materialize, sometimes lasting decades between the first viability studies and the final construction. There are also high financial, social, and environmental costs associated with large dams, coupled with their vulnerability to contamination, high evaporation rates, and the decreasing availability of suitable land and sites for new dams.

Managed Aquifer Recharge (MAR) has been encouraged as a way of advancing nature-based solutions to water scarcity. In fact, according to Bertule et al. (2018), one of the most relevant adaptation responses to climate change in situations of water scarcity interspersed with more infrequent but higher rainfall events is MAR. MAR has been proposed as one of the most effective adaptation measures to reduce the vulnerability of water supply systems under temporary scarcity and climate change conditions (Dillon et al. 2019; Escalante et al. 2019; Guyennon et al. 2017). MAR, as part of broader water storage management planning, and in conjunction with surface water storage, can be used to maximize natural storage during wet periods and increase the resilience of water supply systems during periods of low flow and balance out high seasonal variability. Hence, during periods of abundance, aquifers can be intentionally recharged so that water can later be recovered during periods of scarcity.

The practice associated with using soil or aquifers for water storage for later use dates back more than 3,000 years B.C. in India. Since then, percolation tanks, check dams, and infiltration ponds have been built in arid and semi-arid areas to store water (Gale 2005a, b). The ancient cultures of Peru practiced recharging systems that are still applied in different places; those built in the Sierra de Huarochirí areas are called "amunas" ('channels and pools') (Ancajima 2016a, b).

On a global scale, the application of MAR for diverse uses has notably increased since the1950s (Zheng et al. 2021). In Chile, interest in MAR is more recent, becoming more prevalent since 2012 (CSIRO Chile 2020a, b). Since then, public sector authorities have financed a series of studies and pilot schemes with the objective of acquiring local experience, determining the costs of different techniques, and understanding the associated technical and administrative problems and opportunities. In many cases, these pilot schemes served as demonstration sites to showcase their feasibility and hence were not intended for final commissioning. While these varying experiences most certainly helped instigate momentum for MAR in Chile, most pilot projects to date have been short-lived, uncoordinated, and have been confronted with technical and financial difficulties.

They have also brought about uncertainties, particularly linked to the necessary governance of MAR schemes and the use of water rights associated with the subsequent use of recovered water, aspects which are still being debated at a national level. Thus, despite large efforts, to date Chile has very few continuously operating formal MAR schemes. There have been more recent efforts by the private sector to develop MAR schemes in different regions of the country. However, often these are lacking background information and monitoring data to provide an understanding of their impact on the surrounding socio-hydrological system.

These challenges have not undermined the interest in MAR. It is still very much on the agenda across the public, private, and academic sectors, with new initiatives arising with diverse objectives; from reducing the risk of social conflicts due to droughts or water mismanagement,[6] to minimizing losses in productivity, and even decreasing ecological externalities. Furthermore, MAR is aligned with strategies of the "National Water Table"[7] led by the Ministry of Public Works and coherent with several Regional Water Resource Plans, and the National Aquifer Recharge Plan for Agriculture established in 2019 from the National Irrigation Commission (Comisión Nacional de Riego 2019).

This chapter gives an overview of MAR implementation in Chile and the main challenges experienced from legislative, institutional, technical, environmental, and socioeconomic perspectives. Next, a summary is provided of the Operational Framework for Artificial Aquifer Recharge Projects (MAR Guidelines), published in 2020 by the National Irrigation Commission and elaborated by the CSIRO Chile Research Foundation, with a primary focus on the agriculture sector to encourage the development of simple and small-scale MAR schemes, with limited associated risk.[8] Finally, conclusions and recommendations are given to overcome the obstacles mentioned and further promote sustainable MAR use.

8.2 Methodology

Many of the challenges described in this paper were identified during semi-structured interviews and participatory workshops, conducted as part of the development of the MAR Guidelines. The main objective was to involve key stakeholders, including representatives from public and private enterprise (including water users) and academia to incorporate their local knowledge and experience into the co-development of the MAR Guidelines. This allowed the guidelines to be adapted to local needs and requirements, incorporating stakeholder's uncertainties and suggestions. Fourteen interviews were conducted during April and May 2019 with stakeholders who have a wide variety of experiences with MAR in Chile. Later, in August 2019, a stakeholder workshop with around 40 participants was held with the aim of sharing and discussing the advances in the development of the MAR Guidelines with the key stakeholders and other interested participants and to receive feedback.

---

[6] Examples of social conflicts include: the over-exploitation of groundwater resources close to indigenous communities or environmentally sensitive areas; the dependence of rural water communities on cistern trucks for water supply due to the reduction in groundwater levels.

[7] The *Mesa Nacional del Agua* was established in 2019 with the purpose of proposing a plan to deal with the current water crisis. The latest report from the process identifies 12 key areas to work on, which includes several objectives to deal with water security. It also develops a roadmap for the implementation of a sustainable and long-term water policy (Mesa Nacional del Agua 2022).

[8] To download the MAR Guidelines and for supporting information, see: https://research.csiro.au/rag-chile/en/mar-guides-aquifer-recharge/ and https://www.cnr.gob.cl/recarga-de-acuiferos/

Two smaller workshops were also held during 2019 with an Advisory Panel set-up for the project.

8.3 MAR Implementation in Chile and Its Challenges

In recent years, there has been increasing interest in MAR, or as it is more commonly known as in Chile "artificial aquifer recharge" (here used interchangeably), as an alternative or complementary method to traditional water storage systems, particularly in the agriculture sector. In 2012, after the current drought set in, several public authorities carried out international visits to improve their understanding of good MAR practice around the world and to bring these experiences to Chile. Despite the existence of MAR schemes in the private sector prior to this date, they were relatively unknown and unregulated. Thus, it was this interest from the public sector that promoted the concept more widely in Chile, highlighted by the rapid development of several studies, pilot projects, operational schemes, and dissemination seminars. Table 8.1 gives an overview of the 11 primary MAR projects and pilots conducted in Chile to date, while Fig. 8.1 shows the location of these projects. This figure also shows the location of six cases entered in the Global MAR Inventory of the International Groundwater Resources Assessment Centre (IGRAC). However, these are lacking sufficient information (such as the proponent's name, dimensions of the infrastructure used, water source, etc.) and therefore have not been included in Table 8.1 and are not further discussed in this paper.

A large part of MAR promotion in Chile from the public sector can be attributed to the National Irrigation Commission, part of the Ministry of Agriculture. This is likely due to the importance of the agriculture sector in the country, which accounts for 82% of the total water consumption (Dirección General de Aguas 2016b) and is active primarily in Chile's central zone, the area principally affected by Chile's ongoing mega drought. Traditionally, this sector, as in many arid and semi-arid countries around the world, has looked toward dam construction to solve problems of water availability and use. While this position is still evident, over the last few years there has been increasing discussion at the national level to incorporate more holistic and integrated solutions for water resource management, including the implementation of MAR. Other water infrastructure projects under consideration over the last few years are desalination for potable water and agriculture use in Chile's central zone[9] and water transfer schemes. While desalination and MAR have been backed by the government (Apruebo Dignidad 2021), water transfer schemes have largely been discarded as too expensive, with high external associated costs.

In this regard, in 2018, an inter-ministerial committee was established to focus on MAR development in the public sector, with representatives from the National Irrigation Commission, the Water Authority, the Directorate for Hydraulic Works,

[9] Desalinated water is becoming increasingly used in the mining sector and in some cities for potable water use in the north of the country.

Table 8.1 Summary of primary aquifer recharge projects in Chile since 2012, updated from Page et al. (2020)

N°	Proponent	Location/ River basin	Scale	Status	Dates operated	Recharge method	Source water	Dimensions	Estimated recharge rates (m^3/ day)
1	National irrigation commission	La Ligua/ La Ligua	Pilot	Non-operational	2013	Infiltration ponds	Canal	3 ponds: 250 m^2 each	~7,500
2	The Maipo canal society	Antumapu/ Maipo	Full scale	Operational	2013-present	Infiltration ponds & injection wells	Canal	2 ponds: 3,200 m^2 each; 2 infiltration wells: 0.1 m diameter, 62 m and 73 m deep	~2,200
3	National irrigation commission	Chacabuco/ Maipo	Pilot	Non-operational	2014	Injection well	Canal	1 injection well: 0.15 m diameter and 40 m deep	~300
4	National irrigation commission	Popeta/ Maipo	Pilot	Non-operational	2014	Injection well	Groundwater	1 injection well: 0.15 m diameter and 36 m deep	~400
5	National irrigation commission	Las Cadenas/ Marchigue	Pilot	Non-operational	2014	In-stream dyke	River	1 dyke: 1 m high, 25 m wide, recharge area 37,000 m^2	~1,300
6	National irrigation commission	Lontué/ Mataquito	Pilot	Non-operational	2014	Infiltration pond	Canal	1 pond: 17 m wide, 27 m long and 1.7 m deep	~930

(continued)

Table 8.1 (continued)

N°	Proponent	Location/ River basin	Scale	Status	Dates operated	Recharge method	Source water	Dimensions	Estimated recharge rates (m^3/day)
7	National irrigation commission	La Palma/ Aconcagua	Pilot	Non-operational	2015	Injection well	Canal	1 injection well: 0.15 m diameter and 54 m deep	~20
8	Directorate of hydraulic works	Curimón/ Aconcagua	Pilot	Non-operational	2016	Infiltration ponds	Canal	2 ponds: 2,113 m^2 2,203 m^2	~27,000
9	Groundwater Association: Copiapó—Piedra Colgada—Desembocadura	Toledo/ Copiapó	Full scale	Operational (when water is available)	2017-present	Infiltration ponds	River	3 ponds: 524,000 m^2 463,000 m^2 272,000 m^2	~28,000
10	Copiapó river surveillance board	Cerillos/ Copiapó	Full scale	Operational (when water is available)	2018-present	Infiltration ponds	Canal	3 ponds: 84,000 m^2 125,000 m^2 169,000 m^2	~5,200
11	National irrigation commission	Doñihue/ Cachapoal	Pilot	Operational (when water is available)	2020-present	Infiltration ponds	Canal	2 ponds: 1.250 m^2 each 3 m deep	~1,700

Fig. 8.1 Location of known MAR schemes in Chile, updated from CSIRO Chile (2020a, b). The numbered schemes are described in Table 8.1 and the schemes in italics are included in the Global MAR Inventory of IGRAC but are lacking sufficient information.

the Agricultural Development Institute, and the Corporation to Promote Production from the Ministry of Economics. This committee met every two months, and its principal objective was to align existing MAR initiatives and improve coordination and efficiency of the public institutions involved.

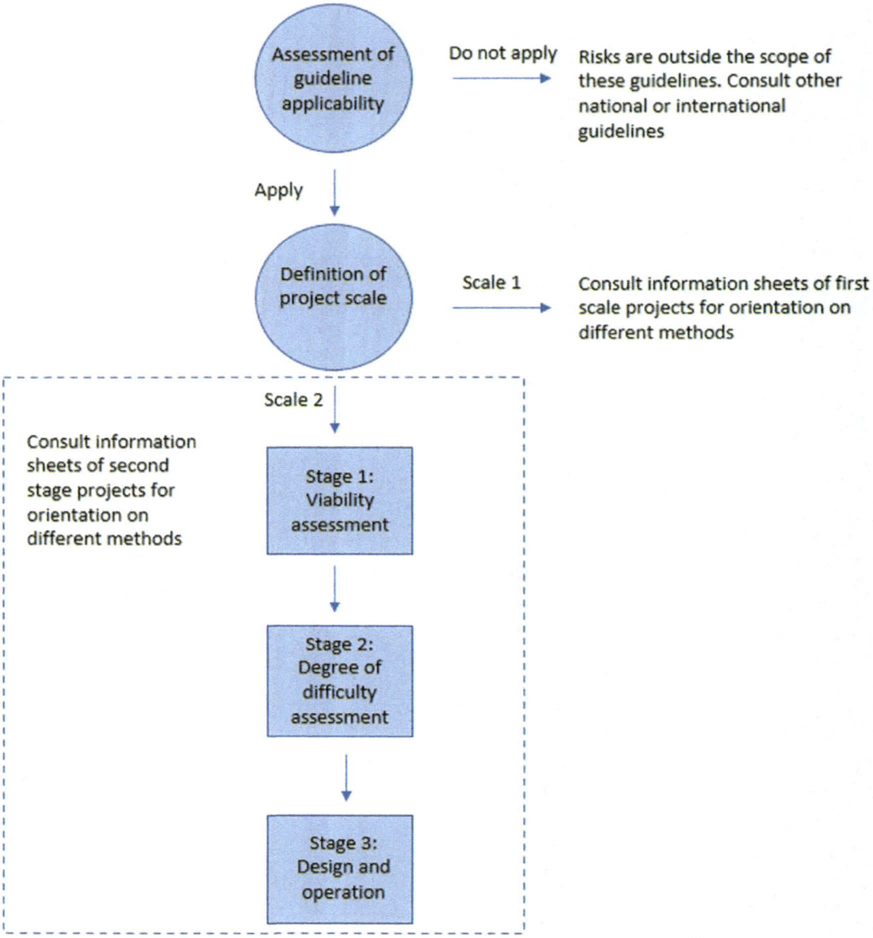

Fig. 8.2 Staged project development and workflow for MAR schemes in Chile (CSIRO Chile 2020a, b). Scale 1 MAR schemes (simple schemes) include: **i** rainwater capture and infiltration systems; **ii** infiltration canals; and **iii** infiltration ditches. Scale 2 MAR schemes (more complex schemes) include: **i** infiltration ponds; **ii** dry wells; **iii** infiltration trenches; and **iv** in-stream infiltration methods

8.3.1 Legislative, Institutional and Financial Challenges

MAR is included in the Chilean legal system via the Water Code (Código de Aguas) established in 1981, based on free market principles. Under this arrangement, the state is responsible for assigning water use rights, which are free of charge and in perpetuity. These water rights are commercial goods, which are tradable between different users, enabled by their separation from the land (Bauer 1998). Once the water rights have been assigned, they are governed by civil legislation related to property rights. Therefore, government agencies are responsible for regulation while the private sector oversees operation (Larraín 2006a, b).

The Water Code refers to the undertaking of MAR works and regulates the water rights associated with MAR projects. MAR is also considered through the regulation standards on Exploration and Exploitation of Groundwater, which establish administrative procedures to request authorisation to develop MAR projects and regulate MAR initiatives. There are also several environmental laws and standards, which are relevant but not specific to MAR, and cover a range of different uses and sectors, such as the law on general environmental considerations, regulation for Environmental Impact Assessment, and standards for liquid discharge to groundwater.

The first reform of the Water Code was established in 2005 and it incorporated the concept of artificial aquifer recharge (Table 8.2). It stated that with prior authorization from the National Water Authority, anyone can carry out MAR projects.[10] This enabled the ability to constitute a provisional water right to use the groundwater derived from the MAR scheme, which could be transformed into a definitive water right after 5 years (Dirección General de Aguas 2012).

In order to enable proponents to present their proposed MAR schemes for evaluation and permit granting, two documents were published between 2014 and 2016 by the Water Authority to assist with this process (Dirección General de Aguas 2014, 2016b). Subsequently, a number of bulletins were released by the Water Authority with the intent of filling in some of the gaps in current legislation and facilitating the development of what they consider relatively low-risk MAR projects (Dirección General de Aguas 2016a, b, c). These bulletins determine that permits from the Water Authority are not required for these low-risk projects, such as rainwater harvesting and infiltration as well as infiltration from hydraulic transport and storage infrastructure (such as canals, ponds, or dams), which do not have impermeable lining (which could also be considered as unintentional infiltration). There are no specific limitations on the size of the schemes permitted or the amount of water infiltrated for these low-risk projects.

However, even for these projects, permits will still be required if new water rights will be requested to recover part of the infiltrated water. Hence, where possible, existing surface water and groundwater rights should be used in conjunction, eliminating the need to request new rights associated with the recovered water. Alternatively, some schemes may aim to improve or maintain groundwater levels and therefore may not require water recovery. In any case, the Water Authority could still request a technical report on the scheme and require the installation of a water quality and quantity monitoring system, if deemed necessary. Also, in the event of water contamination as a result of a MAR scheme, or where third parties have been negatively affected, its operation may be prevented, fined, or shut down.

---

[10] The following water sources considered viable for use in MAR schemes in Chile are: (a) running or standing surface water; (b) groundwater already extracted from an aquifer; (c) rainwater; and (d) treated water, either drinking water, water from a desalination plant, water from sewage treatment plants or water from industrial effluents (Dirección General de Aguas 2016b).

Table 8.2 Historic overview on the legal and institutional progress of MAR in Chile

Year	Development
2005	The first modification of the Water Code incorporated the concept of artificial aquifer recharge
2012	International visits from the public sector to understand more about the application of MAR
2013–2015	First MAR pilot studies carried out by the public sector (National Irrigation Commission)
2014–2016	A methodology to present proposed MAR schemes for evaluation and permit granting published by the National Water Authority
2016 and 2019	Bulletins published by the National Water Authority to clarify some of the gaps in current legislation and to facilitate infiltration from low-risk schemes
2018	An interministerial Committee for MAR development in the public sector established
2019	The existing subsidy for irrigation works extended to include MAR schemes for the agriculture sector
2019	A National Aquifer Recharge Plan for Agriculture published
2019	A proposal to modify the existing Water Code to include a new chapter specifically related to MAR submitted to the Senate
2020	The Operational Framework for Artificial Aquifer Recharge Projects (MAR Guidelines) published
2022	The latest modification to the Water Code clarifies some of the rules for MAR and emphasizes the role that MAR can play in environmental conservation

Subsequently, in 2019, a proposal was submitted by the senate to modify the Water Code, including a new chapter specifically related to MAR, to encourage a more coordinated management between Water User Organizations[11] and the Water Authority. It also establishes the permanent requirement to not affect water quality and to make a detailed record of infiltrated and extracted flows (Escenarios Hídricos 2030 2019).

While the Water Code was updated in 2022, most of the proposals submitted in 2019 were not adopted. However, the new modifications do include some of the rules for the development of MAR projects, stating that the holder of a water right who has carried out MAR works and who wishes to reuse the infiltrated water, whether in

[11] If two or more people have rights to use the waters of the same canal, reservoir, or take advantage of the waters of the same aquifer, they may constitute themselves in a Water User Organisation (irrespective of water use), in order to take the waters from the parent flow, distribute them among the right holders, build, exploit, conserve and improve the catchment works, canals and others that are necessary for their use (Biblioteca del Congreso Nacional de Chile 2022). Water User Organisations are non-mandatory private bodies to coordinate the use of water resources, and are comprised of so-called *Water Communities*, *Canal Associations* in the case of hydraulic works, or *Surveillance Boards* in the case of natural channels. The *Water Communities* can oversee the management of surface water or groundwater (*groundwater communities*) and can be grouped in *Surveillance Boards*, along with *Canal Associations*. Generally, Surveillance Boards are limited to administrative sections of a river and do not operate on a basin-wide scale.

the same or another point of the aquifer, may request the National Water Authority to authorize him/her to exercise his/her right over most of the recharged waters that, according to the technical analysis of the information presented,[12] consider the losses inherent to the recharge, the sustainability of the aquifer and the rights of third parties (Biblioteca del Congreso Nacional de Chile 2022). The new modifications to the Water Code also emphasize the role that MAR can play in environmental conservation such as for ecosystem preservation, including the improvement or maintenance of the sustainability of the aquifer. Finally, they specify that there should be no impact on water withdrawals for human consumption and to water quality.

Despite these efforts, the current legislation regarding MAR in Chile is still considered quite general and confusing. Also, concerns have been raised by a number of stakeholders regarding the complexity of the evaluation and permit process, where new water rights are required for water recovery. They have warned that it is unlikely that this approach becomes widely adopted, particularly in the agriculture sector due to the relatively high costs associated with gathering the required information for evaluation (CSIRO Chile 2020a, b). Finally, fundamental gaps have been identified in the current legislation. In particular, environmental legislation or norms regarding MAR schemes are lacking, with little clarity on requirements around water quality of the source water to be used for recharge (Delgado and Arumi 2019a, b).

With the intent to overcome some of these challenges, a National Aquifer Recharge Plan for Agriculture was articulated by the National Irrigation Commission in 2019 (Comisión Nacional de Riego 2019). The plan sets out a number of short, medium, and long-term measures to facilitate the implementation of MAR, some of which have already begun, such as extending the existing financial subsidy mechanism in place by the National Irrigation Commission to include MAR projects from 2019. This subsidy allows for financing up to 90% of a project after it has been implemented (Biblioteca del Congreso Nacional de Chile 2018). However, the initial investment costs still need to be paid upfront by the proponent of the MAR project and there still exists a risk that after undertaking expensive feasibility studies the project is determined unviable and access to the subsidy is denied.

During the development of the MAR Guidelines with a focus on the agriculture sector (2019–2020), the perceived main barrier to MAR development most often raised by stakeholders during interviews or workshops conducted by the CSIRO Chile Research Foundation was related to a lack of streamlined MAR regulation (CSIRO Chile 2020a, b). The stakeholders go further to state that the current legal framework and associated policy is a major disincentive for MAR development due to it being unclear and too bureaucratic. Furthermore, there were a few cases where proponents seeking to regularize already operational MAR schemes submitted the required documentation to the Water Authority, but they were waiting for several years for the response. Indeed, during interviews conducted with different teams within the

[12] The information presented should include: the technical specifications of the work; information on the hydrogeological sector of the aquifer that justifies the amount of water that is intended to be extracted; the recharge point(s) and the point(s) from which it is intended to extract the water; and a system for measuring and transmitting information at both points.

Water Authority itself, there was confusion regarding where the responsibility lies to evaluate MAR schemes (CSIRO Chile 2020a, b).

Furthermore, there is often a lack of a clear definition of roles and obligations of the stakeholders involved in MAR schemes. Where pilot projects have been established by public authorities, there is often concern about who will be responsible for the operation and maintenance of the subsequent full-scale scheme once it has been implemented. This may explain why many pilot schemes did not proceed to full-scale schemes (Table 8.1). Another challenge mentioned by stakeholders is the lack of capability and resources of the Water Authority to effectively monitor and evaluate (M&E) the performance and compliance of MAR schemes (CSIRO Chile 2020a, b). By some stakeholders, there is a preference to provide more control to Water User Organizations, whereby the Water Authority plays a solely supporting and M&E role. However, for other stakeholders this is concerning due to the large differences in skills and management capabilities of the Water User Organizations and the general lack of knowledge and experience to adequately control MAR schemes.

Finally, the management of surface water and groundwater is separated, with separate Water User Organizations to manage river sections or aquifer sections in a watershed. This separation makes it more difficult to achieve an integrated, collaborative management approach, which is particularly relevant to MAR projects.

8.3.2 Technical and Environmental Challenges

While a number of MAR pilots and projects have been carried out in Chile, it can be appreciated from Table 8.1 that most schemes conducted to date have been short in duration and have been unable to establish as full-scale operations. It is also important to note that most of these schemes, while intentionally implemented to enhance infiltration, are not managed in terms of water quantity and quality, and so are not true MAR sites, according to the technical definition (Page et al. 2020; Dillon 2005).

In terms of the pilot projects carried out by public authorities, there were challenges associated with locating adequate and available sites to use. While there are a few studies on a country-wide scale (Comisión Nacional de Riego 2013; Cabrera Fajardo 2014a, b) or basin-scale (Dirección General de Aguas 2012; Comisión Nacional de Riego 2012; Dirección de Obras Hidráulicas 2012), which identify broad potential recharge zones, according to physical characteristics of the aquifers, there is a need for more localized information on a scale more appropriate to MAR implementation.

Furthermore, in several cases, there was insufficient water available to use to trial the scheme appropriately once it had been implemented (CSIRO Chile 2020a, b). In these cases, the public authority had to depend on the corresponding Water User Organization to provide or transfer water when it was available. However, these pilots were carried out during the mega drought affecting Chile and consequently little or sporadic water was available. It is therefore important that prior to establishing MAR schemes, there is an understanding of water availability, not only of the physical characteristics of the area and potential for recharge as determined on a broad scale in previous studies. This understanding has to also include the knowledge of expected influence of climate change. Also, as these pilots have generally been created under constrained time conditions, to understand the applicability of the infrastructure in the short-term, they have not been designed to capture flood or storm waters which may occur only every 5 to 10 years.

Additionally, a number of technical challenges have been mentioned by stakeholders regarding MAR implementation, mainly related to the lack of information and knowledge about potential aquifers (CSIRO Chile 2020a, b). To evaluate MAR projects, technical information about the aquifers is required on a local scale, which implies the need to understand the local geometry of the aquifer, the piezometric level of the water table and expected changes under the present and future climate, the local stratigraphy of the area, the permeability of the soil and aquifer, etc. Generally, this information is available on a regional scale, where projects use average permeability values to evaluate different hydrogeology sectors.

The hydrogeological modeling undertaken to evaluate the aquifer's behavior is normally carried out at a basin or sub-basin scale, representing large areas. This type of information presents technical limitations to evaluate a MAR project on a local scale. It also limits the confidence of potential proponents to conduct MAR projects, when detailed and expensive studies are required by the Water Authority to evaluate the scheme. While potential proponents of MAR schemes seem willing to invest in the infrastructure required, they seem less inclined to conduct expensive studies to determine the most appropriate location for the scheme, something which often is considered the role of the public authority to carry out.

There is also a lack of confidence in how MAR schemes work because, unlike surface water dams, the water stored in aquifers cannot be seen. In this case proponents are often reluctant to invest in these projects if they don't know exactly where the water is stored and how quickly it can travel through the aquifer system downstream. This highlights the need for continuous monitoring of the surrounding aquifers to understand how the MAR scheme impacts the local area and what benefits can be obtained. This is also important to maintain good relationships with other downstream stakeholders, which could be negatively affected if, for example, flooding or water logging occurs.

Another problem reported is related to there being no official delineation of the river course in some areas. Modification of the river course must only be carried out when prior authorization is obtained. However, when it is not clear where the river delimitation is, particularly in ephemeral rivers which do not have permanent flows, there may be problems when locating MAR schemes close to the river course, such as infiltration ponds which diverts river water. In some parts of Chile, it can be very useful or even necessary to locate MAR schemes close to the river course, particularly in the semi-arid north of the country, which has low and sporadic river flows above underlying aquifers. These areas experience heavy rainfall events every few years and it can be an important local water source if it is adequately captured and treated prior to infiltration. In-stream MAR structures have not been widely implemented and require a permit from the Water Authority to modify the river course.

Despite the fact that there are some good experiences to date in Chile, many schemes do not have continuous measurements of water quality or quantity. Potential reasons for lack of measurements are the additional cost involved, uncertainty around which parameters to measure and how frequently, weak enforcement from authorities to collect the information or lack of a will to share data with neighbours potentially impacted.[13] It has also been reported that in-situ monitoring equipment (such as river flow gauges to measure inflow to a scheme or automatic deep meters for measuring water table depth) are liable to theft, which can provide a disincentive.

Finally, there has been a lot of uncertainty concerning water quality and potential contamination of the aquifers, mainly from agro-chemicals (CSIRO Chile 2020a, b). This is likely due to limited data collection, attributed partly to a lack of guidance regarding relevant information to collect and appropriate parameters to meet. While the Water Authority and Ministry of Environment collect relevant water data (such as meteorological information, river flows, groundwater levels, and water quality), monitoring is often lacking spatial reach and key parameters (particularly for water quality measurements), and there is normally a long delay for the data to be made public (often a 6-month delay).

8.3.3 Socioeconomic Challenges

One of the biggest socioeconomic barriers is related to the uncertainty of costs and benefits of MAR schemes. Unlike other storage infrastructure projects, such as dams, MAR has a relatively short time span of experience, particularly in Chile and other regions of Latin America. This lack of experience, especially in terms of fully operational schemes with adequate monitoring of water volumes and quality, makes it difficult to estimate the costs and benefits of new potential schemes. Even on a global scale, there are only a few studies which look at the financial characteristics and performance of different types of MAR schemes and compare them to other water supply alternatives (Zheng et al. 2021; Ross and Hasnain 2018). What's more, the intrinsic social and environmental values of water are particularly difficult to quantify in monetary terms. At the same time, the value of water varies considerably depending

[13] There is a lot of ambiguity surrounding what and how data is to be reported to local corresponding authorities. According to the bulletins published by the Water Authority, referred to in the previous section, they may request a technical report of a "low-risk" scheme (infiltration of rainwater, infiltration ponds, "leakage" from canals or other hydraulic structures), but this is not mandatory. According to communication with a Regional Water Authority, as part of the development of the MAR Guidelines, pond infiltration schemes on private land with existing water rights do not require a permit from the authority, and only in the case of injection wells was it mentioned that periodic testing of water quality is required, however the procedure for this was not given (CSIRO Chile, 2020a, b).

on the local circumstances, such as water availability, water quality, dependence on the water source, etc.

As previously discussed, current legislation in Chile regarding MAR implies a lengthy and costly evaluation process where permits are required for new water rights to be obtained for water infiltration and/or to extract part of the infiltrated water. Where projects are unable to extract infiltrated water, it can be difficult to perceive the benefits. Furthermore, while some MAR projects may be economically viable, where benefits outweigh the costs, the financial resources may not be readily available. Often water projects compete for limited financial resources against other projects, which also provide social benefits, such as health or transport projects.

8.4 Operational Framework for Artificial Aquifer Recharge Projects

The MAR Guidelines were developed in Chile to reduce some of the uncertainties and challenges previously mentioned concerning the implementation and regulation of MAR projects with a primary focus on the agriculture sector. The guidelines emerged as a result of continuous collaboration between CSIRO Chile and the National Irrigation Commission of Chile, during which the National Aquifer Recharge Plan for Agriculture (Comisión Nacional de Riego 2019) was articulated based on four pillars, suggested by CSIRO Chile: 1. a governance body; 2. regulatory adjustments; 3. promotion instruments; and 4. scientific support. The MAR Guidelines are, in effect, a promotion instrument to encourage farmers and Water User Organizations to implement MAR more widely. The Plan also presents the intention to support the creation of new Groundwater Communities, which could play an important role in the management and/or evaluation of future MAR projects. To date, there have been few established Groundwater Communities and they don't currently have a legal mandate regarding MAR. However, across stakeholders, they are seen as a key institution to support the development and sustainable operation of MAR projects, as well as a move towards a more integrated management of water resources more widely, particularly when coordinated with other Water User Organizations, which manage surface water.

In a broad sense, the guidelines offer potentially interested farmers a practical tool aimed at the development of low-risk MAR projects, which are limited to the following conditions (CSIRO Chile 2020a, b):

- The proponent possesses surface water rights available to infiltrate.
- The source water is restricted to surface water, which has not been previously used, such as rainwater or water extracted from rivers.[14]
- The target aquifer is unconfined and is above the saturated zone.

[14] For example, industrial discharge water or treated wastewater cannot be used.

- The end use of the recovered water is for irrigation purposes and existing groundwater rights will be used for recovery (i.e., no new rights are required).

While the guidelines can also be used for other end uses such as restoring depleted groundwater systems or as a barrier to prevent saline intrusion in coastal areas, they do not cover augmenting drinking water supply, non-potable industrial supply, or non-potable domestic uses, as these present a much higher level of risk and therefore would require more rigorous research, monitoring and treatment approach. It is important to note that the guidelines on their own are not official in nature and therefore compliance with them is not legally required, but their objective is to guide the user and improve their confidence in MAR applications.

The Chilean guidelines were developed based on existing approaches, such as the Guidelines for Drinking Water Quality and the Water Safety Plan (WSP) methods advocated by the World Health Organization (WHO 2006, 2011, 2012)[15] and the Australian risk-based MAR Guidelines (NRMMC–EPHC–NHMRC 2009), however, they were adapted according to the contextual needs, including the requirement for lower-risk usage such as in agriculture. This provided an intermediate framework between current, largely unmanaged MAR projects in Chile and the rigorous existing international frameworks, which rely heavily on extensive data collection and reporting (see Table 8.3 for a comparison between the three international guidelines mentioned). As a result, the Chilean guidelines aimed to use basic information, readily observable at an existing or proposed MAR site to improve the safety of the project and to establish greater accountability as standard practice (see Table 8.4 for further detail). The latter was done using a staged approach and workflow plan, which can be used as a screening tool to identify sites where more rigorous investigations are required.

The workflow for the application of the guidelines presented in Fig 8.2. was developed in collaboration with the National Irrigation Commission and considered the comments raised by stakeholders during the interviews and workshops previously mentioned. As recommended by stakeholders, simple MAR schemes (Scale 1) were included, which do not need to follow the Guidelines' stages (1–3). Instead, information sheets were provided to give examples of the different MAR methods under these schemes, including: (i) rainwater capture and infiltration systems; (ii) infiltration canals; and (iii) infiltration ditches. These simple schemes are likely to be low-cost, low-risk projects and are already being increasingly implemented in Chile. The information sheets provide information about the design and construction of these schemes, any monitoring considerations, examples of cost and typical

[15] Internationally, the general default approach is to adopt the WSP methods, as most countries to date do not have local, country specific MAR guidelines. The WHO has produced three editions of guidelines for water reuse that are relevant to MAR, including the State-of-the-Art Report: Health Risks in Aquifer Recharge using Reclaimed Water published in 2003 (WHO, 2003). These methods recognize that the risk-based approaches that were already well developed in the food industry based on the Hazard Analysis and Critical Control Point (HACCP) system were also needed throughout the water industry, particularly regarding the use of aquifers to recycle wastewater.

Table 8.3 Comparison of coverage of the Chilean MAR Guidelines with the WHO Guidelines (WHO 2011) and the Australian MAR Guidelines (NRMMC–EPHC–NHMRC 2009), in terms of the type of source water and aquifer covered, existing use and conditions of the groundwater, the methodology used, cost of application and confidence in health and environmental protection (CSIRO Chile 2020a, b). A tick implies that the aspect mentioned is considered and guidance is provided

Coverage	Chilean Guidelines	WHO Drinking Water Guidelines	Australian MAR Guidelines
Source Water Type included			
Rural surface water with little prior intervention *	✓	✓	✓
Urban stormwater			✓
Treated sewage effluent			✓
Desalinated water			✓
Industrial wastewater			✓
Aquifer Type included			
Unconfined	✓	✓	✓
Confined		✓	✓
Intended Use of Recovered Groundwater			
Non-potable use**	✓		✓
Drinking locally	✓	✓	✓
Public drinking water supply		✓	✓
Groundwater Conditions Considered			
Sensitive ecosystems dependent on groundwater			✓
Groundwater is brackish	✓		✓
Groundwater is polluted			
Aspects Covered in the Application of the Guideline			
Catchment assessment		✓	✓
Desktop assessment based on available data, even if limited	✓	✓	✓
Risk assessment based on investigations and monitoring data		✓	✓
Monitoring programs after implementation	✓	✓	✓
Management of clogging	✓		✓
Recovery of groundwater after recharge in brackish aquifers	✓		✓

(continued)

Table 8.3 (continued)

Coverage	Chilean Guidelines	WHO Drinking Water Guidelines	Australian MAR Guidelines
Biogeochemical reactions in aquifer			✓
Cost and confidence			
Low cost and ease of implementation	✓	✓	
Improves health protection	✓	✓	✓
High confidence in health protection ***		✓	✓
High confidence in environment protection ***			✓

*Rainwater and river water
**It refers to all water uses except potable water use, e.g., irrigation, industrial, environmental, mining, etc.
***The Chilean guidelines focus on MAR projects in the agriculture sector and therefore do not thoroughly cover these aspects and cannot guarantee 100% protection to health or to the environment. However, the guidelines aim to significantly reduce the risk associated with these aspects.

time frames, and case studies to refer to. The aim is to improve current and future practices of these simple schemes, which are already being implemented in Chile.

The more complex MAR schemes (Scale 2) included are: (i) infiltration ponds; (ii) dry wells;[16] (iii) infiltration trenches; and (iv) in-stream infiltration methods, such as underground dams and recharge releases from existing dams. While they are still relatively simple schemes, which could be implemented in the agriculture sector, they still require further consideration in comparison to Scale 1 schemes. Therefore, a 3-stage assessment approach was given for these schemes, as indicated in Fig 8.2. It is important to note that other MAR schemes such as injection wells were not included in the guidelines, as they would likely require further, more complex considerations, they would need to go through a lengthy evaluation and permit process (as previously discussed), and they generally require a higher investment, which means they are unlikely to be implemented in the agriculture sector.

Scale 2 schemes require the proponent to prepare economic evaluations to demonstrate the value of the project, which is particularly relevant where public funding (subsidies previously described) is used for investment. Table 8.4 provides information regarding each of the three stages of the assessment. The two main advantages of a staged assessment process in MAR development are:

1. Initially, in the first stage, there is a low level of effort, investment, and complex assessments required, which increases incrementally as the project moves through the development stages.
2. Financial risks are minimized by providing early opportunities to terminate projects that are reviewed to be non-viable.

The implementation of the Chilean MAR Guidelines for the recharge of unconfined aquifers in the agricultural sector is an important step closer towards providing

[16] Also known as *Vadose Zone Infiltration Wells* (VZWs) or *recharge shafts*.

Table 8.4 A summary of the development stages for the more complex MAR schemes, based on the Chilean MAR Guidelines (CSIRO Chile 2020a, b)

Development stage	Description	Principal attributes to evaluate
Stage 1: Viability assessment	Evaluates the apparent viability of a MAR project using relevant existing data and information. The objective is to inform the proponents of any fatal flaw in the intended project. At this stage, the proponent should have a basic comprehension of the type of MAR project being considered, such as the source water, type of aquifer, aim of the recharge and/or end use of the recovered water. An understanding of the possible and feasible MAR methods is also required in addition to a preliminary economic evaluation which includes initial estimates of costs and benefits, including a qualitative understanding of likely social and environmental impacts (positive and negative).	• Water demand • Availability of a water source • Hydrogeological evaluation • Space to capture and treat the water • Capabilities and experience • Governance and economic sustainability • Preliminary economic evaluation
Stage 2: Technical assessment	Evaluates technical aspects based on results gathered from fieldwork and as derived from results of the viability assessment (Stage 1), focusing on aspects with higher levels of risk and uncertainty. Fieldwork at this stage could include aquifer pumping tests, permeability tests, water quality testing, soil sampling, and others. Identifies potential risks associated with the project and prepares prevention measures. Includes a sanitary survey to evaluate the potential risks to public health. It is recommended to carry out a detailed economic evaluation, updating the preliminary evaluation developed in the previous stage. This is particularly important when the scheme requires a large investment.	• Quality of the source water • Storage capacity of the aquifer • Potential impacts on neighbours and ecosystems • Presence of reactive or fractured rock (potential to mobilize arsenic) (Fakhreddine et al. 2021) • Sanitary survey (potential sources and routes of contamination) • Scheme approval (where required) • Detailed economic assessment

(continued)

Table 8.4 (continued)

Development stage	Description	Principal attributes to evaluate
Stage 3: Design and operation	A pilot scheme (or at least recharge trials) is recommended at the chosen site before final operation, particularly where it is a relatively large scheme. This ensures that the scaling up of the scheme is more successful. If the pilot or trials are conducted, similar tests to those described in Stage 2 should be carried out to compare and validate the results, over a longer time span and in a more extensive area, providing confidence in the development of a full-scale scheme. These results also allow for the development of a preliminary operation and maintenance plan for the long-term running of the scheme, in addition to a risk management plan. These plans should be periodically reviewed and updated during the subsequent operation of the scheme. Design and construction generally represent the majority of MAR project costs. The design stages can be divided into "functional" and "detailed" design, which is a key input to the final economic evaluation at this stage.	• Pilot or recharge trials • Functional and detailed design • Final economic assessment • Preliminary operation and maintenance plan • Risk management plan • Project management at an operational scale • Water safety plan

a more secured water provision for farmers and a better protection of groundwater, in comparison to current practice in Chile. The guidelines aim to accelerate low-risk MAR projects and at the same time reduce uncertainty and stagnation of other MAR projects, minimizing failures and financial investment.

8.5 Conclusions and Recommendations

The implementation of MAR is part of a larger water security and resilience strategy. In this regard, to ensure that water is used in the most efficient and sustainable way, the authors recommend complementing current land management practices with MAR projects. To achieve this under the current legislative environment, Chile should transition from an individualistic and sectorial water management approach, toward a more systemic and integrated one, acknowledging the interdependence between stakeholders, sectors, and water resources (surface water and groundwater) and looking to pursue water security for the population, the environment, food production, and other productive uses in the long term.

In line with that, MAR has received wide recognition across different sectors in Chile as a promising and necessary component of the response to the current mega drought and future climate change effects. However, there is still a low level of adoption, particularly of managed schemes, which has been assessed to be attributable to at least three principal factors:

1. Firstly, there is large uncertainty due to the technical and knowledge limitations, particularly related to the hydrogeology of the aquifers, water availability for recharge under a future climate, the potential impact on third parties, and the likelihood to achieve required recharge and recovery volumes. This type of uncertainty is very important, and the risks should be systematically managed. There is also a lack of knowledge about the environmental and human health risks associated with MAR, which must be properly managed for its successful operation.
2. Secondly, and partly due to the lack of knowledge just mentioned, there is a lack of operational or demonstrational MAR projects, which has prevented a wider uptake of MAR. Therefore, as MAR is a relatively new technology (in terms of incorporating management as a key consideration), there is still a lack of knowledge and a certain, rightful, distrust regarding its use. On the other hand, dams have been used for decades, the technology is widely known, they provide considerable storage potential, and they are relatively easy to operate and monitor. However, the high financial, social, and environmental costs of dams, as well as their vulnerability to contamination, high evaporation rates, lengthy evaluation and construction times, and decreasing land availability, have triggered the search for alternative storage solutions. In this sense, MAR can complement existing surface water storage infrastructure and improve the conjunctive surface and groundwater management.

3. Finally, as apparent from the interviews with stakeholders (CSIRO Chile 2020a, b), the lack of adequate legislative support and regulation is a source of uncertainty for proponents of MAR projects. This means they face high risks when carrying out potentially expensive research for a prospective project, where they do not have the standards and procedures to guide them, let alone the investment cost, which may or may not be covered by the public subsidy subsequent to implementation. This increases the uncertainty of investors, since there is ambiguity surrounding the life-cycle costs and the long-term economic viability of a project.

The MAR guidelines developed for Chile are expected to increase adoption, reduce uncertainty, accelerate low-risk schemes, and prevent high-risk schemes from proceeding. They also promote projects which do not require the creation of new water rights to extract the infiltrated water. This is likely due to the fact that many river basins are highly over-allocated and therefore the strategy is to promote community-scale projects which improve the health of the aquifer and secure existing water rights, while disincentivizing increased extraction.

As this is the first edition of the MAR Guidelines, it is likely that they will need adjusting after they are applied in practice. It is therefore considered prudent to apply the guidelines in an iterative manner. This will enable understanding how farmers and/or Water User Organizations, as well as authorities, evaluate the application of the MAR Guidelines, allowing time to adjust them to include any new suggestions or to respond to any doubts raised.

It is also recommended that the MAR Guidelines be extended to other purposes and other water sources; for example, incorporating potable water as an end use and using recycled water as source water. For these additions to be complementary to the existing MAR Guidelines, it is essential that they are consistent with its contents and are developed with a similar methodological approach. This would provide more robust guidelines, covering all aspects of MAR, similarly to the Australian MAR Guidelines.

In parallel, these guidelines should be complemented by a wider range of efforts relating to MAR, including:

(1) Encouraging the systematic gathering, exchange and analysis of information and data related to MAR in Chile.
(2) Identifying priority MAR opportunity areas, including at a more local scale.
(3) Generating confidence in the application of MAR projects through the development and promotion of successful projects.
(4) Incentivizing projects, which benefit multiple stakeholders and strengthen Water User Organizations, particularly integrated surface–groundwater ones which are still largely undeveloped in Chile.[17]

[17] As mentioned earlier, there are current efforts to promote the creation of Groundwater Communities which, in coordination with existing Water User Organisations, could improve the integrated management of surface and groundwater.

Concerning the latter, it is crucial to implement adequate project governance. According to the OECD (2017a, b), investment in physical infrastructure is not enough to improve water security. This should be accompanied by governance systems, which allow for the coordination of actions of the different stakeholders. This is particularly relevant for projects, which include groundwater management, as they are potentially more complex and even experts in the field often disagree on the interpretation of the same facts. Therefore, in the case of MAR projects, which mandate an integrated approach to surface and groundwater management, it is important to implement collaborative, legitimate decision-making processes validated by the stakeholders interested or potentially impacted by the decisions.

References

Ancajima R (2016a) Sabiduría ancestral, un camino en la adecuación al cambio climático. XXVII Congreso Latinoamericano de Hidráulica. Lima, Perú

Ancajima R (2016b) Ancestral wisdom, a path to climate change adaptation. XXVII Latin American Congress of Hydraulics. Lima, Peru

Bauer C (1998) Against the current: privatization, water markets, and the state in Chile. Springer, US

Bertule M, Appelquist LR, Spensley J, Trærup SL, Naswa P (2018) Climate change adaptation technologies for water: A practitioner's guide to adaptation technologies for increased water sector resilience. UNEP DTU Partnership

Biblioteca del Congreso Nacional de Chile (2018) Principales contenidos y alcances de la Ley N° 18.450 (Ley de Riego). Biblioteca del Congreso Nacional de Chile (BCN), Santiago

Biblioteca del Congreso Nacional de Chile (2022) Ley 21435. Reforma El Código de Aguas. Ministerio de Obras Públicas. Biblioteca del Congreso Nacional de Chile (BCN), Santiago

Bozkurt D, Rojas M, Boisier J, Valdivieso J (2017) Climate change impacts on hydroclimatic regimes and extremes over Andean basins in central Chile (preprint). Hydrology and Earth System Sciences Discussions

CSIRO Chile (2020) Marco operativo para proyectos de recarga artificial en acuíferos: informe final. Comisión Nacional de Riego, Santiago

CSIRO Chile (2020) Operatonal framework for artificial aquifer recharge projects: final report. National Irrigation Commission, Santiago

Cabrera Fajardo G (2014a) Análisis desde Arica hasta el Maule: Dónde Hay Condiciones para la Recarga Artificialde Acuíferos en Chile. RevistaAIDIS 47:32–39

Cabrera Fajardo G (2014b) Analysis from Arica to Maule: where are there conditions for artificial aquifer recharge in Chile? AIDIS Magazine 47:32–39

Center for Climate and Resilience Research (2020) La megasequía 2010–2019: Una lección para el futuro. Accedido el 09 de 08 de 2022, de https://www.cr2.cl/megasequia/

Centre for Climate and Resilience Research (2020) The mega drought 2010–2019: a lesson for the future. Accessed 09 Aug 2022, from https://www.cr2.cl/megasequia/

Comisión Nacional de Riego & Gobierno Regional de Atacama (2012) Estudio Prefactibilidad: Mejoramiento del Sistema de Aguas Subterráneas para su Utilización en la Cuenca del Río Copiapó, Región de Atacama. Elaborado por: Jorquera Ingeniería, Santiago

Comisión Nacional de Riego (2013) Estudio diagnóstico de zonas potenciales de recarga de acuíferos en las regiones de Arica y Parinacota a la región del Maule. GCF Ingenieros Ltda, Elaborado por

Comisión Nacional de Riego (2019) Plan Nacional de Recarga de Acuíferos para la Agricultura. Comisión Nacional de Riego, Santiago

Comisión Económica para América Latina y el Caribe (CEPAL) (2012) La Economía del Cambio Climático en Chile. Naciones Unidas, Santiago
Correa-Parra J, Vergara-Perucich J, Aguirre-Nuñez C (2020) Water privatization and inequality: gini coefficient for water resources in Chile. Water. https://doi.org/10.3390/w12123369,12(3369)
Delgado V, Arumi J (2019) La recarga artificial de aguas subterráneas en el derecho chileno parte I. 03 de 05 de 2019 en El Mercurio
Delgado V, Arumi J (2019) Artificial recharge of groundwater in Chilean law part I. 03/05/2019 in El Mercurio
Dillon P, Stuyfzand P, Grischek T, Lluria M, Pyne RDG, Jain RC, Bear J, Schwarz J, Wang W, Fernandez E, Stefan C, Pettenati M, van der Gun J, Sprenger C, Massmann G, Scanlon BR, Xanke J, Jokela P, Zheng Y, Rossetto R, Shamrukh M, Pavelic P, Murray E, Ross A, Bonilla Valverde JP, Palma Nava A, Ansems N, Posavec K, Ha K, Martin R, Sapiano M (2019) Sixty years of global progress in managed aquifer recharge. Hydrogeol J 1(30)
Dillon P (2005) Future management of aquifer recharge. Hydrogeol J 13
Dirección de Obras Hidráulicas (2012) Análisis de potencialidad de recarga artificial acuíferos primera y tercera sección valle del Aconcagua. GeoHidrología, Elaborado por
Dirección General de Aguas (DGA) (2019a) Circular N°1 2019. Dirección General de Aguas
Dirección General de Aguas (DGA) (2019b) Circular N°2 2019. Dirección General de Aguas
Dirección General de Aguas (2012) Investigación recarga artificial de acuíferos cuencas del Río Choapa y Quilimarí, Región de Coquimbo. Dirección General de Aguas. Realizado por: AC Ingenieros Consultores Ltda, Santiago
Dirección General de Aguas (2014) Diagnóstico de metodología para la presentación y análisis de proyectos de recarga de acuíferos. Dirección General de Aguas. Realizado por: AMPHOS 21 Consulting Chile Ltda, Santiago
Dirección General de Aguas (2016a) Atlas de Agua—Chile 2016. Dirección General de Aguas, Santiago
Dirección General de Aguas (2016b) Guía metodológica para la presentación de proyectos de recarga artificial. Dirección General de Aguas. Realizado por: el Departamento de Administración de Recursos Hídricos, Santiago
Dirección General de Aguas (2016c) Circular N°4 2016. *1–4*.Dirección General de Aguas
Donoso G, Lictevout E, Rinaudo J (2020) Groundwater Management Lessons from Chile. In: Sustainable groundwater management: a comparative analysis of French and Australian policies and implications to other countries, pp 481–509
Economic Commission for Latin America and the Caribbean (CEPAL) (2012) The economy of climate change in Chile. United Nations, Santiago
Escalante EF, Sauto JSS, Gil RC (2019) Sites and indicators of MAR as a successful tool to mitigate climate change effects in Spain. Water 11(1943)
Escenarios Hídricos 2030 (2019) Transición hídrica: el futuro del agua en Chile. Portafolio de medidas, acciones y soluciones. MAS seguridad hídrica. Fundación Chile, Santiago
Fakhreddine S, Prommer H, Scanlon B, Ying S (2021) Mobilization of arsenic and other naturally occurring contaminants during managed aquifer recharge: a critical review. Environ Sci Technol 55(4)
Foro Consultivo Científico y Tecnológico (FCCyT) (2012) Diagnóstico del agua en las Américas. Red Interamericana de Academias de Ciencias. Distrito Federal: Foro Consultivo Científico y Tecnológico, México
Gale I (2005a) Estrategias para la gestión de recarga de acuíferos (GRA) en zonas semiáridas. United Nations Educational, Scientific and Cultural Organization (UNESCO), Paris
Gale I (2005b) Strategies for managed aquifer recharge (MAR) in semiarid zones. United Nations Educational, Scientific and Cultural Organization (UNESCO), Paris
Garreaud R, Boisier J, Rondanelli R, Montecinos A, Sepúlveda H, Veloso-Aguila D (2019) The central chile mega drought (2010–2018): a climate dynamics perspective. Int J Climatol 6:74

General Directorate of Water/National Water Authority (2012) Artificial aquifer recharge research Choapa and Quilimarí river basins, Coquimbo Region. General Directorate of Water. Prepared by: AC Ingenieros Consultores Ltda, Santiago
General Directorate of Water/National Water Authority (2014) Diagnosis of methodology for the presentation and analysis of aquifer recharge projects. General Directorate of Water. Prepared by: AMPHOS 21 Consulting Chile Ltda, Santiago
General Directorate of Water/National Water Authority (2016a) Water Atlas—Chile 2016. General Directorate of Water, Santiago
General Directorate of Water/National Water Authority (2016b) Methodological guide for the presentation of artificial recharge projects. General Directorate of Water. Prepared by: Department of Water Resources Administration, Santiago
General Directorate of Water/National Water Authority (2016c) Circular N°4 2016. 1–4. General Directorate of Water
General Directorate of Water/National Water Authority (2019a) Circular N°1 2019. General Directorate of Water
General Directorate of Water/National Water Authority (2019b) Circular N°2 2019. General Directorate of Water
Guyennon N, Salerno F, Portoghese I, Romano E (2017) Climate change adaptation in a Mediterranean semi-arid catchment: Testing managed aquifer recharge and increased surface reservoir capacity. Water 9(689)
Hernandez K, Madeira C (2022) The impact of climate change on economic output across industries in Chile. PLoS ONE 17(4). https://doi.org/10.1371/journal.pone.0266811
Hydraulic Works Directorate (2012) Analysis of the potential for artificial aquifer recharge in the first and third sections of the Aconcagua valley. Elaborated by: GeoHidrología
Jiménez JP (2012a) Plan maestro de gestión de recursos hídricos Región de O'Higgins. Dirección General de Aguas, Santiago
Jiménez JP (2012b) Master plan for water resources management O'Higgins Region. General Directorate of Water, Santiago
Larraín S (2006a) El agua en Chile: entre los derechos humanos y las reglas del mercado. Revista de la Universidad de Bolivariana, Polis
Larraín S (2006b) Water in Chile: between human rights and the rules of the market. Polis, J Boliv Univ
Library of the National Congress of Chile (2018) Main contents and scope of the Law N° 18.450 (Irrigation Law). Library of the National Congress of Chile (BCN), Santiago
Library of the National Congress of Chile (2022) Law 21435. Reform of the Water Code. Ministry of Public Works. Library of the National Congress of Chile (BCN), Santiago
Mesa Nacional del Agua (2022) Mesa Nacional del Agua. Informe final. Ministerio de Obras Públicas, Santiago
Ministerio de Medio Ambiente (2015) Atlas de cambio climático de la zona semiárida de Chile. Accedido el 02 de 03 de 2022, de https://educacion.mma.gob.cl/wp-content/uploads/2015/09/Atlas-CC-ZonaAridas-Semiaridas-Chile.pdf
Ministry of the Environment (2015) Climate change atlas of Chile's semi-arid zone. Accessed 2 March 2022, from https://educacion.mma.gob.cl/wp-content/uploads/2015/09/Atlas-CC-ZonaAridas-Semiaridas-Chile.pdf
National Irrigation Commission and Regional Government of Atacama (2012) Prefactibility Study: Improvement of the Groundwater System for its use in the Copiapó River Basin, Atacama Region. Elaborated by: Jorquera Ingeniería, Santiago
National Irrigation Commission (2013) Diagnostic Study of potential aquifer recharge zones in the regions of Arica and Parinacota to the region of Maule. Elaborated by: GCF Ingenieros Ltda
National Irrigation Commission (2019) National plan of aquifer recharge for agriculture. National Irrigation Commission, Santiago
National Water Table (2022) National Water Table. Final report. Ministry for Public Works, Santiago

Natural Resource Management Ministerial Council, Environment Protection and Heritage Council & National Health and Medical Research Council (NRMMC–EPHC–NHMRC) (2009) Australian Guidelines for Water Recycling (Phase 2): Managed Aquifer Recharge. Canberra
Organisation for Economic Co-operation and Development (OECD) (2017a) Brechas y estándares de gobernanza de la infraestructura pública en Chile: análisis de gobernanza de infraestructura. OECD Publishing, Paris
Organisation for Economic Co-operation and Development (OECD) (2017b) Gaps and standards of public infrastructure governance in Chile: analysis of infrastructure governance. OECD Publishing, Paris
Page D, Gonzalez D, Bennison G, Burrull C, Claro E, Jara M, Valenzuela G (2020) Progress in the development of risk-based guidelines to support managed aquifer recharge for agriculture in Chile. Water Cycle 1:136–145
Polade SD, Pierce DW, Cayan DR, Gershunov A, Dettinger MD (2014) The key role of dry days in changing regional climate and precipitation regimes. Sci Rep 4(4364)
Ross A, Hasnain S (2018) Factors affecting the cost of managed aquifer recharge (MAR) schemes. Sustain Water Resour Manag 4:179–190
Scientific and Technological Consultive Forum (FCCyT) (2012) Water diagnosis in the Americas. Inter-American Network of Science Academies. Federal District: Scientific and Technological Consultive Forum, Mexico
Sheffield J, Wood EF (2008) Projected changes in drought occurrence under future global warming from multi-model, multi-scenario, IPCC AR4 simulations .Climate Dyn 31(79)
Superintendencia de Servicios Sanitarios (SISS) (2021a) Informe de gestión del sector sanitario 2020. Superintendencia de Servicios Sanitarios, Santiago
Superintendent of Sanitary Services (SISS) (2021b) Report of the management of the sanitary sector 2020. Superintendent of Sanitary Services, Santiago
Vázquez R (2019a) Meteochile Blog: Lo que sabemos (y lo que no) sobre el cambio climático en Chile. 12 de 09 de 2019. Accedido el 02 de 03 de 2022, de: https://blog.meteochile.gob.cl/2019/09/12/lo-que-sabemos-y-lo-que-no-sobre-el-cambio-climatico-en-chile/
Vázquez R (2019b) Meteochile Blog: What we know (and what we don't) about climate change in Chile, 12/09/2019. Accessed 2 March 2022, from: https://blog.meteochile.gob.cl/2019/09/12/lo-que-sabemos-y-lo-que-no-sobre-el-cambio-climatico-en-chile/
Water Scenarios 2030 (2019) Water transition: the future of water in Chile. Portfolio of measures, actions and solutions. MORE water security. Chile Foundation, Santiago
World Health Organisation (WHO) (2003) Health risks in aquifer recharge using reclaimed water: state of the art report. In: Aertgeerts R, Angelakis A (eds) World Health Organization. Regional Office for Europe: https://apps.who.int/iris/handle/10665/83800
World Health Organisation (WHO) (2006) Protecting groundwater for health; managing the quality of drinking-water sources available. World Health Organisation and IWA Publishing
World Health Organisation (WHO) (2011) Guidelines for drinking water quality, 4th edn. World Health Organization, Geneva
World Health Organisation (WHO) (2012) Water safety planning for small community water supplies: step-by-step risk management guidance for drinking-water supplies in small communities. World Health Organisation, Geneva
World Resources Institute (2015) Ranking the world's most water-stressed countries in 2040. Accessed 01 March 2022, from: https://www.wri.org/blog/2015/08/ranking-world-s-most-water-stressed-countries-2040
Zheng Y, Ross A, Villholth K, Dillon P (eds) (2021) Managing Aquifer recharge: a showcase for resilience and sustainability. United Nations Educational, Scientific and Cultural Organization (UNESCO), Paris, France, p 379. ISBN: 978-92-3-100488-9

Chapter 9
From Managed Aquifer Recharge to Managing Aquifer Recharge: Developing a Strategic Approach to Artificial Recharge in India

Himanshu Kulkarni, Uma Aslekar, Siddharth Patil, Neha Bhave, Jayesh Desai, and Imran Siddique

Abstract The rapid transition from a narrow focus on developing groundwater resources to wider thinking on groundwater governance has been a result of many different developments in the South Asian Region, particularly in India. Policies and practices of groundwater recharge, especially through engineered mechanisms have also evolved from artificial recharge to the concept of managed aquifer recharge (MAR). The large potential for MAR in India prompts strategic thinking on enabling projects on groundwater recharge, bearing in mind the diversity of India's aquifers and the status of groundwater extraction in different regions of India. MAR is not just about adopting and adapting artificial recharge to different geographies but is also about understanding the contexts for MAR. Understanding groundwater conditions, making correct choices of recharge and conservation measures and gauging impact are an integral part of MAR, explained through case studies in this paper. Embedding groundwater recharge planning within the larger framework of participatory groundwater management and governance will provide a further impetus to change the very meaning of MAR—from managed aquifer recharge to managing aquifer recharge.

Keywords Aquifers · Managed aquifer recharge · Hydrogeology · Scale · Participation

H. Kulkarni (✉) · U. Aslekar · S. Patil · N. Bhave · J. Desai · I. Siddique
Advanced Center for Water Resources Development and Management (ACWADAM), Pune, India
e-mail: acwadam@gmail.com

D. Saha et al. (eds.), *Managed Groundwater Recharge and Rainwater Harvesting*, Water Resources Development and Management,
https://doi.org/10.1007/978-981-99-8757-3_8

9.1 Introduction

Groundwater resources have become the lifeline of India's water supply—in rural-agrarian situations and the growing urban-industrial context. More than 85% of India's rural population depends upon groundwater for its domestic water requirements (World Bank 2010). Nearly 70% of the water for India's agriculture comes from groundwater sources (Ministry of Agriculture 2013), at least half of the urban water supply is groundwater based (Narain 2012) and even the industry uses a significant amount of groundwater, although the exact figures at the national level remain elusive. A study of a sample of industrial sectors indicated that some 55% of the 27 sectors studied used groundwater either as a standalone resource or in conjunction with other sources (Parveen et al. 2012). Put together, these statistics indicate that at least 100 billion Indians use groundwater in India daily, one way or the other. On one hand, this dependency underscores the importance of groundwater but on the other, it also demands a careful synthesis of vulnerabilities from over extraction and contamination and the articulation of groundwater management strategies.

The last few national-level groundwater assessments (Central Ground Water Board 2006; 2011, 2014, 2017, 2019) indicate India's annual groundwater extraction to be about 250 billion cubic metres (BCM) or 250 km^3. These numbers are further ratified through a global synthesis of the growing trends in groundwater extraction, especially for agriculture (Shah 2009; Margat and van der Gun 2013). India's annual groundwater extraction is perhaps the largest in the world today with the intensity of groundwater extraction being in the range of 50–100 mm/year (Margat and van der Gun 2013). Groundwater over-abstraction and contamination together have led to a high level of vulnerability of resources and population across large swaths of India's landscape, with nearly 60% of the districts being affected by depleting or contaminated groundwater or both (Kulkarni and Vijayshankar 2009). The latest assessments available at the district and block levels show just how widespread the dual challenges of groundwater depletion and contamination are (Fig. 9.1).

Stressed groundwater resources are obviously reflected in the depletion of groundwater in aquifers and the contamination of groundwater sources (Saha et al. 2022). However, there are deeper, invisible implications of these problems, stemming from the fact that aquifers, a common pool resource come under increasing, often intensive competition, leading to the potential conflict between different users (Kulkarni and Vijayshankar 2014). Intra-sectoral competition is observed in agriculture domestic, industrial, and ecosystem users, clearly representing the deepening of the groundwater crisis. What is even more serious is the inter-sectoral competition, suggesting that the groundwater crisis is also widening and is, therefore, now at a widespread scale (Patil et al. 2017). The conflict between farming communities and the city of Chennai over filling up tankers from wells used for agriculture is one clear case that is well-documented (Janakarajan 2008). Groundwater competition has many dimensions but is usually a result of a race between supply and demand wherein the stocks of groundwater (availability and quality) are often presumed to be infinite.

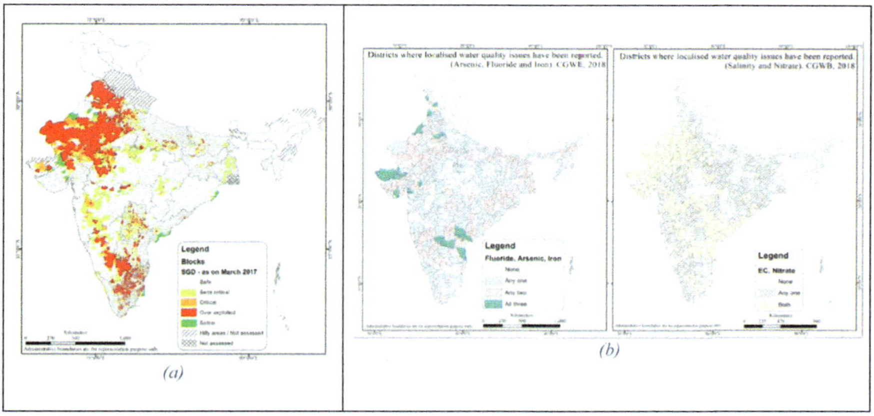

Fig. 9.1 **a** Stage of groundwater development in blocks and tehsils **b** Localised water quality issues in different districts of India. (Central Ground Water Board 2019) and (Central Ground Water Board 2018). *Note* Maps with administrative divisions and the colour coding is for representative purposes only

It is interesting to note how artificial recharge, aquifer storage and recovery (ASR) and managed aquifer recharge (MAR) evolved in India during the period from the 1970s to the 2020s. While the legacy of harvesting water with a focus on groundwater recharge dates to much older periods (Agarwal and Narain 1997; Joji et al. 2021), the early 1970s saw a lot of activity, marked by the drought of 1971–1972. This was a period when many rural employment guarantee efforts linked to the development of water harvesting structures were launched in states like Maharashtra.

At the same time, the rapid development in technology, techniques and analytics behind recharge has dominated global discussions on groundwater recharge. Such development has included understanding specific issues that required interventions of engineered recharge, the need for systematic planning and the socio-economic challenges and opportunities around these interventions (Pyne 1995). For instance, the efficacy of widespread watershed development in India, particularly during the 1990s, showed varying benefits from artificial recharge, across a wide range of agro-climatic and hydrological zones (COMMAN 2005). This variability was as much a consequence of the poor understanding of groundwater science (Kulkarni 1998) as much of the lack of a perspective in dealing with the imbalance between (augmented) supplies and increasing demand (Batchelor et al. 2000).

The diversity of India's aquifer settings has been described in detail in many documents (Kulkarni 2005; Central Ground Water Board 2012; Kulkarni et al. 2018; Bonsor et al. 2017; Mukherjee et al. 2015; Saha et al. 2017).

Aquifer Storage and Recovery (ASR) is defined as the storage of water in a suitable aquifer during times when water is available and recovery during times when it is needed (Pyne 1995). Dillon (2005) and Dillon et al. (2009) make a distinction between ASR and aquifer storage transfer and recovery (ASTR). ASR involves injection to and recovery from the same well while ASTR includes mutually exclusive wells for injection and recovery. Further, Gale (2005) described Managed Aquifer Recharge (MAR) as, "the intentional storage and treatment of water in aquifers"; the term 'artificial recharge' has also been used to describe such activities but adverse connotations of (the term) 'artificial' in a society where community participation of water resources management is becoming more prevalent, suggesting that a new name is required. More recently, a global narrative on MAR explains how, over a period of 60 years, MAR has captured the interest of various stakeholders with a growing awareness of its potential in replenishing over-allocated aquifers, restoring brackish aquifers and even enabling energy recovery fully, at the same time acknowledging that the rate of growth of MAR is not apace of the global rate of groundwater depletion (Dillon et al. 2019).

While MAR holds a lot of promise in the Indian context with the potential of addressing both, the problem of depleting aquifers and that of contaminated sources of groundwater, it is important to note that a systematic and strategic approach is required. It is particularly required in the absence of aquifer-level information at local scales, where much of the effort on MAR is concentrated. This paper attempts to highlight certain issues, particularly in the evolution of the recharge programme in India while drawing attention to key cases where ACWADAM and some of its partner organisations have been involved in more comprehensive planning and management of groundwater. The need to focus on demand-side management and groundwater governance is the compelling aspect behind the paradigm for 'Managing Aquifer Recharge' in India.

9.2 The Trajectory of Artificial Recharge: From Groundwater Resource Development to Groundwater Governance

Groundwater usage in India took off unprecedentedly in the 1970s and 1980s. Even as far back as the late 1980s and the mid-1990s, researchers reported early signs of groundwater over-extraction marked by the expansion in well numbers and competitive drilling of deeper wells from different regions of Western India (Rushton 1990; Moench 1992; Macdonald et al. 1995). As far back as the 1970s, especially after the drought years of the early nineteen seventies, three initiatives emerged that were to define the trajectory of groundwater resource management in large regions of India.

1. Technologies like the down-the-hole hammer (DTH) technique of drilling bore wells in the large tracts of hard-rock regions of the country were introduced
2. Large-scale technology uptake that enabled farmers to access deeper groundwater to fulfil irrigation demands. Deeper wells meant that larger pump capacities were needed to lift groundwater to higher levels from the aquifers below

3. The third element of the technology was achieved through the advent of the submersible pump that could be lowered to any depth for extracting groundwater, particularly for agriculture in rural areas and also in growing urban India.

The rising numbers of sources, pumps, and the shift to irrigated cropping marked this phase (1970 and 1980s) of groundwater resources development, which was incidentally accompanied by the genesis of the watershed management programme in larger areas of western and southern India. Large-scale soil and water conservation activity began in India in 1951 (Tideman 1996). What began essentially as a programme to control the rapid erosion of topsoil soon also became a programme to integrate the programmes on augmentation of groundwater.

The 1990s and early 2000s were periods of developing large-scale programmes on water conservation, mainly through integrated watershed management. By the mid-1980s, conferences on groundwater recharge began to be hosted across India followed by debates in the 1990s that focused on the social and environmental impacts of groundwater overdraft, during which the potential role of recharge in mitigating these impacts was highlighted (Moench et al. 1993). This period also brought out the need to infuse a strong hydrological, and significantly systematic hydrogeological input, into watershed management (Kulkarni 1998).

The institutional context of groundwater recharge, until the late 1990s and early 2000s was quite limited. Interest in supply augmentation of water resources increased significantly with a dramatic rise in the number and variety of organisations undertaking recharge activities (Farrington et al. 1999; Kerr and Chung 2002). While the overall approach to watershed guidelines attempted to customise the focus based on areas covered and activities implemented (James and Robinson 2001; Gale et al. 2002), most approaches continued to focus on dry land, rain-fed regions of India. At the same time, newer programmes were being adopted in the moister and forested regions like the Western Ghats and the Himalayas (Joy and Paranjape 2004).

The Central Ground Water Board's replenishable resource assessments on replenishable groundwater resources, reveal that within the span of fewer than 10 years–1996 to 2004—the proportion of districts in the semi-critical, critical, and over-exploited categories has grown from 9 to 31%, the areas in these categories from 5 to 33% and the population affected from 5 to 35% (CGWB 2006, 2011, 2014, 2017). This jump shows that the groundwater crisis had deepened considerably and there was an urgent need to expand the mandate of groundwater management to that of groundwater governance (Shah 2009; Kulkarni and Vijayshankar 2014). The emphasis on groundwater governance was also a consequence of the global push for groundwater governance. Broadly speaking, the Indian discussion on groundwater governance has focused on four broad areas:

1. Augmenting Recharge
2. Organisational reform in groundwater
3. Groundwater regulation through legislation
4. Socially driven participatory groundwater management and governance.

The Master Plan for Artificial Recharge to Groundwater in India (CGWB 2005) stated that about 14% of the total land area of the country (about half a million km^2) is suitable for artificial recharge, with estimated potential equalling 18% of the annual groundwater extraction (CGWB 2002). This plan was preceded by a Guideline on Artificial Recharge to Groundwater in India (CGWB 2000, 2002), updated since to incorporate more detailed guidelines and designs (CGWB 2007). The magnitude of MAR as a percentage of groundwater use (in five representative states of India) was found to be only 7.7%, although this figure could be somewhat conservative (Dillon et al. 2019).

Policy and legislation on groundwater have also focused on the need for resource augmentation through systematic groundwater recharge programmes (CGWB 2002; Shah 2009; Shah 1993). Measurement and documentation of MAR structures are quite limited (Dashora et al. 2018) despite the fact that India has the potential for several millions MAR structures (Dillon et al. 2019). Moreover, a practical and experiential perspective on the diverse nature of MAR is often missing from literature. The next section attempts to throw light on the contextual experiences regarding MAR as part of the paradigm of aquifer-based, community groundwater management in India, experiences that ACWADAM has documented as part of its pursuit of participatory groundwater governance in partnership with different organisations. There could be numerous others but only a limited number of various dimensions are presented in the next section, through four experiential examples.

9.3 Experiential Lessons from MAR: Geographies and Contexts

The decentralised, often organic expansion of groundwater access, the rapid progression from rain-fed agriculture to irrigated cropping, the urban sprawl, and the need for coping with climate vagaries have contributed to an ever-increasing dependency on groundwater in India as in many other regions of the world. Technology has enabled individual access to groundwater, an invisible common pool resource (CPR), in many ways. In the trade-off between food and income security, water security not only became complex but now has serious implications in furthering various Sustainable Development Goals, particularly SDG6—Sustainable Water and Sanitation for All. The last hundred years have witnessed a groundwater story unlike anything before, leading to a paradox of solving problems while creating some more. Cutting across these time scales, India also possesses a long history of water conservation that is quite well-documented, particularly in publications such as 'Dying Wisdom' (Agarwal and Narain 1997). Many of these include sustainable practices of harvesting rainwater while also addressing water spreading for infiltration to the subsurface.

Managed Aquifer Recharge (MAR) programmes must be designed and executed through integration with larger programmes like Participatory Groundwater Management (PGWM) with strategic convergence with efforts like the Mahatma Gandhi National Rural Employment Guarantee Scheme (MGNREGS) and watershed programmes, so as to replenish groundwater, especially in the context of sustaining drinking water supplies. They would also help in mitigating the impacts of salinity ingress in coastal areas. Natural recharge areas identified on the basis of such efforts must be protected, in rural and urban areas. There are many ongoing efforts, both by the Government like, National Aquifer Mapping Programme (NAQUIM and Atal Bhujal Yojona (ABHY) and by the Civil Society Organisations that are attempting to integrate MAR into larger efforts in water management and governance from local to regional scales.

We present four case studies that highlight the dimension of understanding and quantifying groundwater recharge, implementing MAR and gauging impacts that ACWADAM has been involved in.

9.4 Case 1: The Dynamics of Recharge—The Case of Local Crystalline Aquifers of East India

The ambitious programme on NAQUIM is a comprehensive programme covering the entire country and includes both demand and supply side measures in addressing groundwater stress and overexploitation of aquifers (Saha et al. 2019). However, it is necessary to further sharpen the focus on the periodic national assessment and aquifer mapping efforts with the following key features:

1. Comparing groundwater recharge and discharge against the potential aquifer storage at scales of operation (often hyper -local scales involving a single village or a micro watershed)
2. Estimating groundwater recharge through natural infiltration and artificially engineered methods
3. Increasing the periodicity of assessments to an annual basis (from the current periodicity of two to three years).

We summarise, how, at scales of operating groundwater management projects (villages or watersheds are often the units of project management), estimating recharge on an annual basis becomes important. Eastern India, particularly regions of Jharkhand and the western regions of West Bengal, is underlain by crystalline aquifer systems typically ranging between 8 and 15 m in thickness. With annual average rainfall varying between 1000 and 1300 mm, these shallow aquifer systems are discharged and recharged on an annually cyclic basis. However, the rate and quantum of recharge is highly variable. So are the potential groundwater storages, especially in the shallow, weathered-fractured aquifers. The study brings out the differences in recharge cycles in similar aquifer settings during an inter-annual estimation of storage, recharge, and discharge (Patil et al. 2015) (Fig. 9.2).

The aquifers under these two villages represent different capacities of groundwater storage. The shallow unconfined aquifer system in Bhubhui village in Jharkhand has nearly twice the potential groundwater storage (507 mm) than that of the aquifer system in Churinsara (249 mm). Both aquifers show annual recharge and discharge cycles implying that recharge from rainfall is capable of saturating or refilling the

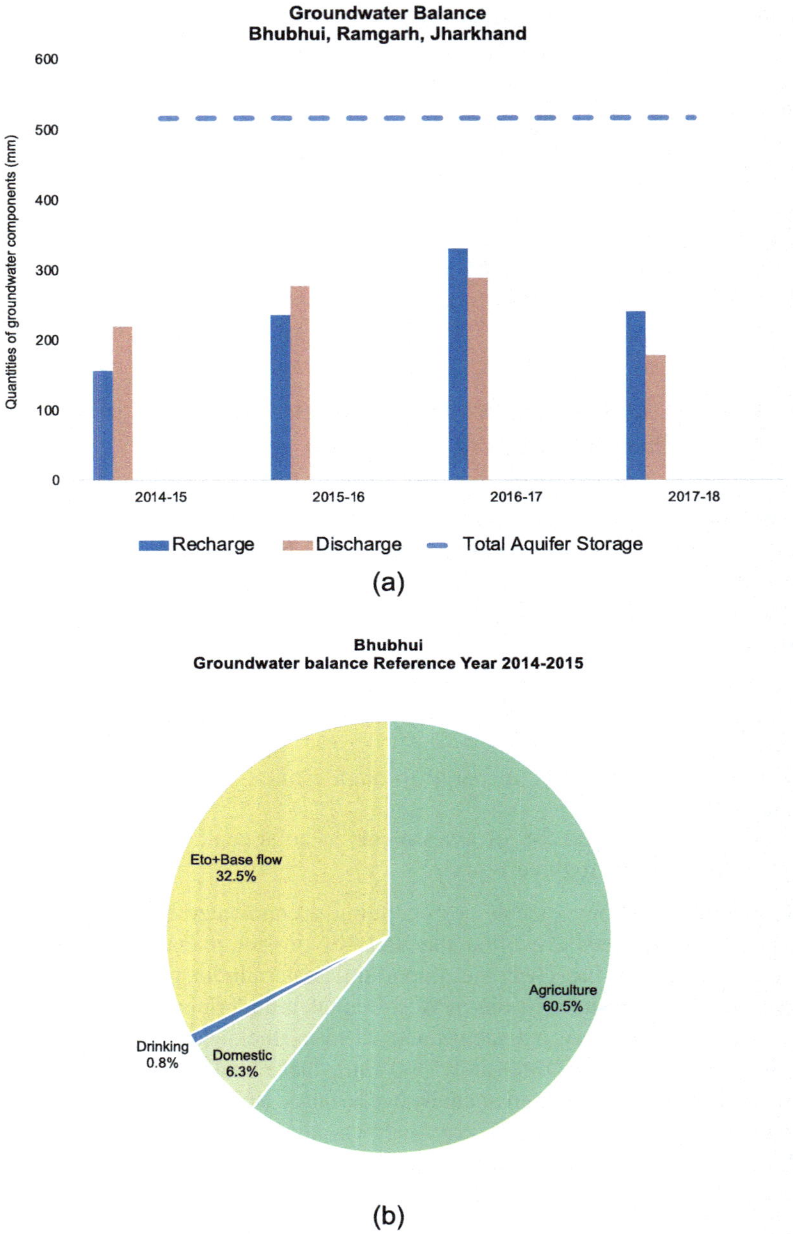

Fig. 9.2 Metrics for the Bhubhui aquifer system (modified after (Patil, Bhave, & Kulkarni, 2015)) – **a** A groundwater balance for multiple years against the potential aquifer storage; **b** Consumption (demand-side) water balance estimates for 2014–15

aquifer to its initial saturation level. However, the inter-annual relationship between groundwater recharge and extraction is so dynamic that without reference to the potential aquifer storage, it is not possible to place groundwater recharge in the proper context especially when comparing the two sites (Fig. 9.3).

In Bhubhui, groundwater pumping from the shallow aquifer is more than that in Churinsara. The demand-side water balance indicates that nearly 40% of the groundwater discharge in 2014–2015 was used in agriculture—2 to 3 crops per year—as compared to only 2.3%. In Churinsara, where agriculture is mainly rainfed. However, the main difference between the two systems is the gap between the potential aquifers storage and the annual recharge and discharge from the aquifer. In Bhubhui, the difference between the maximum recharge and the potential aquifer storage is 185 mm, while in Churinsara it is only 50 mm. The groundwater discharge ranges for Bhubhui are also greater than that from Churinsara. Although the region experienced normal annual rainfall, the aquifer system does not get completely saturated in Bhubhui, providing the potential and opportunity for focused MAR. On the other hand, the Churinsara aquifer system is thinner, with limited potential storage, limited extraction, and reasonably good recharge. Every year the rainfall saturates the aquifer to its full capacity. Here, even though 2015 was a deficit rainfall year, the aquifer system was recharged to its full capacity. Hence, focusing on conserving available groundwater in a limited storage aquifer system rather than a compulsive argument for MAR would be a strategic way forward in Churinsara.

The data, especially from Churinsara shows that it is possible that the Stage of Ground Water Extraction (SGE) is beyond 100% but the water levels do not show a declining trend. Although our data is not over a period of more than 10 years, it can be observed that the hyper-local nature of the aquifers means that even limited rainfall can completely re-saturate the aquifers even when the SGE is more than 100%. On the contrary, a cursory look at the SGE index of Bhubhui village gives the perception that the ground water situation has greatly improved over the years given that the SGE has reduced from 139 to 73%. However, when we consider groundwater availability, this ratio does not provide the entire picture as the availability across all the years does not show a significant change (Fig. 9.4).

This research has revealed that recharge of local shallow aquifers is governed not only by the amount of rainfall but by the inherent aquifer storage capacity and its intermittent saturation condition. In Bhubhui, the actual recharge is lower than the potential recharge as the system never attains full saturation despite receiving normal rainfall. In Churinsara, even 50% deficit rainfall (as compared to the long-term annual average rainfall) can completely saturate the shallow aquifers. Thus, there is an opportunity for MAR in Bhubhui aquifers, but limited potential for the same in Churinsara aquifers.

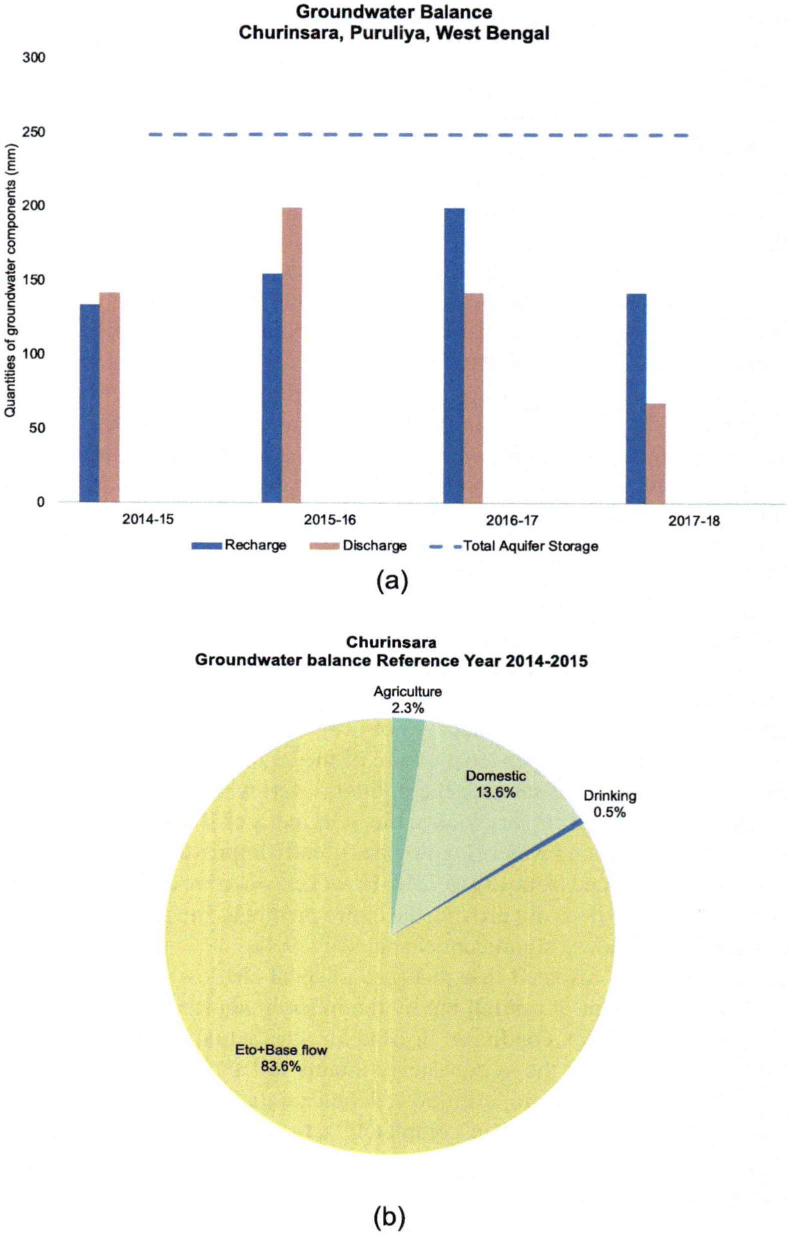

Fig. 9.3 Metrics for the Churinsara aquifer system (modified after (Patil et al. 2015))—**a** A groundwater balance for multiple years against the potential aquifer storage; **b** Consumption (demand-side) water balance estimates for 2014–2015

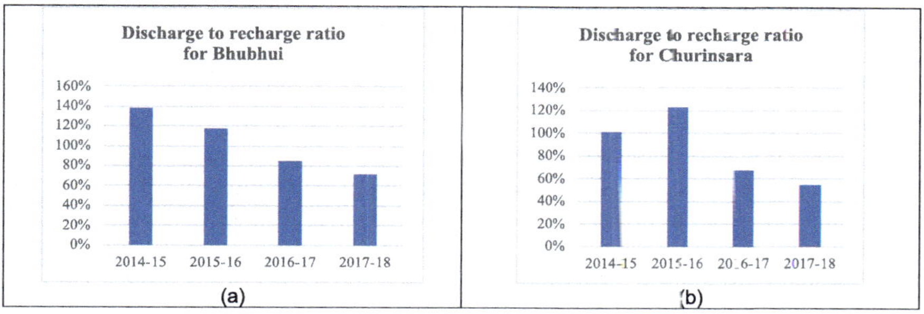

Fig. 9.4 Ratio of annual discharge to recharge for **a** Bhubhui and **b** Churinsara

9.5 Case 2: Understanding the Impact of Check Dams on Groundwater Recharge

Check dams are constructed in large parts of India, mainly with the purpose of water harvesting either for storage or for recharge. They are either in the form fear than structures, stone masonry, or concrete structures. Some of these are gated structures while others are in the form of earthen structures, stone masonry, or concrete structures. Some of these are gated structures while others are in the form of weirs with top-up design or separate water weirs for routing peak discharge downstream (Samaj Pragati 2007. In many cases, it is difficult to find a correlation between the local hydrogeology and the location of such structures, particularly with the purpose of artificial recharge as the plannings are often devoid of information regarding hydrogeological, drainage and geomorphological aspects (Kulkarni 2018).

Augmenting Groundwater Resources through Artificial Recharge - The AGRAR Project - captured the details of such structures in three locations in India (Gale et al. 2006). Of these, the Kolwan valley site was studied in detail (Kulkarni 2005) To begin with, it is relevant to state that watershed projects remain one of the most popular and successful models of decentralised, conservation practices in India. The hydrogeological nuances resulting from highly variable rainfall and geological factors are seldom considered while assessing recharge to groundwater, especially in areas of watershed development (Kulkarni 1998, 2005). In the1970s under rural employment guarantee schemes of various State Governments, earthen bunds were constructed to harvest runoff, especially in the Western and Southern parts of India. Percolation tanks were built to harvest and impound water at scales of micro-watersheds in villages. However, the lack of hydrogeological understanding at these scales implied that aspects like groundwater recharge strategies were decided based on slopes, soils, land-use etc.

The Deccan Volcanic Province in central-west India offers a unique hydrogeological setting for the study of artificial recharge. The type of basalt, its degree of weathering, and the nature and intensity of fracturing control the occurrence and movement of groundwater resources in the Deccan basalt (Adyalkar and Mani 1972;

Deolankar 1980; Saha and Agrawal 2006; Kulkarni et al. 2000). The Kolwan valley in Pune district of Maharashtra state is a typical river basin located in the Deccan basalt setting and offers a unique opportunity to study groundwater in its various dynamics: physical, social, institutional, and economic.). Three check dams were constructed along the mainstream in Chikalgaon watershed as a part of GOMUKH's (local NGO) watershed development programme implemented in 1997–1998 and funded under the Drought Prone Area Programme (DPAP). Out of these, CD1 is an earthen check dam whereas CD2 and CD3 are masonry check dams (Fig. 9.5). CD1 has a top width of 33.5 m with a separate concrete masonry spillway that has a top width of 6 m. CD2 is a masonry structure with a top width of 25.8 m and is located at 200 m upstream of CD1. CD3 is also a masonry structure with a top width of 22 m. It is located some 500 m upstream from CD2. The masonry check dams CD2 and CD3 are 'top-over' structures with their spillways designed as rectangular weirs on their crests.

Intensive monitoring of water levels was done. All monitoring equipment like AWS, automatic water level recorder, observation boreholes, staff gauges, inflow and overflow weirs were located within a maximum distance of 500 m from the recharge sites (Fig. 9.5).

The area of Chikhalgaon watershed is 1.98 km^2, with the three structures located within a length of 1 km from each other. The catchment areas for the three

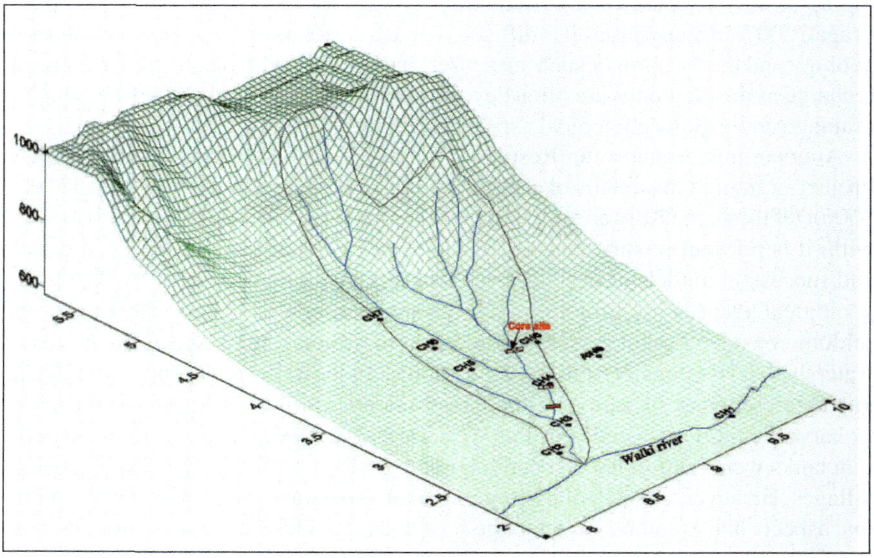

Fig. 9.5 Wire frame for Chikhalgaon watershed, indicating the location of the three recharge structures CD1, CD2, and CD3. Figure also shows the location of the AWS and Chikhalgaon village

check dams vary, with CD3 having the smallest catchment size and CD1 catchment nearly equalling the total watershed area. The structures are relatively small in the volume of water they hold, the total bank full volume in the three check dams equals only 3.24 mm of water, with reference to the Chikhalgaon watershed area, where the rainfall recorded for the year 2004–2005 was 1840 mm. However, at local scales, this water has significance for recharge to the local, unconfined basalt aquifer. A summary of the structural comparison of the three check dams and their water balances is given in Table 9.1.

The check dam CD3 is located on a second-order stream in the natural recharge area of a shallow unconfined basalt aquifer system mapped in Chikhalgaon village (Kulkarni 2005). Badarayani et al. (2005) provide a detailed account of the hydrogeology of the site and the estimation of recharge from the percolation tank, i.e., CD3. We provide a further brief of this summary with the relevance of such estimation in the overall canvass of MAR in the Deccan Basalt aquifer system of West Central India.

The research made an attempt to separate the recharge resulting from the floor of the check dam through direct infiltration. The quantification of the recharge is based on two phases in the annual water cycle for the site during which water is retained by CD3. These were:

1. Monsoon (wet season) period between June and October 2004, i.e., during the period of overflow from CD3. This period can be called the period of flow through the structure after its full storage capacity has been reached
2. Post-monsoon (dry season) period between October and December 2004, i.e., the period during which there was neither any inflow nor any overflow from CD3.This period can be called the period of the pond age.

A combination of groundwater level data from a set of automated piezometers drilled on the basis of systematic aquifer mapping, weather data including high-frequency monitoring of rainfall events and the changes in the stocks, inflows to and outflows from CD3 was used to calculate infiltration from CD3. Table 9.2 summarises the infiltration events from CD3.

We can clearly note that the cumulative infiltration from CD3 is of the order of 33 mm, one order of magnitude lower than the natural infiltration (105 mm). While results were not conclusive, it may not be wrong to state that once the aquifer reaches a certain degree of saturation, infiltration is routed as interim recharge into the aquifer

Table 9.1 Details of check dams in Chikhalgaon watershed

Structure	Type	Top, in m above MSL	Height at the centre of the structure, in m	Capacity of the structure as bank full volume, in m^3	Catchment area of structure in km^2
CD3	Masonry	617.05	2.05	1530	0.79
CD1	Masonry	599.2	1.2	583	1.6
CD1	Earthen	598.5	4.5	2070	1.72

Table 9.2 A summary of the analysis of infiltration events from CD3 (after Badarayani et al. 2005)

Period of event (in 2004)	Rainfall in mm	Estimated infiltration from the structure in mm	Natural infiltration estimated from water level behaviour in a controlled piezometer outside the influence of CD3, in mm
June	392	22	58
July	289	3	12
August	896	2	15
September–October	211	4	21
November–December	52 (only in November)	2	0
Total	1788	33	105

only to be slowly discharged to base flows downstream. Hence, apart from a benefit to wells (improved sustenance), despite their seemingly low volumes, such structures play a role in increasing the travel time for rainwater to reach streams and rivers through the shallow basalt aquifers.

The most significant takeaways from the AGRAR case study (particularly from the detailed study of a single water harvesting/recharge structure) are:

1. For low storage aquifers from western portions of the Deccan Basalt Region, infiltration of natural water remains a significant contribution to groundwater recharge, particularly in rural settings
2. When structures are planned for facilitating groundwater recharge, particularly based on the conceptual framework of MAR, it is noted that the water table mound created in the unconfined aquifer in the immediate vicinity of a water-filled check dam, leads to an augmentation of recharge
3. The response of a shallow basalt aquifer to both, the natural and artificial infiltration of water is quick and occurs over a few weeks, with the maximum impact during the early season, when the groundwater table in wells is deep
4. Given the heterogeneity of the basalt aquifer, the figures of natural and artificial recharge vary from place to place.

9.6 Case 3: Improving the Efficiencies of Existing Recharge Interventions: Desilting of Pokharni Tank, Randullabad, Satara District, Maharashtra

Randullabad is a typical village located in Western Maharashtra on the upland portions of the Deccan Plateau, east of the Western Ghat scarp line. The village is located within a watershed of 854 hectares (ha) and has been one of the few promising examples of participatory groundwater assessment and recharge methods in India (Zwarteveen et al. 2021). The aquifer system in Randullabad consists of

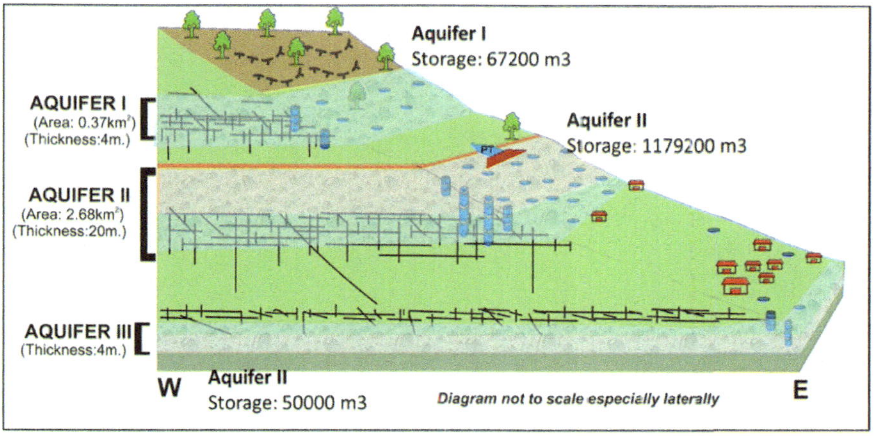

Fig. 9.6 Conceptual diagram of the aquifer system in Randullabad (Aslekar et al. 2013). The location of the Pokharni percolation tank is indicated as PT in the diagram

three shallow unconfined basalt aquifers, illustrated through a conceptual diagram in figure (Fig. 9.6).

Pokharni 'percolation tank' is one of the oldest watershed structures constructed in Randullabad. It is located on a tributary of the main stem representing a natural recharge area (identified through a combination of detailed geological mapping and water level contour maps) for the main shallow unconfined aquifer system in the village (Aslekar et al. 2013). The dam wall is 5 m in height and its width is 140 m. A spillway 3 m wide is constructed at the end of the earthen bund of the tank. The total water spread area of this tank is approximately 4 hectares (ha). The total catchment area of the dam is 1.22 km^2, i.e., 122 ha.

There are 30 wells immediately downstream from the dam that tap the unconfined aquifer. The tank is located at the boundary of Randullabad and Karanjkhop villages and therefore the benefits of this tank are shared by the two villages through the 'shared' unconfined aquifer (Fig. 9.7).

The storage capacity of the tank was around 200,000 m^3 of water at its full capacity (Aslekar et al. 2013). A lot of silt that got accumulated at the base of the check dam over a period of over 38 years had reduced the water-holding capacity and also the vertical percolation of the stored water into the aquifer below. Detailed surveys yielded estimates of an average of 1500m^3/year as the rate of silt deposition in the tank, largely eroded from the upstream slopes in the watershed. Watershed activities including soil conservation and water infiltration structures were first suggested and executed in the catchment of the tank. The deposit of silt was as thick as 2 m in some parts of the storage area of the tank, reducing the water storage capacity by almost 30–40% of its originally designed stage.

In a normal rainfall year (the average annual rainfall is between 700 and 800 mm), the tank has the capacity to recharge the aquifer through about three episodes of rain

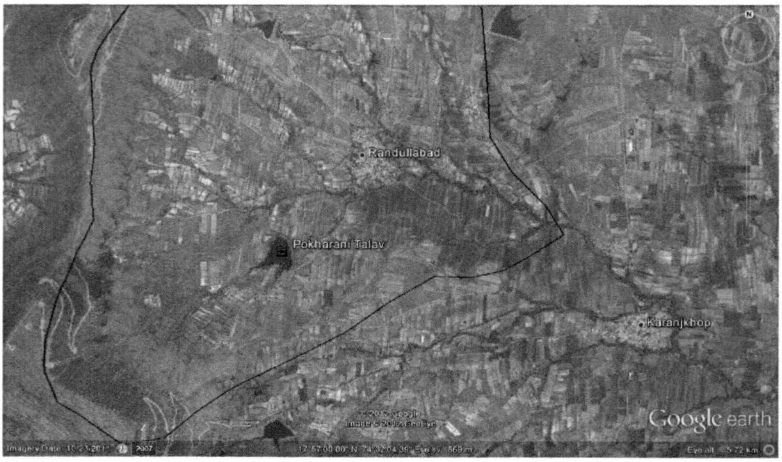

Fig. 9.7 Location of Randullabad village and Pokharni tank on Google Earth. The image also includes an overlay of the watershed boundary in which the tank and the village are located

during each monsoon. This implies that an equivalent of at least 40–50 mm of this water is likely to infiltrate from the storage in the percolation tank during every such episode. With the presence of silt, most of the potential recharge water from the check dam was either lost to evaporation or flowed through the spillway as runoff. In some portions of the tank, the level of silt was at the same elevation as the exit point of the spillway.

Water level data from the 30 dug wells downstream indicate direct recharge benefit from the tank. Even with the present recharge, the wells had enough water throughout the year and farmers were growing 2 or 3 crops every year. However, during drought years, 30% of these wells dried up during the summer months. A gridded contour survey of the area of the tank and its surroundings—over 8.86 ha area using a 10 m × 10 m grid to generate 900 data points—was conducted (Fig. 9.8). The survey was conducted for the tank, its spillway, and the portion just downstream of the tank.

The grid survey gave useful information about the silt accumulated in the check dam. It also gave information about the subsidence of the dam wall. On the basis of the data, various cross-sections were prepared, indicating the amount of silt to be removed for each contour level (Fig. 9.9).

The survey showed that 78,350 m^3 of silt was deposited in the tank during the last 40 years. The storage capacity of the dam before desilting was 117,000 m^3 at a 4 m level from the base. At a conservative figure of 12% of rainfall-recharge for basalt (rainfall-recharge is 20% for Randullabad), the recharge from this dam was calculated to be 14,000 m^3 at full capacity. If the dam was desilted completely, the storage capacity would increase upto 124000 m^3 and the recharge would increase by 40% (due to the removal of the clay deposited) and would go up to 50,000 m^3. It was thought that if the dam wall was repaired, it would allow the dam to store the

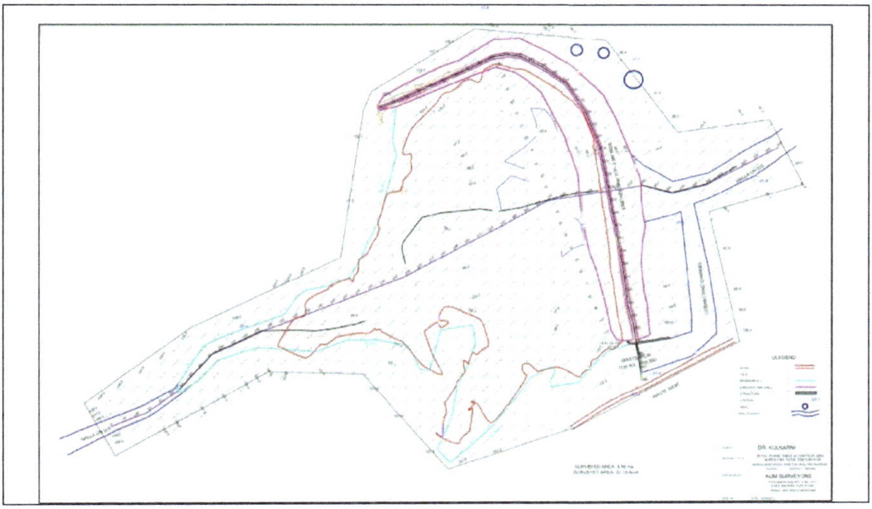

Fig. 9.8 Grid survey for Pokharni tank in Randullabad

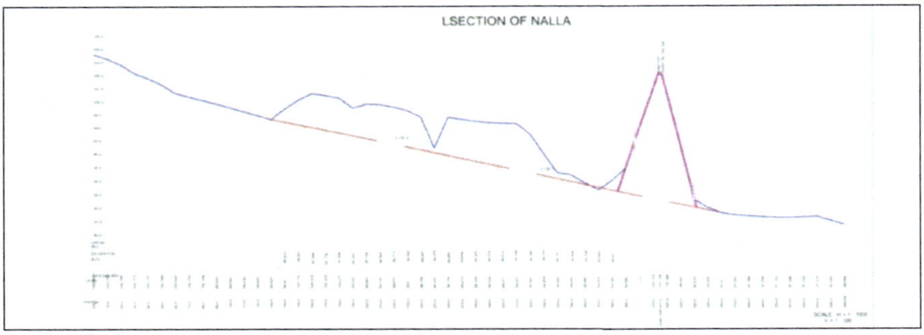

Fig. 9.9 A cross section along the stream profile (Lsection of Nalla) in the percolation tank (purple), showing amount of silt deposited at different parts along the section (blue)

water to the fullest capacity of about 200,000 m^3 leading to potential percolation of 80,000 m^3 which would lead to a significant recharge to the aquifer (Table 9.3).

Aquifer II, the main aquifer in Randullabad has a total storage capacity of 1179000 m^3, estimated based on detailed aquifer mapping and characterisation including a comprehensive groundwater balance (Aslekar et al. 2013). In a normal rainfall year, about 280,000 m^3 is the estimated groundwater recharge for the aquifer. *Pokharni* dam would contribute to almost 29% of this total recharge, also leaving scope for improving base flows to the downstream portions of aquifer II.

At Randullabad, the hydrogeological investigations leading up to the development of a community based water governance system also included a protocol for aquifer-based groundwater management (Aslekar et al. 2013). The effort involved

Table 9.3 Pre and post-desilting estimates for Pokharni dam

Pre-desilting			Post-desilting		
Water column in the dam (m)	Water storage (m^3)	Potential percolation leading to recharge (m^3)	Water column in the dam (m)	Water storage (m^3)	Potential percolation leading to recharge (m^3)
1	11,650	1368	1	18,750	8880
2	32,500	3885.6	2	44,600	21,350.4
3	64,500	7725.6	3	78,450	37,598.4
4	117,000	14,025.6	4	124,700	59,798.4
5	Leakage	Leakage	5	200,750	96,342

water security impacts such as drinking water security throughout the year, improved groundwater recharge, protection of recharge areas for the three aquifers and ban on bore well drilling, all through resolutions of the gram panchayat in the gram sabha.

Hence, the desilting of the tank could be considered a good example of MAR as it was embedded into the larger effort of aquifer-based PGWM. More specifically, improving the recharge efficiencies of existing structures such as old percolation tanks hold a lot of promise in India, especially in regions underlain by hard-rock aquifers. The social process in doing so in Randullabad can also be seen as an effort to forge a convergence between MAR and water governance, a feature that is being increasingly highlighted as an important aspect in MAR literature (Dillon et al. 2019).

9.7 Case 4: 'Springshed' Management: A Customised Strategy to MAR in the Himalayan Region

A large proportion of the Himalayan population depends on spring water for their domestic needs and livelihoods, whether for livestock or agriculture. Almost 64% of the cultivable area in the Himalayas is fed by natural springs that often constitute the only source of irrigation in the Indian Himalayan Region (NITI Aayog 2018). The Himalayan ecosystem is fragile and susceptible to several changes caused due to both natural factors and anthropogenic interventions. The erratic rainfall pattern, seismic activity, and ecological degradation associated with land-use change are posing huge pressures on mountain aquifer systems. It is reported that half of the perennial springs have already dried up or have become seasonal resulting in acute water shortages for drinking and other domestic purposes across hundreds of Himalayan villages. The science of hydrogeology that governs the occurrence and movement of water in the underground aquifers is not as well understood or applied in spring water management as it is in the relatively plain regions where aquifers are tapped by wells. Most mainstream approaches to water conservation in the mountains have focused on augmenting groundwater recharge through conventional watershed management

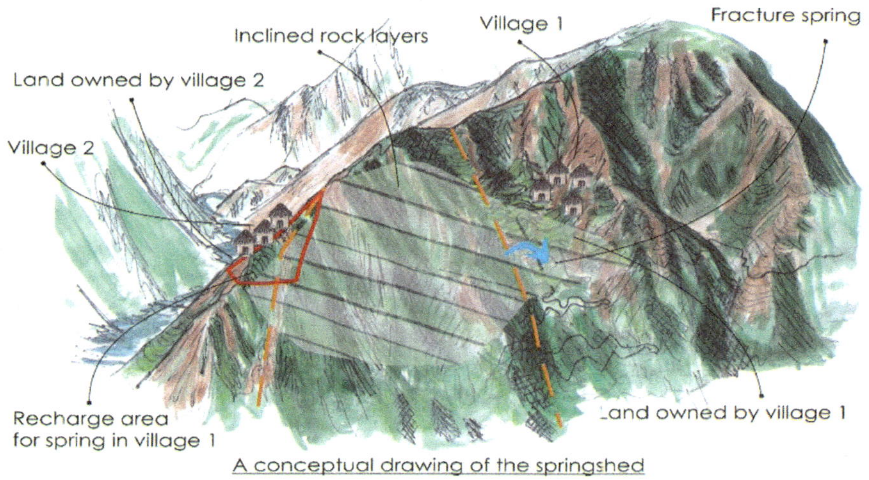

Fig. 9.10 A conceptual depiction of the concept of a 'springshed' that integrates watersheds, the underlying aquifer system and habitations within a landscape of ridges and valleys

without taking 'aquifers' into consideration (Kulkarni et al. 2021). The most important factors that go unnoticed in conventional conservation approaches in this region are spatial heterogeneity and anisotropy in geologic parameters. Challenges posed by the topography of the region further guarantees that the conventional approach of watershed development in the Himalayas is not enough (Fig. 9.10).

Advances in knowledge of Himalayan springs are limited. The concept of a 'springshed' in the Himalaya evolved in response to various field-level initiatives in conserving water in the Himalayan Region to revive springs that form the lifeline of water for mountain communities (Rural Management and Development Department 2014; Shrestha et al. 2018; Khadka et al. 2019).

Spring water conservation and augmentation represent an important aspect in the shift away from standardised, straight-jacketed conservation efforts that consider only slopes, soils, and runoff as the main parameters. The assumption that there is little groundwater in mountainous regions dominated our water praxis and policy. It was only when the crisis of spring depletion emerged in areas that depended upon spring water did we wake up to the fact that there are complex mountain aquifers that discharge groundwater naturally in the form of springs. While watershed management prevailed in many of India's mountainous regions, it lacked any integration with the system of aquifers. Infusion of hydrogeology has led to the development of the concept of 'springshed management' that is able to include an understanding of the relationship between watersheds and the underlying aquifers (Rathod et al. 2021; Siddique et al. 2019). At the same time, springshed management offers a unique dimension for MAR and the forward linkage to managing mountain aquifers. An action-research-based spring revival programme was implemented in Dailekh

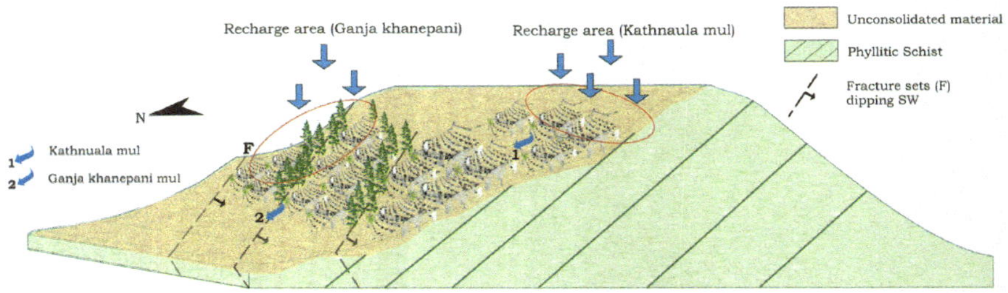

Fig. 9.11 A conceptual 3-D layout for the Ganjakhanepani and Kathnaula Springs (mul = spring)

district, which is a part of Bheri zone in mid-western Nepal in collaboration with ICIMOD (Khadka et al. 2019) (Fig. 9.11).

Kathnaula Spring emerges along a slope, fed by a local aquifer system comprising colluvial deposit, with its recharge zones along the ridgeline area. The Ganjakhanepani Spring emerges further down along the same slope and receives water from the fractured schists with its recharge area falling along the fracture zone in the same watershed. This implies that the two springs receive water from two different aquifer systems with their individual recharge areas within a single watershed. As part of the programme, recharge measures were implemented in the recharge zone of Kathnaula Spring while the Ganjakhanepani Spring was treated as a controlled spring (Shrestha et al. 2018).

The Ganjakhanepani Spring (controlled spring) shows a decrease in discharge during the lean season of 2017 as compared to 2016 in the absence of any treatment measures. Kathnaula Spring (treated spring—a type of MAR intervention), which has very low discharge during the lean season of year 2016 shows improvement in its lean season discharge in the year 2017, especially from a steep drop in discharge to a gradual decline. This illustrates the impact of the MAR effort through springshed management even over a short period (Fig. 9.12).

There are many such examples of improved spring discharges although long-term data (in the absence of resources for measurement) is not easily available. Another example is from the four monitored springs under the Dhara Vikas Programme of RM & DD (now RDD), Government of Sikkim. In the case of three such springs in West Sikkim District, the lean season discharge improvement over 5 years after the implementation of a recharge programme in 2011 was in the range of 28–31% (Tambe et al. 2019).

9.8 Synthesis and Discussion

India's groundwater narrative is particularly interesting because of its ever-increasing reliance on groundwater on one hand and the compounding crises surrounding groundwater resources on the other. The problem is further complicated by the

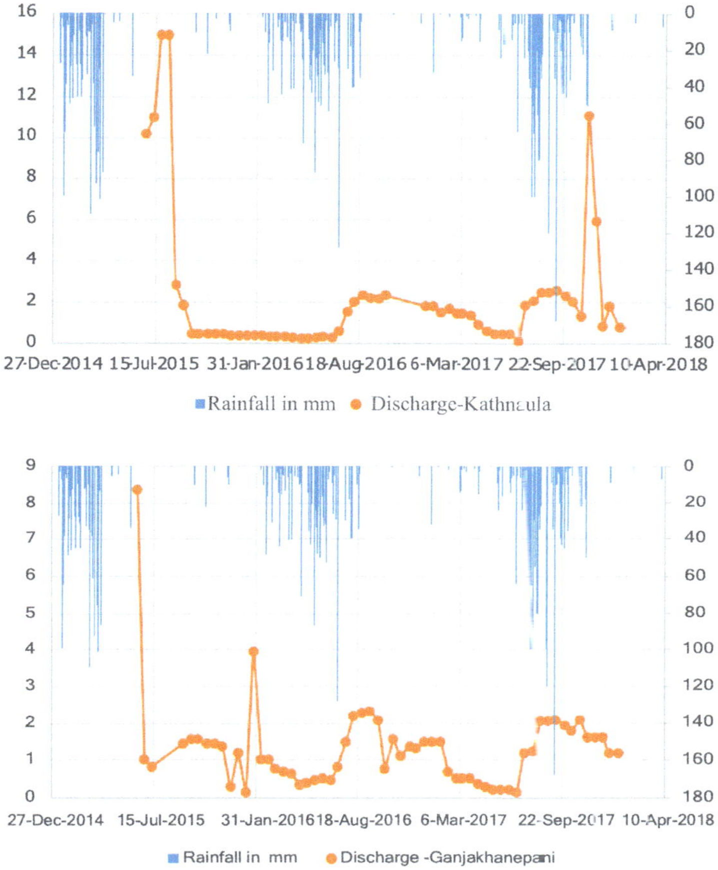

Fig. 9.12 A 40-month hydrograph for **a** Kathnaula and **b** Ganjakhanepani Springs

sheer numbers—of people, of sources, of the dimensions in socio-economic and socio-ecological consequences, and by the lack of an integrated approach to groundwater management and governance. While national water programmes like Atal Bhujal Yojana and the Jal Jeevan Mission have attempted to incorporate the concept of participatory groundwater management, it is necessary to develop a perspective on the protection, management, and governance of aquifer systems in India by involving community participation. Aquifer depletion, groundwater contamination and the conflicts created between aquifers and ecosystems, especially rivers, often go together.

The shrinking size of landholdings, the decentralised nature of governance required to address the largely atomistic nature of groundwater extraction and the diverse socio-hydrological aquifer settings makes it difficult to implement top-down approaches to groundwater management, including the often-obvious instrument of groundwater legislation.

Local participatory groundwater management and governance fosters a paradigm shift on many fronts. While we need to deal with the dichotomous nature of groundwater management and governance—which one of these is a subset of the other?—the paradigm of groundwater governance attempts to move the focus from sourcing of groundwater to that of resource (aquifer) management using the principles of managing groundwater as a common pool resource through a protocol that is based on social hydrogeology, a concept that is gaining importance across the globe. It is within this framework that MAR must be perceived to make the shift from managed aquifer recharge to managing aquifer recharge.

The high degree of groundwater vulnerability in India calls for groundwater management strategies that include supply and demand-side interventions. At the same time, in order to strike a balance between supply and demand, resource understanding becomes imperative. Hence, acknowledging, recognising, and understanding aquifers must all become part of the groundwater management strategy. Conservation practices, particularly artificial groundwater recharge, become an important component of such a strategy. Watershed projects remain the most popular and often the most successful models of decentralised, hydrology-based conservation practices in India. However, the lack of hydrogeological understanding at the scales of micro-watersheds (units of 100–1000 hectares), implies that aspects like groundwater recharge strategies are decided merely on the basis of slopes, soils, and land-use without due consideration to hydrogeological factors.

MAR requires focusing on strategies of augmenting recharge on one hand and integrating such strategies with addressing the larger challenges in aquifer-based groundwater management and governance. Managing aquifers is a function of science, communities, and regulation. MAR strategies, therefore, must further consider certain key elements of attempting the questions of where one locates strategies based on an understanding of the scientific aspects of recharge, community wisdom, and acceptance and regulation (e.g., protection of recharge sites from external interference). Specific questions on how one makes the choice of an appropriate intervention and how one goes about implementing measures of recharge, keeping the safety and sustainability of such measures in mind during the planning of MAR, are significant to this process. The key elements of implementing MAR at different scales may differ and the basic questions to be answered during such interventions might vary from scale to scale (Fig. 9.13).

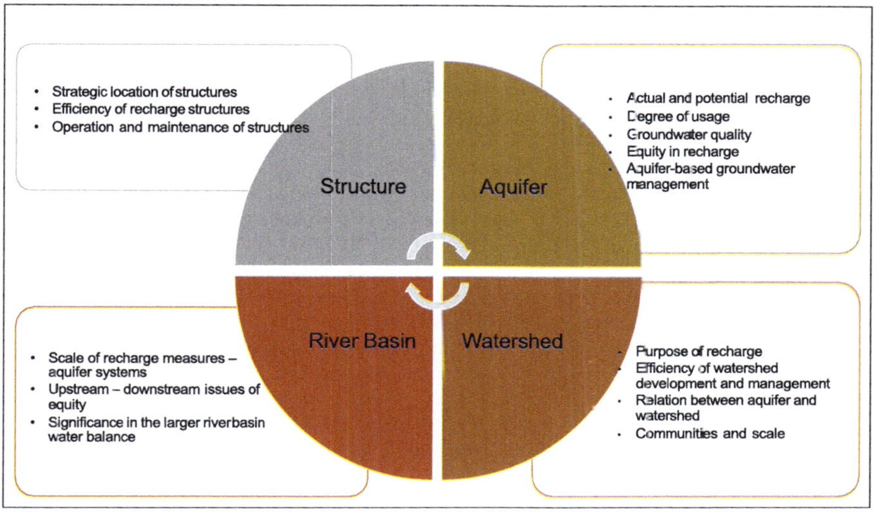

Fig. 9.13 Four salient elements of MAR at different scales (Note: these are not exhaustive and are only indicative based on the limited experiential context of integrating MAR with PGWM)

Acknowledgements ACWADAM was the lead agency in all the four case studies presented. The authors acknowledge the ACWADAM teams who were part of these projects. We would also like to thank:

1. For case study 1, Australian Centre for International Agricultural Research, Western Sydney University, Charles Sturt University, University of Adelaide, University of Queensland, CSIRO, WorldVeg, PRADAN, BCKV, and the smallholder farmers from the project sites for support during ACWADAM's research on groundwater as part of the larger collaboration.

2. For case study 2: British Department for International Development (DFID), the British Geological Survey, UK, and Gomukh Trust, Pune, who supported and were collaborators with ACWADAM for the Kolwan Valley research site.

3. For case study 3: The Randullabad participatory groundwater management (PGWM) initiative was undertaken as a collaborative project involving the Village Watershed Committee (VWC) of Randullabad Gram Panchayat, BAIF Development Research Foundation who planned and implemented the watershed component of the programme under the Indo-German Watershed Development Programme of the National Bank for Agriculture and Rural Development (NABARD);ACWADAM, the knowledge partner, who developed aquifer-related information and communication worked with the financial support of Arghyam Trust.

4. For case study 4: The springshed management effort in Sindhupachok and Dailekh districts of Nepal was anchored through the CGIAR (WLE) project on spring water studies. ACWADAM's component was supported by ICIMOD for which they are gratefully acknowledged. Helvatas undertook the socio-economic dimension of the project and helped in the fieldwork. We thank them as well.

References

Adyalkar P, Mani V (1972) Geohydrological study in parts of Sholapur district and its bearing on the feasibility of tubewells in the Deccan Trap. Curr Sci 42(12):441–445. Accessed from http://www.jstor.org/stable/24074449

Agarwal A, Narain S (1997) Dying wisdom: rise, fall, and potential of India's traditional water harvesting systems. Centre for Science and Environment, New Delhi

Aslekar U, Upamanyu A, Kulkarni H (2013) Participatory groundwater management in Randullabad. ACWADAM Technical report: ACWA/Hydro/2013/H28, Pune

Samaj Pragati Sahyog (2007) National Rural Employment Guarantee Act. Watershed Works Manual. Baba Amte Loksashaktikaran Centre, SPS. Accessed from https://doi.org/samajpragatisahayog.org/wp-content/uploads/2014/12/watershed_manual_english.pdf

Badarayani U, Kulkarni H, Phadnis V (2005) Groundwater recharge from a percolation tank to a Deccan basalt aquifer: a case study from western Maharashtra, India. In: Recharge systems for protecting and enhancing groundwater resources: proceedings of the 5th international symposium on management of aquifer recharge ISMAR5, vol 13. Berlin, Germany, pp 180–185. UNESCO International Hydrological Programme, UNESCO-IHP, Paris. Accessed from https://recharge.iah.org/ismar/ismar5

Batchelor CH, Rama Mohan Rao MS, James AJ (2000) Karnataka Watershed Development Project: Water Resources Audit : KAWAD report 17. Karnataka Watershed Development Society, Bengaluru

Bonsor H, Macdonald A, Ahmed KM, Burgess W, Basharat M, Calow R, Zahid A (2017) Hydrogeological typologies of the Indo-Gangetic basin alluvial aquifer, South Asia. Hydrogeol J 25:1377–1406. https://doi.org/10.1007/s10040-017-1550-z

COMMAN (2005) Community Management of Groundwater Resources in Rural India. British Geological Survey Commissioned Report CR/05/36N

CGWB (2005) Master plan for artificial recharge to groundwater in India. Central Ground Water Board, Ministry of Water Resources, Government of India. https://doi.org/cgwb.gov.in/documents/MASTER%20PLAN%20Final-2002.pdf

Central Ground Water Board (2000) Guide on Artificial Recharge to Groundwater. Ministry of Water Resources, Government of India, Faridabad. Accessed from http://cgwb.gov.in/documents/ArtificialRecharge-Guide.pdf

Central Ground Water Board (2006) Dynamic groundwater resources of India (As on March,2004). Ministry of Water Resources, Government of India, Faridabad

Central Ground Water Board (2007) Manual on Artificial Recharge of Groundwater. Ministry of Water Resources, Government of India. Accessed from http://cgwb.gov.in/documents/Manual%20on%20Artificial%20Recharge%20of%20Ground%20Water.pdf

Central Ground Water Board (2011) Dynamic Ground Water Resources of India (as on March 2009). Ministry of Water Resources, Government of India, Faridabad

Central Ground Water Board (2012) Aquifer Systems of India. Ministry of Water Resources, Government of India, Faridabad. Accessed from http://cgwb.gov.in/AQM/India.pdf

Central Ground Water Board (2014) Dynamic Groundwater Resources of India (As on 31st March2011) Ministry of Water Resources, River Development and Ganga Rejuvenation, Government of India, Faridabad

Central Ground Water Board (2017) Report of the groundwater resource estimation committee (2015). Ministry of Water Resources, River Development & Ganga Rejuvenation, Government of India. Accessed from http://cgwb.gov.in/Documents/GEC2015_Report_Final%2030.10.2017.pdf

Central Ground Water Board (2018) Groundwater Quality in Shallow Aquifers of India. Ministry of Water Resources, River Development and Ganga Rejuvenation, Government of India, Faridabad

Central Ground Water Board (2019) National Compilation on Dynamic Groundwater Resources of India (2017) Ministry of Jal Shakti, Department of Water Resources, River Development and Ganga Rejuvenation ,Government of India, Faridabad

Central Ground Water Board (2002) Master plan for artificial recharge to Groundwater. Ministry of Water Resources, Government of India, Faridabad. Accessed from https://www.indiawaterportal.org/sites/default/files/iwp2/MASTER_PLAN_Final_2002.p df

Dashora Y, Dillon P, Maheshwari B, Soni P, Dashora R, Davande S, Mittal HK (2018) A simple method using farmers' measurements applied to estimate check dam recharge in Rajasthan, India. Sustain Water Resour Manag 4:301–316. https://doi.org/10.1007/s40899-017-0185-5

Deolankar S (1980) The Deccan Basalts of Maharashtra, India–their potential as aquifers. Groundwater 18(5):434–437. https://doi.org/10.1111/j.1745-6584.1980.tb03416.x

Dillon P (2005) Future management of aquifer recharge. Hydrogeol J 13(1):313–316. https://doi.org/10.1007/s10040-004-0413-6

Dillon P, Pavelic P, Page D, Beringen H, Ward J (2009) Managed aquifer recharge: an introduction. National Water Commission, Commonwealth of Australia, Canberra. Accessed from https://recharge.iah.org/files/2016/11/MAR_Intro-Waterlines-2009.pdf

Dillon P, Stuyfz P, Grischek T, Lluria M, Pyne RG, Jain RC, Sapiano M (2019) Sixty years of global progress in managed aquifer recharge. Hydrogeol J 27:1–30. https://doi.org/10.1007/s10040-018-1841-z

Farrington J, Turton C, James AJ (1999) Participatory watershed development: challenges for the twenty-first century. Oxford University Press, New Delhi

Gale I (2005) Strategies for Managed Aquifer Recharge (MAR) in semi-arid areas. United Nations Educational, Scientific and Cultural Organization (UNESCO), Paris. Accessed from https://assets.publishing.service.gov.uk/media/57a08c8340f0b6497400127e/R8169-MAR-Strategies.pdf

Gale I, Macdonald D, Calow R, Neumann I, Moench M, Kulkarni H, Palanisami K (2006) Managed Aquifer Recharge: an assessment of its role and effectiveness in watershed management. British Geological Survey Commissioned Report CR/06/107N

Gale I, Neumann I, Calow M, Moench M (2002) The effectiveness of Artificial Recharge of groundwater: a review. British Geological Survey Commissioned Report, CR/02/108N, Keyworth. Accessed from https://assets.publishing.service.gov.uk/media/57a08d4aed915d622c0018d7/R8169-AGRAR_Review.pdf

James AJ, Robinson E (2001) Water and sustainable rural livelihoods in Andhra Pradesh: Background paper. Natural Resources Institute, University of Greenwich, Greenwich, UK. Accessed from https://www.ircwash.org/sites/default/files/James-2001-Water.pdf

Janakarajan S (2008) Unequal power, unequal contracts and unexplained resistance: the case of peri-urban areas of Chennai. In: Joy KJ, Paranjape S, Gujja B, Goud V, Vispute S (eds) Water conflicts in India: amillion revoltsin the making. Routledge, New Delhi, pp 54–58

Joji VS, Gayen A, Saha D (2021) Harvesting of water by tunnelling: a case study from lateritic terrains of Western Ghats, India. J Earth Syst Sci 130:202. https://doi.org/10.1007/s12040-021-01687-y

Joy KJ, Paranjape S (2004) Watershed development review: issues and prospects. Centre for Inter disciplinary Studies in Environment and Development, Bengaluru. Accessed from https://www.indiawaterportal.org/sites/default/files/iwp/WatershedDevelopment_FullRep ort.pdf

Kerr J, Chung K (2002) Evaluating watershed management projects. Water Policy 3(6):537–554. https://doi.org/10.1016/S1366-7017(02)00016-8

Khadka K, Pokhrel G, Dhakal M, Desai J, Shrestha R (2019) Spring-shed management: an approach to revive drying Springs in the Himalayas. In: Pant B, Singh A, Rawal D (ed) Proceedings of the Seminar on "Leaving No One Behind". Nepal Academy of Science and Technology, Kathmandu, pp 10–22

Kulkarni H, Deolankar S, Lalwani A, Joseph B, Pawar S (2000) Hydrogeological framework of the Deccan basalt groundwater systems, west-central India. Hydrogeol J 8(4):368–378. https://doi.org/10.1007/s100400000079

Kulkarni H, Vijayshankar PS (2014) Groundwater resources in India: an arena for diverse competition. Local Environ: Int J Justice Sustain 19(9):990–1011. https://doi.org/10.1080/13549839.2014.964192

Kulkarni H (1998) Watershed development and management a movement seeking inputs in earth sciences. Geol Soc India 52(2):239–241
Kulkarni H (2005) Groundwater overdraft: a physical perspective. In: Calow R, Macdonald D (eds) Comman 2005. Community management of groundwater resources inrural India–research report. British Geological Survey Commissioned ReportCR/05/36N, pp 1–13
Kulkarni H (2018) Hydrogeology in strategic water conservation. Water OwnShip Account 18(2–113). Accessed from https://geographyandyou.com/people-and-ownership/hydrogeology-in-strategic-water-conservation
Kulkarni H, Vijayshankar PS (2009) Groundwater: towards an aquifer management framework. Econ Polit Wkly 44(6):13–17. https://doi.org/10.2307/40278472
Kulkarni H, Aslekar U, Patil S (2018) Groundwater management in India: status, challenges and a framework for responses. In: Mukherjee A (ed) Groundwater of South Asia. Springer, Singapore, pp 615–642. https://doi.org/10.1007/978-981-10-3889-1_36
Kulkarni H, Desai J, Siddique M (2021) Rejuvenation of Springs in the Himalayan Region. In: Pandey VP, Shrestha S, Wiberg D (eds) Water, climate change, and sustainability. Wiley, New York, pp 97–107. https://doi.org/10.1002/9781119564522.ch6
Macdonald DM, Kulkarni H, Lawrence AR, Deolankar SB, Barker JA, Lalwani AB (1995) Sustainable groundwater development of hard-rock aquifers: the conflict between irrigation and drinking water supplies from the Deccan basalts of India. British Geological Survey (WC/95/052). Accessed from https://nora.nerc.ac.uk/id/eprint/21393/
Margat J, vander Gun J (2013) Ground water around the world: a geographic synopsis. CRC Press, London. https://doi.org/10.1201/b13977
Ministry of Agriculture (2013) State of Indian Agriculture: 2012–2013. Department of Agriculture and Co-operation, Ministry of Agriculture, Government of India, New Delhi
Moench M (1992) Drawing down the buffer-science and politics of ground water management in India. Econ Polit Wkly 27(13):7–14. Accessed from https://www.epw.in/journal/1992/13/review-agriculture-review-issues-specials/drawing-down-buffer-science-and-politics
Moench M, Turnquist S, Kumar D (eds) (1993) Proceedings of the workshop on water management: India's groundwater challenge. Ahmedabad, India: Pacific Institute Collaborative Groundwater Project, VIKSAT
Mukherjee A, Saha D, Harvey CF, Taylor RG, Ahmed KM, Bhanja SN (2015) Groundwater systems of the Indian Sub-Continent; J Hydrol: Reg Stud, 4, 1–14
Narain S (2012) Excreta Matters, Vol. I: How Urban India is Soaking Up Water, Polluting Rivers and Drowning in Ins Own Waste. New Delhi: Centre for Science and Environment
NITI Aayog (2018) Report of Working Group I: Inventory and Revival of Springs in the Himalayas for Water Security. NITI Aayog. Accessed from https://www.niti.gov.in/writereaddata/files/document_publication/doc1.pdf
Parveen S, Sen R, Ghosh M (2012) India's deepening water crisis? water risks for Indian industries: a preliminary study of 27 industrial sectors. Federation of Indian Chambers of Commerce and Industry (FICCI) and Columbia University Water Center (CWC). Accessed from https://water.columbia.edu/sites/default/files/content/publications/Indias-Deepening-Water-Crisis.pdf
Patil S, Bhave N, Kulkarni H (2015) Value and challenges of embedding hydrogeology-research in agriculture, livelihoods and NRM programmes on East India Plateau (EIP). In: Research and education for rural development and food security to build resilient rural environments. Australian and Indian perspectives, New Delhi. https://doi.org/10.6084/m9.figshare.9249473
Patil S, Kulkarni H, Bhave N (2017) Groundwater in the Mahanadi Basin. Forum for Policy Dialogue on Water Conflicts in India, Pune. https://doi.org/10.13140/RG.2.2.11561.95846
Pyne RD (1995) Groundwater recharge and wells: aguide to aquifer storage recovery. CRC Press, Boca Raton. https://doi.org/10.1201/9780203719718
Rathod R, Kumar M, Mukherji A, Sikka A, Satapathy KK, Mishra A, Khan M (2021) Resource book on springshed management in the Indian Himalayan Region: guidelines for policy makers and development practitioners. International Water Management Institute (IWMI); New Delhi,

India: NITI Aayog, Government of India; New Delhi, India: Swiss Agency for Development and Cooperation (SDC). https://doi.org/10.5337/2021.230
Rural Management and Development Department (2014) Dhara Vikas Handbook: A user manual for springshed development to revive Himalayan Springs. In: Tambe S (ed) Government of Sikkim
Rushton K (1990) Recharge in Mehsana alluvial aquifer, India. In: Lerner DN, Issar AS, Simmers (eds) Groundwater recharge. A guide to understanding and estimating natural recharge, vols 8/ 1990, International Contributions to Hydrogeology). Verlag Heinz Heise for the International Association of Hydrogeologists (IAH), Hannover, pp 297–312
Saha D, Agrawal AK (2006) Determination of specific yield using a water balance approach–case study of Torla Odha watershed in the Deccan Trap province, Maharashtra State, India. Hydrogeol J 14:625–635
Saha D, Sikka AK, Goklani R (2022) Artificial recharge endeavours in India: a review. Water Secur. https://doi.org/10.1016/j.wasec.2022.100121
Saha D, Marwaha S, Dwivedi SN (2019) National aquifer mapping and management programme: a step towards water security in India. In: Singh A, Saha D, Tyagi A (eds) Water governance: challenges and prospects. Springer Water. Springer, Singapore, pp 49–66. https://doi.org/10.1007/978-981-13-2700-1_3
Saha D, Marwaha S, Mukherjee A (2017) Groundwater resources and sustainable management issues in India. In: Saha D, Marawaha S, Mukherjee A (eds) Clean and sustainable groundwater in India. Springer, Berlin. ISBN 978-981-10-4551-6, https://doi.org/10.1007/978-981-10-4552-3_1
Shah T (1993) Groundwater markets and irrigation development: political economy and practical policy. Oxford University Press
Shah T (2009) Taming the anarchy: groundwater governance in South Asia. Resources for the Future, Washington D.C., United States of America. https://doi.org/10.4324/9781936331598
Shrestha RB, Desai J, Mukherji A, Dhakal M, Kulkarni H, Mahamuni K, Bajracharya S (2018) Protocol for reviving springs in the Hindu Kush Himalayas: Apractitioner's manual. ICIMOD Manual 2018/4. International Centre for Integrated Mountain Development (ICIMOD), Kathmandu
Siddique MI, Desai J, Kulkarni H, Mahamuni K (2019) Comprehensive report on Springs in the Indian Himalayan Region. ACWADAM Tech. Report ACWA/Hydro/2019/H88; Swiss Agency for Development Cooperation (SDC), ACWADAM. https://doi.org/10.13140/RG.2.2.12104.06408
Tambe S, Rawat GS, Bhutia NT, Sherpa PN, Dhakal S, Pradhan S, Arrawatia ML (2019) Building sustainability in the Eastern Himalaya: linking evidenceto action. Environ Dev Sustain 22:5887–5903. https://doi.org/10.1007/s10668-019-00456-8
Tideman EM (1996) Watershed management: guidelines for Indian conditions. Omega Scientific Publishers, New Delhi
The World Bank (2010) Deep wells and prudence: towards pragmatic action for addressing groundwater overexploitation in India. World Bank. Accessed from https://doi.org/documents.worldbank.org/en/publication/documents-reports/documentdetail/272661468267911138/deep-wells-and-prudence-towards-pragmatic-action-for-addressing-groundwater-overexploitation-in-india
Zwarteveen M, Kuper M, Olmos-Herrera C, Dajani M, Kemerink-Seyoum J, Cleaver F, Beckett L, Lu F, Kulkarni S, Kulkarni H, Aslekar U, Borjeson L, Verzijl A, Dominguez Guzman C, Ore MT, Leonardelli I, Bossenbroek L, Ftouhi H, Chitata T, Hartani T, Saidani A, Johnson M, Peterson A, Bhat S, Bhopal S, Zakaria K, Deshmukh R, Joshi D, Komakech H, Joseph K, Mlimbila E, De Bont C (2021) Transformations to groundwater sustainability: from individuals and pumps to communities and aquifers. In: Fisher E, Boyd E, Brondizio E (eds) Transformations to sustainability: critical social science perspectives, vol 49. Elsevier B.V., pp 88–97. https://doi.org/10.1016/j.cosust.2021.03.004

Chapter 10
Managed Aquifer Recharge Projects in the Western Karoo, South Africa: Progress and Challenges

D. Hohne, F. Fourie, S. Esterhuyse, H. Gericke, and M. Butler

Abstract Managed aquifer recharge (MAR) in Southern Africa is nothing new and various projects have been completed with great success since the 1950s. In the Western Karoo, South Africa, climate change is predicted to cause prolonged droughts. Recent droughts have had devastating effects on the water supply to dispersed communities, affecting their lives and livelihoods. By using relatively simple and cost-effective MAR schemes (annual recharge of between 0.18 and 1.1 Mm^3 per scheme), depleted aquifers have been replenished locally, restoring and improving water supplies. In this chapter, four sites (Saaipoort, Smouskolk, Williston, and Sutherland) in the Western Karoo are described in terms of the MAR schemes implemented and their progress and ongoing challenges. The first scheme was started in 2000, but most of them are built on or expand existing water supply infrastructure, basically augmenting capture and harvesting of occasional rainfall for recharge in the vicinity of production boreholes. The rationale for expanding the systems was in most cases that the existing systems could not keep up with population growth

D. Hohne · F. Fourie · H. Gericke
Department of Water and Sanitation, Pretoria, South Africa
e-mail: HohneD@dws.gov.za

F. Fourie
e-mail: FourieF@dws.gov.za

H. Gericke
e-mail: GerickeH@dws.gov.za

S. Esterhuyse (✉)
University of the Free State, Bloemfontein, South Africa
e-mail: EsterhuyseS@ufs.ac.za

M. Butler
Environmental Isotope Laboratory, iThemba LABS, Cape Town, South Africa
e-mail: mj.butler@ilabs.nrf.ac.za

S. Esterhuyse
Centre for Environmental Management, Natural and Agricultural Sciences, PO Box 339, Bloemfontein 9300, Republic of South Africa

D. Saha et al. (eds.), *Managed Groundwater Recharge and Rainwater Harvesting*, Water Resources Development and Management,
https://doi.org/10.1007/978-981-99-8757-3_9

and the higher variability in rainfall. Despite the progress, the MAR schemes have had varying success due to factors like limited and very erratic rainfall and geological factors such as impermeable clay layers that restrict recharge. Based on the findings, and per the Water and Sanitation Masterplan for South Africa, additional schemes are being investigated for more towns in the Western Karoo. This chapter showcases the progress and experiences of current MAR schemes in South Africa in terms of performance and impacts, with a focus on the Western Karoo cases, and discusses, apart from the biophysical and performance aspects, the socioeconomic and institutional aspects. The aspiration is that new MAR projects can be planned and implemented with high confidence in their success, to help with assured future water supply in the Western Karoo, South Africa, and beyond.

Keywords Managed aquifer recharge · Climate change · Groundwater · Drought · Western karoo · South Africa · Performance and impact evaluation · Socioeconomic and institutional aspects

10.1 Introduction

In South Africa, settlements and towns historically formed where water was found. These settlements were almost always near rivers and springs,[1] while farming land usually got water from groundwater resources through wells and boreholes. Groundwater played a major role in reducing the backlog of domestic water supply as towns grew, and it currently supplies >60% of communities in towns and villages across the country (DWS 2016a). Twenty-two per cent of towns[2] and practically all villages use groundwater as a sole water supply source, while 34% of towns and villages use a combination of surface water and groundwater, and 44% use surface water (DWS 2016a). Groundwater is used extensively because surface water resources are limited in South Africa, especially in the interior of the country. When considering sectoral groundwater use, 78% is used for irrigation in the agriculture sector, 7% is used for domestic water supply, 6% for livestock watering, 5% for mining, and 4% for industrial use (Nel 2017). The semi-arid Western Karoo,[3] which includes the Hantam, Kareeberg, Karoo Hoogland, and Ubuntu Local Municipalities (Fig. 10.1) has non-perennial rivers, and around 95% of the communities depend on groundwater as their main water source (DWS 2016b).

[1] In South Africa, according to the National Water Act, springs are considered surface water (RSA, 1998).

[2] Towns are defined as an area with a population ranging from 3,000–5,000 people, where municipal services provide infrastructure and services such as water supply and sanitation (Lehohla, 2016). Villages are defined as less populous areas where such structured municipal services are not provided.

[3] 'Western Karoo' is not a geographically uniquely defined term. It is mostly located in the Northern Cape Province, and Fig. 10.1 gives the definition as used in this chapter.

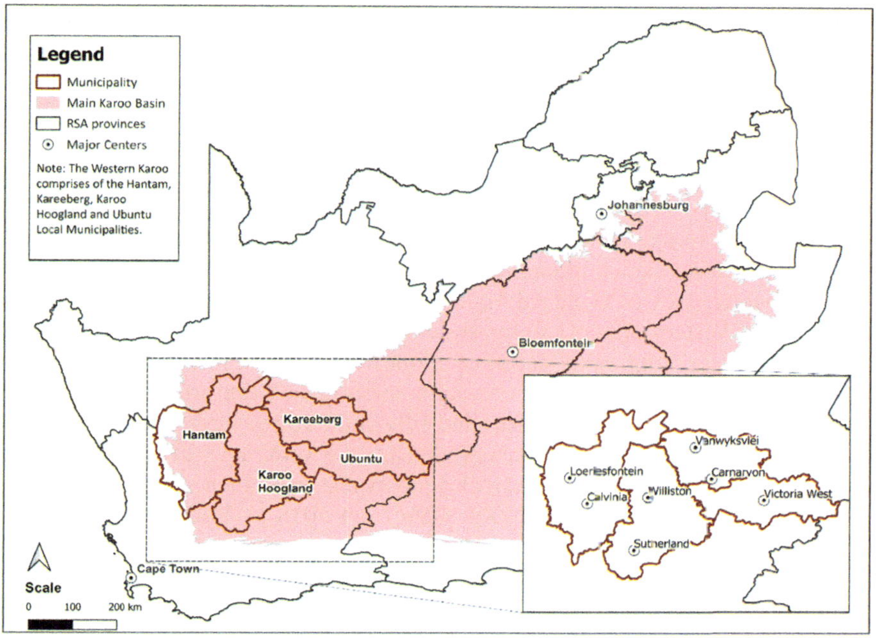

Fig. 10.1 An overview map of the Western Karoo, here defined as the local municipalities of Hantam, Kareeberg, Karoo Hoogland, and Ubuntu (an area of 106,493 km^2)

Boreholes drilled in the late 1800s to early 1900s could only reach a depth of around 30 m (100 ft) due to the available technology. In the 1960s, boreholes reached around 60–80 m depth and were later drilled to a depth of 120 m. With technological improvements and the demand for more water, boreholes were drilled to a depth of 150 m up until 2016 (Hohne 2016), and from 2018 they were drilled to a depth of up to over 200 m in places like Calvinia (Fig. 10.1) (Murray 2021). An ongoing drought from 2013 to 2022, however, spurred the drilling of even deeper boreholes to provide water for communities (Murray 2021). In Victoria West (Fig. 10.1), boreholes that were drilled to around 300 m in 2001 (Chevallier et al. 2001), are now being investigated for more intense water provision. In addition to providing water for domestic and agricultural purposes, the running of renewable energy plants (solar and wind plants) in central South Africa also requires groundwater resources. These energy plants put the groundwater resources under pressure, and original projections of risks to water supply are failing because these water uses were not accounted for in the 10-year plans (Fourie 2021a). Importantly, experience shows that groundwater resources are limited, and without seriously considering augmenting groundwater supplies through enhanced recharge, these resources may fail in the future, as deepening wells is not a long-term solution.

The Groundwater Resource Assessment (GRA II) project estimated that South Africa has a stock of 235,500 Mm^3 of groundwater, and, not including droughts and failing systems, 19,000 Mm^3/yr makes up the utilisable (renewable) groundwater potential (Woodford et al. 2005). Groundwater is therefore an invaluable resource, and with climate change knocking on the door, it will only become more valuable (Murray 2016; Fourie 2021a). Managed aquifer recharge (MAR) can further enhance the potential and sustainable use of South Africa's groundwater will become ever more critical. Apart from MAR schemes in other areas in Southern Africa, such as the Western Cape where they serve larger communities (Braune and Israel 2021), MAR schemes in the Western Karoo are especially important for socioeconomic development in smaller communities.

This chapter assesses Managed Aquifer Recharge (MAR) schemes in the Western Karoo. The Western Karoo aquifer systems occur in the fractured rock systems (igneous dolerite, kimberlite, and breccia intrusions in sedimentary rocks) of the Karoo Supergroup. The Karoo Supergroup consists of the Beaufort, Ecca, and Dwyka Groups with intrusive dolerite dykes and sills in the Karoo large Igneous Province (Neumann et al. 2011). The dolerite intruded into the weak zones of these formations creating locally delimited fractured rock aquifers that are influenced by these dolerite dykes and sills (Johnson et al. 2006). Data from 2018 to 2022 were assessed to determine if these MAR schemes recharge effectively during periods of rainfall after droughts. The chapter also considers why these schemes are successful or fail.

10.2 Climate Change Impacts on Water Availability in South Africa and the Western Karoo

Climate influences water availability in South Africa. The location of South Africa's perennial rivers and water bodies is determined by the prevailing climate, with rainfall being the primary hydrological factor (Avenant et al. 2016). The climate also largely influences the occurrence and recharge of groundwater (Mvandaba et al. 2019). The Western Karoo climatic conditions range from semi-arid to arid and are characterised by a strong east–west gradient, with the west being drier than the east. The Karoo Basin's mean annual precipitation is relatively low, ranging from 500 mm/yr in the east to less than 100 mm/yr in the west (Avenant et al. 2016). Evaporation is high with potential evaporation values ranging from 2000 to >2600 mm/yr (Nkosi et al. 2021). The temperature fluctuates significantly, both daily and seasonally (Avenant et al. 2016). The day temperature ranges from 26 °C to more than 30 °C in summer (October to April) and from 10 °C to 18 °C in winter (May to September) (Schulze 2005).

Climate change predictions indicate that the western part of South Africa will become drier, and it is expected that the dry periods will last longer (Shepard 2019). Drought conditions will become the norm in the western parts of the Karoo, interspersed by more intense rainfall events (Dennis and Dennis 2012). Rainfall data from the area confirm fears of extended dry periods linked to climate change—with

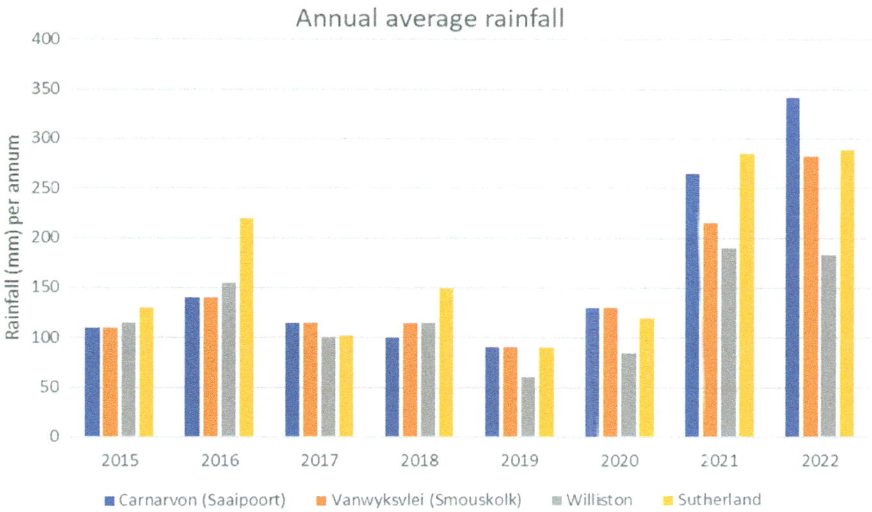

Fig. 10.2 Observed average annual rainfall from January 2015 to December 2022 for the four Western Karoo towns where running MAR schemes were investigated (data from the South African Weather Service 2022)

a reduction in summer wet-season rainfall over the interior of South Africa (Roffe et al. 2020).

Figure 10.2 shows the average rainfall for the four towns in the Western Karoo where the investigated MAR schemes are operating. In Williston, from 1964 to 2012, the maximum number of consecutive years with below-average rainfall was three. However, since 2013, all annual rainfall figures have been below the long-term average value for the area. This below-average rainfall likely means less groundwater recharge and a continuation of the drought conditions (Fourie 2021a).

The baseline groundwater recharge potential in South Africa is mainly influenced by the rainfall distribution of climate zones across the country, influencing the amount of rainfall that is available to recharge aquifers, and to a lesser extent by the distribution of aquifer systems across the country (Dennis and Dennis 2012). Hence, low groundwater recharge potential zones occur in the west of the country, correlating with high aridity regions, while high recharge potential areas occur in the east of the country and coincide with wet and humid climates (Mvandaba et al. 2019). With increasing predicted aridity in the western parts of the country linked to climate change, natural baseline groundwater recharge is also projected to decrease in the coming years (Mvandaba et al. 2019). Managed artificial recharge (MAR) is, therefore, important to aid in water supply security.

Some aquifers in the western parts of South Africa have still not recovered from the 2014 drought, despite subsequent above-average rainfall (Roffe et al. 2020). The decrease and variability in rainfall, which is linked to climate change, leads to a decrease in, or more intermittent, surface water supply (Dennis and Dennis 2012;

Vogel and van Zyl 2016; Cavalho Resende et al. 2019; Edokpayi et al. 2020). The level of groundwater recharge may be more sensitive to the intensity of rainfall events (Cuthbert et al. 2019), where groundwater may benefit from the projected higher intensity of rainfall events providing disproportionately more recharge (Cuthbert et al. 2019). The Saaipoort (Carnarvon), Smouskolk (Vanwyskvlei), and Williston MAR sites have ‘extremely high’ inter-annual variability in water availability (Ebrahim et al. 2020). This high variability in climate and water availability makes groundwater resource management particularly important in this region (Hohne et al. 2021).

10.3 Brief History of Managed Aquifer Recharge in South Africa

Before the commissioning of formal MAR schemes by the South African government in 1971, many farmers in the Western Karoo were running informal MAR schemes, which included building earth dams and rock walls in river channels to slow down intermittent water flow to allow for better infiltration into the riverbed and underlying aquifers (Fourie 2021b). However, with the initiation of systematic water research at the time when the Water Research Commission (WRC) was established with the promulgation of the Water Research Act in 1971 (RSA 1971), more modern and science-based MAR approaches were developed in Southern Africa.

South Africa has developed and implemented various formal MAR schemes by the Department of Water and Sanitation (DWS) to supply communities with water (DWA 2010a). The first was in Loeriesfontein in the Western Karoo (Fig. 10.1), which initiated a scheme around 1958 where dams with spillways were developed in a river system and boreholes were drilled downstream of this network of dam systems. This system has been providing water to the town from 1958 to 2016, at which time the boreholes, dams, and rivers ran dry because of a decline in rainfall since 2011. The system was not used from 2016 to 2020. Based on records from the National Groundwater Archive, relatively significant rainfall during this period, however, occurred in the winter months. Boreholes recovered so much that all of them are now in use again and providing water to the town in conjunction with a new well field. This example illustrates the effect that climate change uncertainties have on MAR systems as well as the potential time lag between rainfall and its impact on groundwater systems. MAR is not a silver bullet solution, and the large variability and uncertainty of climate must be considered during the planning of MAR systems.

In 1963, a MAR scheme called ‘Waterskema Williston’ was commissioned at Williston in the semi-arid Karoo (Fig. 10.1) (Seward 1983). This scheme was an infiltration system, where water from the Sak River was pumped to ‘saaidamme’[4]

[4] Saaidamme: or “planting dams” are used by the farmers to infiltrate water in the alluvial sands. Flood waters are diverted into a series of large, flat basins between 1 and 100 ha in size, which are ringed by low earth walls forming shallow dams or infiltration basins. The water is allowed to stand in the basins up to 1 m deep for between 1 and 3 days to saturate the alluvial soils and is then

and then onto an earth dam immediately northeast of Williston, from where the water was allowed to recharge the aquifer (Seward 1983). Three different MAR schemes are still in operation today in Williston. Two of them are infiltration schemes that make use of infiltration dams next to the production boreholes. One of the schemes has been there since before 1984 (Louw 2022), while another was developed in 2019 (Esterhuyse 2019a). The third MAR scheme is a water transfer scheme from the early 2000s where water is pumped from one aquifer compartment to another (Murray and Harris 2010).

Research commissioned by the WRC in the late 1990s and early 2000s (Murray and Tredoux 1998, 2002) laid a foundation for the practical and knowledge-based implementation of MAR in Southern Africa (Murray 2002, 2016). In 2007, the South African Department of Water Affairs (the predecessor of the DWS), supported by the WRC, developed and rolled out a national Artificial Recharge Strategy (DWAF 2007) to increase awareness of MAR and develop MAR demonstration sites. A comprehensive list of potential site-specific artificial recharge areas was produced for all the Water Management Areas (WMA's)[5] as a follow-up to the strategy (DWAF 2009). Since the establishment of the WRC, several MAR schemes have been piloted and commissioned in South Africa. Table 10.1 provides a historical summary of MAR schemes in South Africa (Tredoux et al. 2002; Murray 2007; DWA 2010a, 2010c). Figure 10.3 explains the different types of MAR schemes.

10.4 How MAR Schemes Fit into South Africa's Policies on Water Supply and Security

Groundwater in South Africa was considered 'private water' until the promulgation of the National Water Act (NWA) in 1998 (RSA 1998). Today, groundwater is considered a national asset and an integral part of South Africa's water resources. It is therefore included in national water supply planning (DWS, 2018), with a specific groundwater strategy developed in 2010 (DWA 2010e) and revised in 2016 (DWS 2016a). Specific national MAR strategy and guideline documents have also been developed (DWAF 2007, 2009; DWA 2010a).

MAR can be licensed under Section 21(e) of the NWA as a controlled activity, and it refers to MAR as '*the intentional recharging of an aquifer with any waste or water containing waste*' (Fourie et al. 2016). Any water that is artificially recharged is considered wastewater or water-containing waste. Most of the new projects are seen

released. The stored soil water is used by deep rooted crops like lucerne that is planted directly in the basin (Murray et al., 2009).

[5] Water Management Areas (WMAs): The National Water Act defines a WMA as: "an area established as a management unit in the national water resource strategy within which a catchment management agency will conduct the protection, use, development, conservation, management and control of water resources" (RSA, 1998).

Table 10.1 A historical timeline of MAR schemes commissioned in South Africa

Date	Scheme name	Aquifer type	Water Source	Scheme type	Recharge capacity (Mm^3/yr)	Status	Additional information	References
1958	Loeriesfontein[a]	Alluvium	Ephemeral stream	Infiltration basin	No data	Operational	Enhance water supply to the town	DWA (2010a)
1960–1979	Cape Flats	Sand	Treated wastewater and stormwater	Infiltration basin	No data	Pilot tested, not operational	This formally commissioned pilot-scale MAR scheme augmented freshwater supplies of the Cape Peninsula and operated from 1973 to 1979	Braune and Israel (2021), DWA (2010b)
1963	Williston[a]	Sand	Surface water from the Sak River	Infiltration basin	No data	Not Operational	The Sak River is not perennial, therefore the option of groundwater use was more sustainable	Seward (1983, 1986)

(continued)

Table 10.1 (continued)

Date	Scheme name	Aquifer type	Water Source	Scheme type	Recharge capacity (Mm^3/yr)	Status	Additional information	References
1970 and 1982	Atlantis	Sand	Urban stormwater and treated wastewater	Infiltration basin	2.7–4.2	Operational	The MAR scheme, which uses indirect recycling of water, was inspired by the Cape Flats pilot site. The scheme started shortly after the development of Atlantis town commenced in the mid-1970s. The scheme is currently successfully managed by the local municipality	Braune and Israel (2021), Ebrahim et al. (2020), DWA (2010b), Tredoux et al. (2002),
1984	Williston[a]	Fractured rock	Ephemeral stream	Infiltration basin	No data	Operational	Enhance water supply to the town	Esterhuyse (2019a, b)
1998	Williston[a]	Fractured rock	Groundwater	Injection cum water transfer scheme	0.18	Temporarily operational	This scheme is currently recovering from drought	Esterhuyse (2019a, b)

(continued)

Table 10.1 (continued)

Date	Scheme name	Aquifer type	Water Source	Scheme type	Recharge capacity (Mm^3/yr)	Status	Additional information	References
2000	Calvinia[a]	Breccia pipes	Treated surface water	Injection cum water transfer scheme	0.1	Not Operational	The scheme has not been in operation because of the mineralised nature of the host rock. The recharged and abstracted water is unfit for human consumption and would need to be blended with the dam water before consumption, which the municipality is not able to do	DWA (2010d)
2001	Kharkams	Fractured rock	Surface runoff	Injection scheme	0.005	Not Operational	Was implemented and failed a year later due to the municipality's lack of management	Braune and Israel (2021), Ebrahim et al. (2020), Tredoux et al. (2002)

(continued)

Table 10.1 (continued)

Date	Scheme name	Aquifer type	Water Source	Scheme type	Recharge capacity (Mm^3/yr)	Status	Additional information	References
2007	Plettenberg Bay	Fractured rock	Raw and treated water from the wastewater treatment works blended with water from the Keurbooms River	Injection scheme	0.75	A borehole feasibility assessment has been done	Has not been implemented yet	DWA (2010c), Murray (2007)
2010	Sedgefield	Sand aquifer	Wastewater	Dune infiltration	0.5 on the pilot scale	Not operational	Has not been implemented yet	Braune and Israel (2021), DWA (2010c), Ebrahim et al. (2020)
2010	Hermanus	Sand	Stormwater	Subsurface infiltration dams	0.004	Not operational	Has not been implemented yet	Braune and Israel (2021)
2017	Saaipoort[a]	Primary aquifer	Surface water from Carnarvon Leegte River		No data	Operational	Enhance water supply to surrounding towns	Hohne et al. (2021)
2017	Smouskolk[a]	Fractured rock	Rainwater	Infiltration scheme	1.1	Operational	Enhance water supply to the town	Ebrahim et al. (2020), Hohne et al. (2021)

(continued)

Table 10.1 (continued)

Date	Scheme name	Aquifer type	Water Source	Scheme type	Recharge capacity (Mm^3/yr)	Status	Additional information	References
2019	Williston[a]	Sand	Rural stormwater	*Infiltration basin*	No data	Operational	Enhance water supply to the town	Esterhuyse (2019a, b))
2021	Sutherland*	Fractured rock	Rural stormwater River Dam	Infiltration boreholes	No data	Operational	Enhance water supply to the town	Mulder (2021)

[a]Schemes located in the Western Karoo

Types and Examples of Managed Aquifer Recharge schemes		
Infiltration basin		**Example**
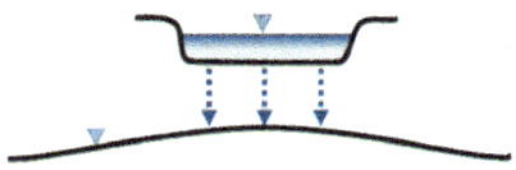	Basins are constructed in unconfined sand or gravel aquifers. Surface (or other) water is diverted to basins and infiltrates through an unsaturated zone to the underlying aquifer. • For sandy unconfined aquifers – not suited for clayey soils. • Treatment: infiltration is faster with clean water than with turbid water; treatment lengthens the "runs" before having to scrape the basins and remove fine material. The unsaturated zone provides natural treatment. • Cost: moderate – can recharge up-gradient of existing boreholes.	Atlantis (Western Cape), Loeries-fontein and Williston (Western Karoo)
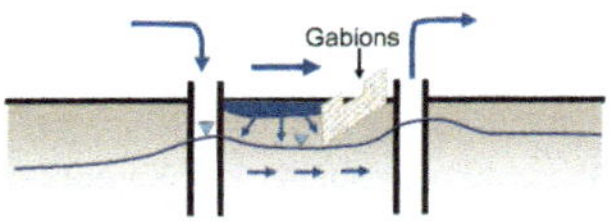	Gabions can be used to allow better infiltration of surface water, which is diverted to basins and allowed to infiltrate into the unconfined aquifer. • For sandy unconfined aquifers. • Treatment: The unsaturated zone provides natural treatment. • Cost: moderate – can recharge up-gradient of existing boreholes, gabions aid recharge.	Sutherland (Western Karoo)
Dune infiltration		
	The infiltration of water through sand dunes and abstraction from boreholes/wells/ponds downstream • For unconfined sedimentary aquifers. • Treatment: minimal pre-treatment; source water should be reasonably clear to prevent clogging. • Cost: low – only shallow abstraction required.	Sedgefield (Western Cape)
Sand dam		
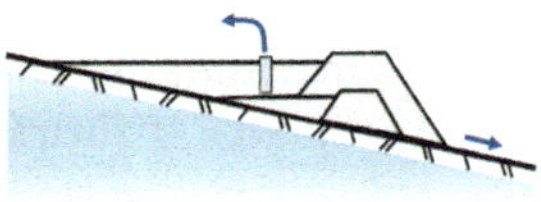	Built in ephemeral streams in arid areas on low permeability lithology. They trap coarse sediment when flow occurs. The detention wall is established to create an artificial "aquifer" behind it. Often built on fracture features in the landscape to speed up the recharge of the natural aquifer underneath. Abstraction can be from the sand aquifer or the aquifer below. It takes years to create the artificial aquifer, but the approach allows for a phased implementation. • Cost: low to moderate	Kituï District (Kenya)
Borehole injection (Aquifer Storage and Recovery - ASR)		

Fig. 10.3 Different types of MAR schemes and examples from Africa (adapted from DWA 2010a; Fourie et al. 2016)

Types and Examples of Managed Aquifer Recharge schemes		
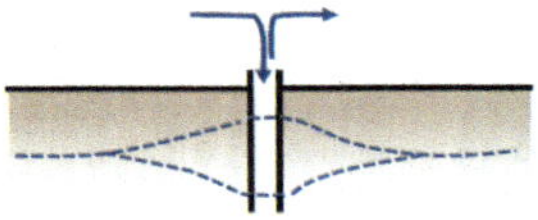	The injection of water into a borehole for storage and recovery from the same borehole • Suitable in both confined and unconfined aquifers. Optimal in areas where groundwater has a shallow hydraulic gradient. • Cost: usually high.	Kharkams (Northern Cape)
Borehole injection (Aquifer Storage, Transfer and Recovery - ASTR)		
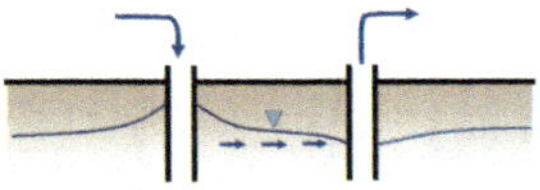	The injection/infiltration of water into a borehole for storage and recovery (mostly from different boreholes) • Suitable in both confined and unconfined aquifers. • Treatment: usually high treatment required – need to remove sediment/debris to prevent borehole clogging. • Cost: moderate to high.	Williston, Smouskolk, Sutherland (Western Karoo)
Infiltration boreholes		
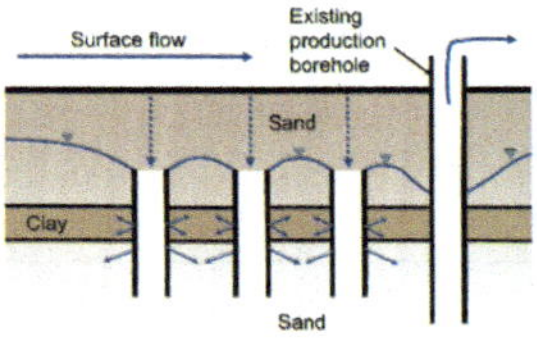	• Clay layers are broken, and subsurface boreholes filled with gravel and pebbles to enhance the recharge of the underlying aquifer when the intermittent river is flowing. • Cost: low to moderate. There is a discontinuous localized clay layer forming an unconfined or semi-confined aquifer. The clay layer needs to be broken to ensure better recharge of the underlying aquifer.	Saaipoort (Western Karoo)
ASR and ASTR combined		
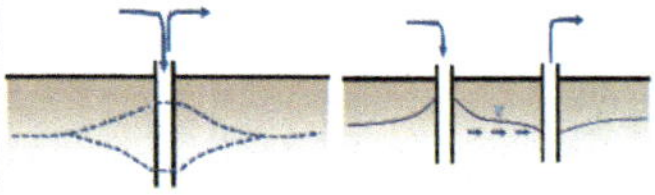	A combination of ASR and ASTR. Treatment: Full treatment to potable standards before injection. • Cost: usually high.	Windhoek (Namibia) Cape Flats (Western Cape)

Fig. 10.3 (continued)

as pilot projects and do not need to be authorised when working in stormwater channels that are manmade (Ackerman 2022), but where changes in natural streams are proposed, authorisation would be needed under Section 21(c) and (i). If abstraction is taking place, Section 21(a) of the NWA would be triggered.

Based on the MAR strategies, site-specific MAR schemes have been implemented. MAR schemes in the western Karoo, where placement is supported by a good understanding of the geology, have been effective in providing water to communities. Several new schemes are being planned and developed to help provide water to Karoo towns.

10.5 Case Studies of Mar Schemes in the Western Karoo

Each case study addresses the biophysical aspects of the MAR scheme, including the geology, hydrogeology, climate, isotopic characteristics, and site conditions, as well as the successes and challenges of each MAR scheme in terms of whether it can provide water during droughts and whether the water quality is within the acceptable range for domestic use. The location of the MAR case study sites can be seen in Fig. 10.1. The Saaipoort MAR site is located 10 km northwest of Carnarvon and the Smouskolk MAR site is located 15 km south of Vanwyksvlei. Both sites are located in the Kareeberg Local Municipality within the Pixley Ka Seme District Municipality. The Williston MAR scheme is located 5 km east of Williston and the newly developed scheme in Sutherland is located 3 km south of Sutherland. Both towns are in the Karoo Hoogland Municipality within the Namaqua District Municipality in the Northern Cape Province.

10.5.1 Saaipoort: Optimising MAR in an Area with a Discontinuous Clay Layer

Site Information

Carnarvon is in a summer rainfall area (Roffe et al. 2020). The mean annual precipitation for Carnarvon is 204 mm/yr, which occurs during the late summer months. Most surface runoff and recharge of groundwater occur between November and May (Esterhuyse 2019a). Carnarvon is in quaternary catchment D54B, which contains various non-perennial rivers. The most important of these is the Carnarvon Leegte River, in which the Saaipoort MAR scheme is located (Fig. 10.4). The Saaipoort scheme is a strategic water source and is a regional bulk scheme that provides two towns (Carnarvon and Vanwyksvlei) with water. Because of its importance, the possibility of enhancing recharge to ensure a sustainable water source was investigated.

The geology of the Saaipoort MAR scheme is the Ecca Group of the Karoo Supergroup, which consists of sandstone and siltstone with less shale (Esterhuyse 2019a). Dolerite intrusions are predominantly targeted for the groundwater they hold (Chevallier et al. 2001). Exploration boreholes were drilled in 1992 up to a depth of 30 m in gravel and sand (Hohne et al. 2021). The Saaipoort MAR scheme has a primary aquifer with a discontinuous clay layer at the top. Recharge is facilitated through clay-penetrating sand- and gravel-filled boreholes in the riverbed, while recovery occurs via existing boreholes close to the river (Fig. 10.4).

Water Levels and Stable Isotopes

Figure 10.5 shows rainfall and water levels from production boreholes (G39891 and G39895) and monitoring boreholes (G39885, G39887, G39888, and G39896). The two production boreholes have been providing the town of Carnarvon with water 24/

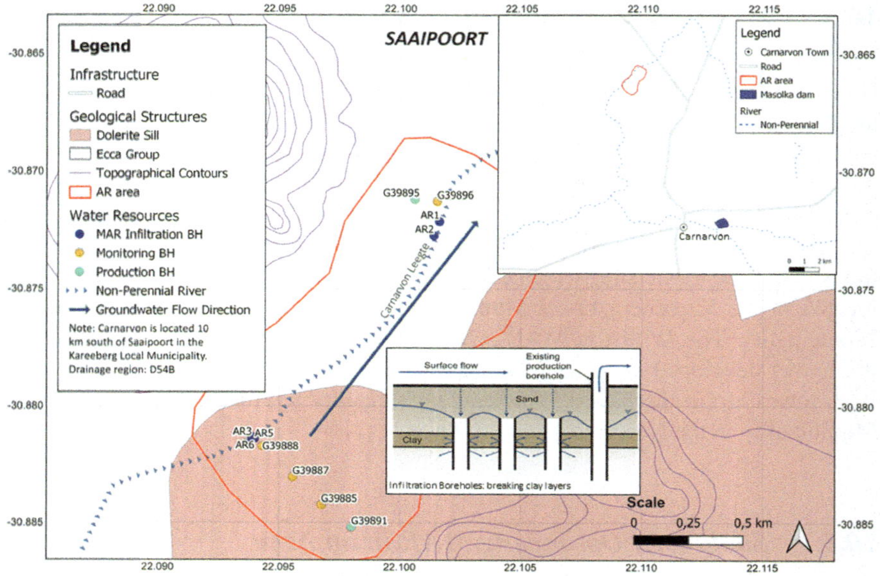

Fig. 10.4 Detailed locality map showing the MAR infiltration boreholes and production boreholes of the Saaipoort MAR scheme

7 for two years (2017–2019). The scheme is currently being monitored monthly and water levels for April 2022 indicated that recharge is taking place in all the boreholes after good rainfall (Fig. 10.5).

Flash floods recharged the aquifer in 2019, and from December 2021 to April 2022, when above-normal rainfall occurred. A few rainfall events of more than 70 mm took place in February 2022, which were recorded shortly after in observation borehole G39896 equipped with a continuous datalogger. The associated recharge wetted the clay layer and increased the water levels in all the boreholes in the system from March 2022 (Fig. 10.5). This aquifer is therefore probably being recharged when all the clays are wet, and water is perched on top.

The isotope data have been plotted for deuterium against oxygen 18 (Fig. 10.6). The Global Meteoric Water Line ($\delta D = 8.0 \cdot \delta^{18}O + 10.0$) (Clark and Fritz, 2013) and the Beaufort West local meteoric water line ($\delta D = 5.4 \cdot \delta^{18}O + 2.6$), developed by Van Wyk (2013) are indicated in the figure. The stable isotope signatures of monitoring boreholes G39888 and G39885 plot along the Beaufort West local meteoric water line (Fig. 10.6). The Beaufort West line is specifically relevant to drought conditions and indicates isotope samples taken during droughts (Van Wyk 2013). The relatively depleted stable isotope $\delta^{18}O$ and δD signatures of borehole G39888 indicate that this borehole is recharged by a source with depleted isotopes, possibly direct infiltration by rainwater depleted in $\delta^{18}O$ and δD. Borehole G39885 shows an increase in water level in Fig. 10.5, and its $\delta^{18}O$ and δD isotopes show a relative enrichment, indicating possible mixing between older stagnate water and

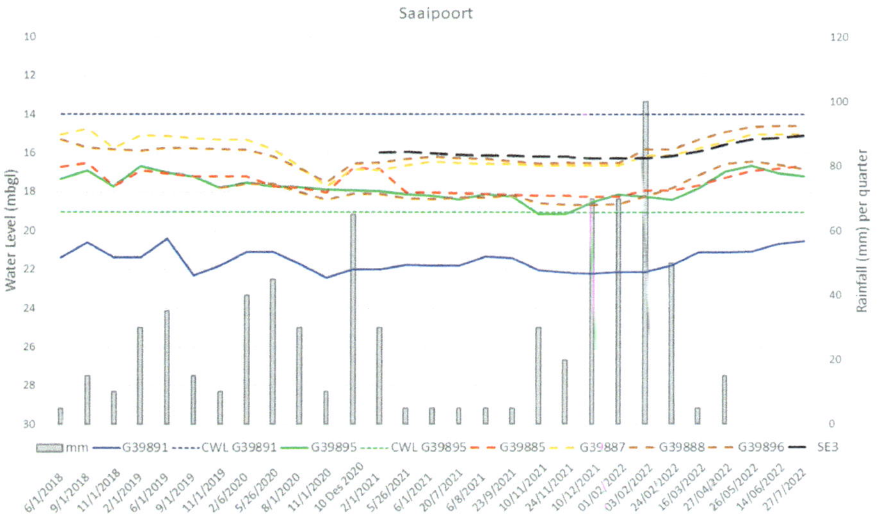

Fig. 10.5 Quarterly rainfall and groundwater levels for monitoring boreholes (thick dashed lines: G39885, G39887, G39888, and G39896) and production boreholes (solid lines: G39891 and G39895) and critical water levels (CWL) (thin dashed levels: G39891 and G39895) for the Saaipoort MAR scheme from 2018 to 2022

newer more humid recharged water (Xi 2014), or the infiltration of standing water. This confirms the theory that this aquifer is being recharged when all the clays are wet and there is old water pushed laterally towards the production borehole which augments recharge (focused recharge).

Rainfall from three rainfall events (one on 2 February and two on 3 February 2022) were also sampled and their isotopic signatures were analysed for comparison with the isotopic signatures of the borehole water samples. A complete annual record of rainfall is required to compare these samples to identify if the rainfall events were responsible for the recharge at Saaipoort. Samples from the Carnarvon Leegte River and the Masolka Dam were also taken for isotopic analysis.

Successes and Challenges

The water supply scheme successfully supplied water to Carnarvon from 2004, while the MAR scheme was implemented in 2017. A flood event in 2019 and recent large rainfall events from December 2021 to April 2022 recharged the aquifer to some extent. The water level data of monitoring boreholes G39896 and G39888 close to the river showed an increase in water levels. These monitoring boreholes are located close to the MAR boreholes. Breaking the clay layers, therefore, appears to increase recharge at this MAR site. However, even though the MAR scheme boreholes have been rested from February to October 2021 when the new pipeline to Vanwyksvlei was commissioned (Sect. 10.6), the recharge of the production boreholes was poor. Alien vegetation (*Prosopis* species (Fabaceae)) takes 200 L/d per tree (Esterhuyse

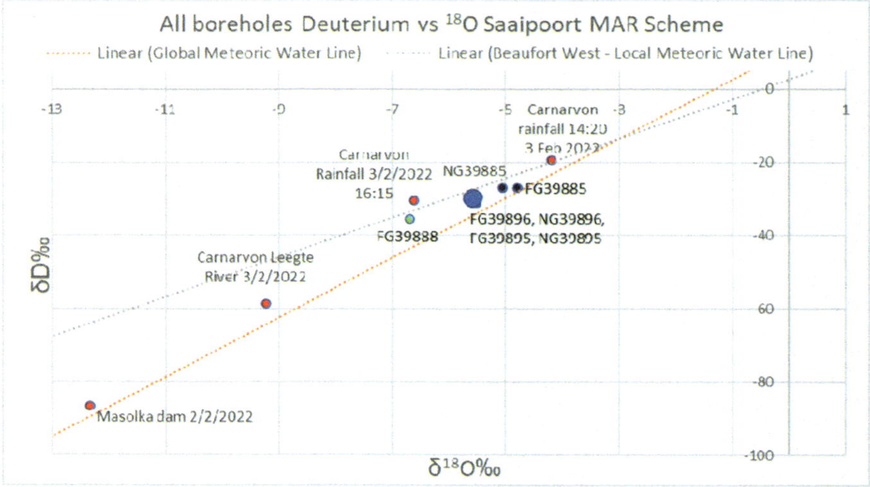

Fig. 10.6 Isotope data for the Saaipoort MAR scheme, November 2021–February 2022. Samples of rainfall, groundwater (Gxxxxx), Carnarvon Leegte River, and Masolka Dam are indicated. The larger dot represents the isotope signature of four samples that all plot in the same locality on the graph. (Note: Sample names starting with an N indicate a November 2021 sample and sample names starting with an F indicate a February 2022 sample)

2014a) and is presumed to play a large role in depleting the aquifer in the area. The groundwater levels relative to the critical water levels in Fig. 10.5 show that limited rainfall and the over-abstraction that took place from 2019 to 2020 also put the aquifer under pressure. The critical levels are set out in the water license conditions, indicating minimum and maximum groundwater levels that should not be exceeded (DWAF 2004). The case shows that rainfall and recharge after 2019 were too little relative to demand and that invasive species and the increasing rainfall variability with climate change could be an issue to sustain this MAR scheme in the future unless further MAR is introduced.

10.5.2 Smouskolk: Augmenting the Recharge of a Fractured System that Supports Existing Production Boreholes

Site Information

Vanwyksvlei receives rain in the late summer from March to May with an average rainfall of 221 mm per annum (Visser et al. 2004; Roffe et al. 2020). Vanwyksvlei town is located in quaternary catchment D54B and its MAR scheme (Smouskolk) is in quaternary catchment D54C. The MAR scheme is in the non-perennial Smous

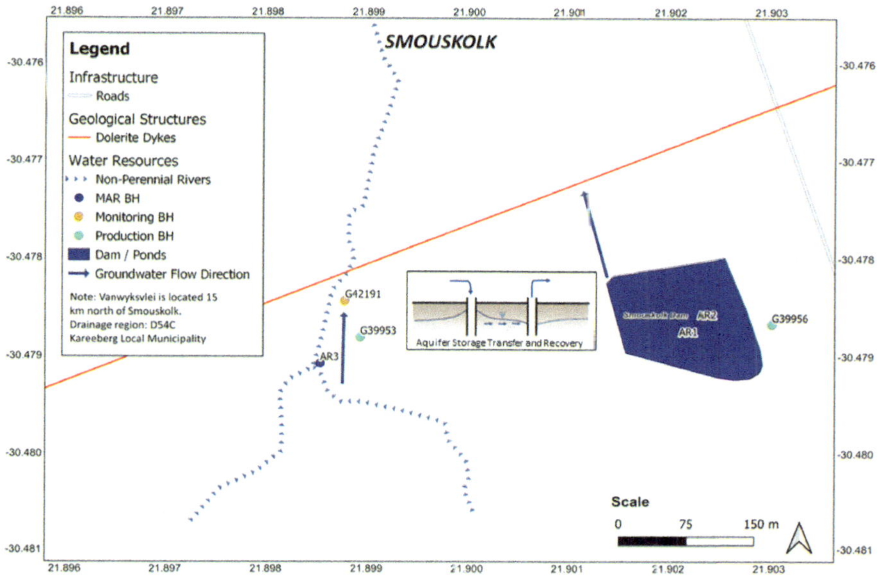

Fig. 10.7 Detailed locality map showing the MAR boreholes and production boreholes of the Smouskolk MAR scheme. AR1 and AR2 are located close to G39956, while AR3 is located close to G39953 and borehole G42191

River that drains into the Vanwyksvlei Dam close to Vanwyksvlei (Visser et al. 2004) (Fig. 10.7).

The Smouskolk MAR scheme is a combination of recharge and abstraction wells and bank infiltration that targets a dry river and a small dry dam (Hohne et al. 2021). Seasonal excess surface water recharges naturally via the infiltration boreholes that are located in the dam (AR1 and AR2) and the riverbed (AR3) and is abstracted via the ASTR abstraction boreholes located off the dam and river. The Smouskolk MAR site is in the Ecca group, which consists of weathered alluvium and calcrete at the surface followed by dolerite sills and Tierberg and Whitehill shales sandwiched between the sills (Chevallier et al. 2001). The production boreholes of the Smouskolk MAR scheme were drilled between 1982 and 1996 to depths between 120 and 180 m. In 2017, three boreholes that serve as recharge enhancement boreholes (AR1–AR3) were drilled. The recharge boreholes have slotted casings sticking out of the ground to ensure better natural water infiltration to the aquifer system. The boreholes were drilled to a depth of 80 m to ensure that the recharge of water only takes place in the fractured rock between two dolerite sill layers. It is believed that these infiltration boreholes are on the same fracture system as, and support, the production boreholes (Chevallier et al. 2001).

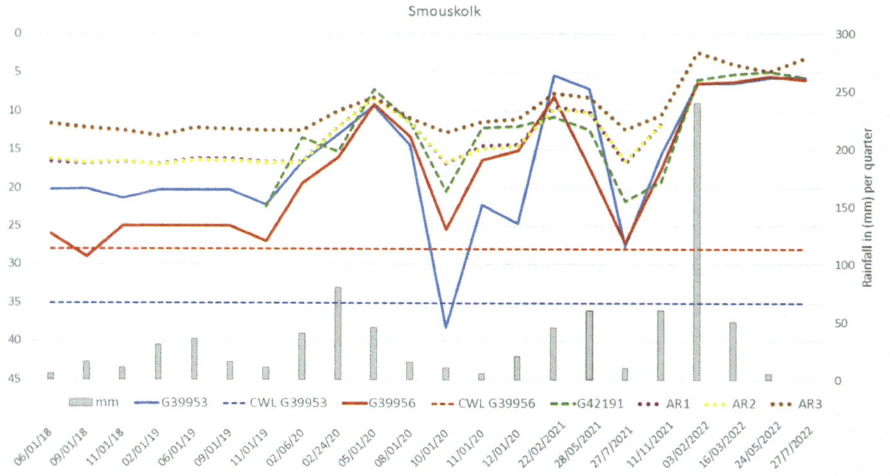

Fig. 10.8 Quarterly rainfall and groundwater levels for production boreholes (solid lines: G39953 and G39956), critical water levels [thin dashed lines (CWL): G39953 and G39956] and MAR infiltration boreholes (thick dashed lines: (AR1-3) for the Smouskolk MAR scheme from 2018 to 2022

Water Levels and Stable Isotopes

The increase in water levels in the MAR boreholes (AR2 and AR3) and the production boreholes (G39953 and G39956) can be seen in October 2020, July 2021, and February 2022 after rainfall events (Fig. 10.8). The first activation of the MAR system occurred after it received rain from 18 to 20 February 2020 at a volume of 50–75 mm (ARC, 2020) (Fig. 10.8).

Stable isotope samples were collected and analysed for δD and δ^{18}O for the river water on 3 February 2022, while isotope samples were collected from two boreholes (observation borehole G42191 and production borehole G39953) in June 2021, October 2021, and February 2022 (Fig. 10.9). The production borehole G39953 showed enrichment in δD and δ^{18}O from June 2021 to October 2021 and depletion in isotopes from October 2021 to February 2022. The enrichment may be a result of static 'old' water that is being pushed out of the aquifer with the new recharged water where isotopically enriched winter rainfall is recharging the aquifer in the late winter months, while major events during summer (like in December 2021 and February 2022), which are isotopically depleted, replenish the river water. The latter is the main recharging source of the system. Observation borehole G42191 is not located next to a MAR borehole and is more depleted in isotopes than G39953, which can mean the water in this part of the aquifer is still some of the older recharged water. The fresher recharged water is located closer to the production wells. The depleted isotope signal of the water in the Smouskolk River confirms the intense rainfall event on 1 February 2022, when 110 mm of rainfall was recorded in the area.

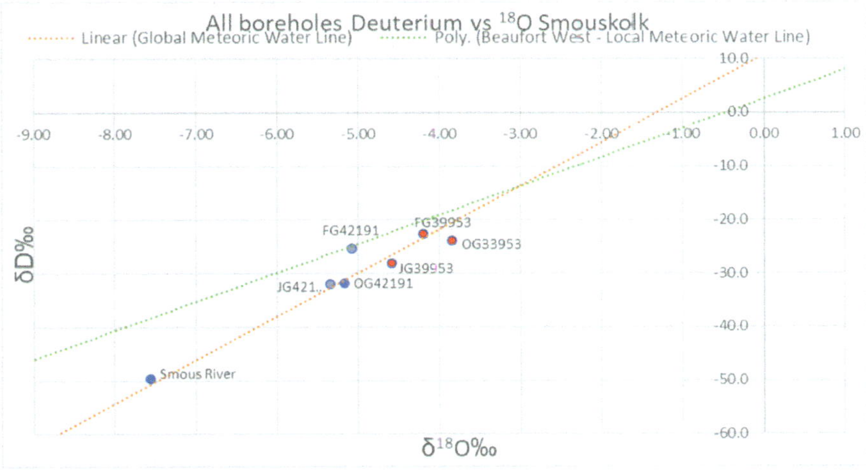

Fig. 10.9 Isotope data for the Smouskolk MAR scheme for June 2021, October 2021, and February 2022. *Note* The months of sampling are indicated with J, O, and F, at the beginning of the borehole number. The Beaufort West local meteoric water line (Van Wyk 2013) is specifically relevant to drought conditions and indicates isotope samples taken during the drought

Successes and Challenges

This MAR scheme, which has been operational since 2017, successfully and quickly recharges the aquifer. The aquifer also recharges during very small rainfall events. Large rainfall events recharged the aquifer in December 2021 to full capacity within 10 days. Data collected in September 2022 indicated that the aquifer is still fully recharged and that its water levels are at 5 mbgl. A positive aspect is that the chemical water quality improves when the aquifer recharges, while iron bacteria also disappear during the recharge of this system. Challenges include that the aquifer is not being managed and is over-abstracted, as indicated by the CWLs (Fig. 10.8) (Van der Westhuizen et al. 2007). If demand is managed properly, the scheme, enhanced by the infiltration boreholes, could provide water to the town for many years, especially in the drought years like was seen from 2014 to 2019.

10.5.3 Williston: Moving Groundwater Across Confined Aquifer Compartments to Equilibrate Storage and Enable Sustainable Use

Site Information

Williston is classified as a late summer rainfall area (Esterhuyse 2014b) and is in a transitional zone between summer and winter rainfall (Roffe et al. 2020). A few non-perennial river channels are found close to town with the Sak River located in the south

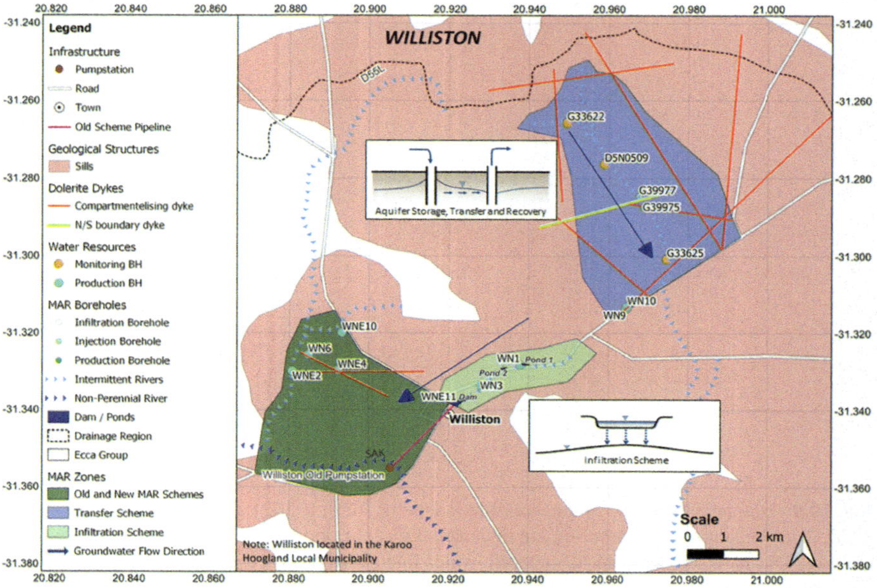

Fig. 10.10 Detailed locality map showing the MAR boreholes and production boreholes of the Williston existing and planned MAR schemes

of town. Williston has three different MAR schemes, located in quaternary catchment D55L (Fig. 10.10). The geology of this area is the Ecca group sedimentary sequence of shale, sandstone, and mudstones. These formations are intruded by dolerite dykes and sills of up to 100 m thick, also called the Williston Dolerite Ring Complex (WDRC). The WDRC is most often targeted for the groundwater it holds (Chevallier et al. 2001; Esterhuyse 2014b). The dolerite dykes of the WDRC compartmentalise the aquifer at Williston (Fig. 10.10).

The largest MAR scheme, established in 1998, is a transfer and recovery scheme located in the northeast of Williston in one of the non-perennial rivers (purple area in Fig. 10.10). This scheme transfers water from one aquifer compartment (northern) to another (southern) and is abstracted 4 km further south to provide the town with water (Fig. 10.10). Abstraction and observation boreholes have been drilled to depths ranging between 70 and 270 m. These boreholes were drilled in 1983 (WN9/G33076), 1994 (WN10; G33625 and G33623), and 2007 (reconstruction of G39977). The aquifer in this scheme is confined and consists of fractured rock. Before the scheme was implemented, the groundwater level of the northern compartment was 20 m higher than the southern compartment (Chevallier et al. 2001).

Under ideal operating conditions, the groundwater should be abstracted from a depth of 70 m from the northern compartment borehole (G39977) with a water level of 10 mbgl, and should be transported via a closed pipeline for 50 m across the dolerite dyke (which is a hydraulic boundary), and should then be injected at a constant rate of 3 L/s (0.26 ML/d) at a depth of 70 m into the southern injection borehole, which has a water level of 30 mbgl (Hohne et al. 2021).

The town has two other MAR schemes, implemented later, in 1983 and 2019, which are infiltration schemes at localities WN1 and WNE11 on the western side of the town, respectively (Fig. 10.10). This area will be the focus of the new MAR schemes for WN6, WNE2, WNE4, and WNE10. Additional MAR schemes will be developed in the near future.

Water Levels and Stable Isotopes

A gradual drop in groundwater levels was visible in the boreholes being monitored higher up in the catchment (G33622, D5N0509 and in the MAR extraction borehole itself (G39977) until 15 May 2021 when it was switched off, after which very slow recovery took place until the end of the observation period (Fig. 10.11a). The monitoring boreholes (G33625) and the production boreholes (WN9 and WN10) in the southern injection compartment showed a relatively constant water level for about a year until 15 May 2021, after which they started to decline.

From 14 February 2022, the production boreholes (WN9 and WN10) were switched off and no recovery in the water level was seen even with 100 mm of rainfall recorded in March 2022. This indicates that throughout extended periods of drought (from 2013 to 2022), water levels can stay constant with the help of the MAR system, by transferring water from the northern compartment to the southern compartment. However, after 7 years of poor rainfall (2013–2022) and with constant pumping from both the northern and southern compartments, the aquifer seems to be exhausted and purposeful MAR schemes must be developed to harvest and put water directly underground in the southern compartment, as natural percolation from rainfall is either insufficient or slow.

At the other two infiltration schemes, water levels for production boreholes (WN1, WN3, and WNE11) were declining, but on 24 October 2021, 47 mm rainfall occurred, and the intermittent streams started to flow, which filled the three dams in this water course, and which created a groundwater level rise (Fig. 10.11b).

Isotope analyses were done on the groundwater of the old MAR scheme in July and October 2021 of the water of boreholes G36622, G33625, G39977, WN9, and WN10 (Fig. 10.12). The October 2021 samples are slightly more enriched in δD and δ^{18}O compared to the July 2021 samples. This can be due to the mixing of recently recharged groundwater with older groundwater with some recharge from small rainfall events. The plots can also be more enriched because surface water evaporates before infiltration, becoming more enriched in summer than in winter with more arid conditions.

The samples for the February 2022 isotope analyses of the production boreholes WN9 and WN10 indicate very little change (data not shown). This can be because the system was rested in May 2021, and WN9 and WN10 only draw old water that is not recharged out of the southern compartment. It is hoped that new data will reveal that recharge took place in the northern compartment (Xi 2014). To determine if these samples indicate recharge, tritium and carbon analyses must be included. Boreholes G33625 and G33622 plot differently from the other boreholes WN9 WN10, G39977, and D5N509. G33625 is located in the southern compartment and seems not to be

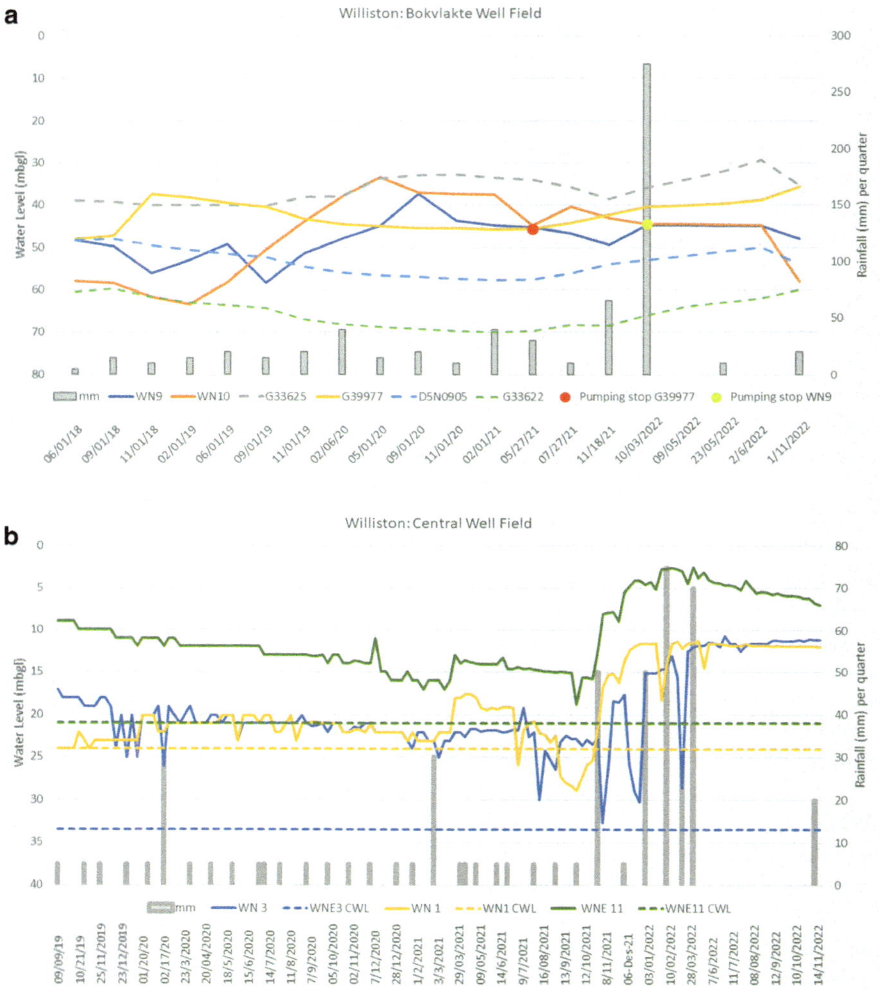

Fig. 10.11 **a** Quarterly rainfall and groundwater levels for production boreholes (solid lines: WN9 and WN10), observation boreholes (dashed lines: G33625, D5N0905, and G33622), and MAR injection borehole (thicker dashed line: G39977) for the old Williston MAR scheme from 2018 to 2022. The pumping stops refer to G39977, which stopped pumping in May 2021, and WN9, which stopped pumping in February 2022. **b** Quarterly rainfall and groundwater levels for production boreholes (solid lines: WN1, WN3, and WNE11) and CWLs (dashed lines: WN1, WN3, and WNE11) in the new Williston infiltration scheme

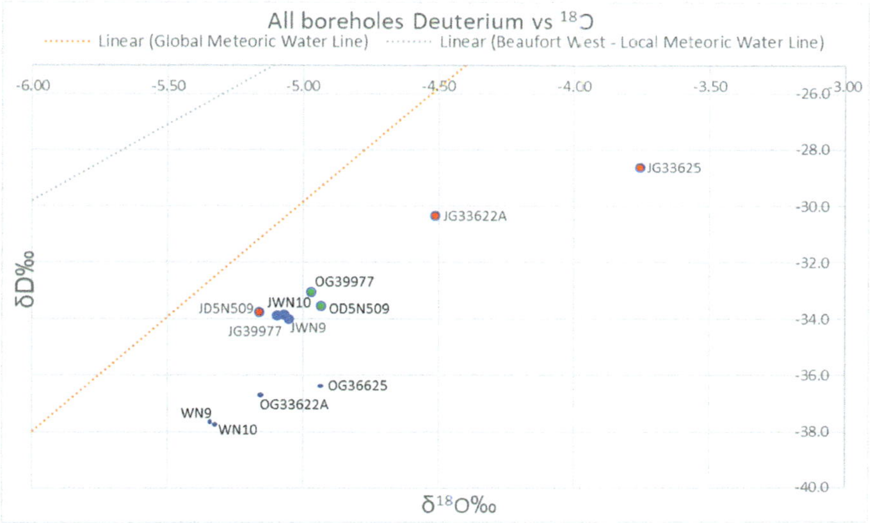

Fig. 10.12 Isotope data for the Williston MAR scheme for June and October 2021. *Note* The months of sampling are indicated with J and O at the beginning of the borehole number

influenced by the MAR scheme. Borehole G33622 is located north of the scheme in the northern compartment and might not be hydraulically connected due to the site being compartmentalised.

Successes and Challenges

Good ambient groundwater quality means that water can be abstracted and used directly. This system has maintained good water quality from the outset of the MAR scheme in 1998.

Water levels in the northern compartment have dropped to 45.8 mbgl due to the continuous drought from 2013 to 2022 despite that the management of the system was sharply enhanced in 2018 to try to get back to ideal conditions as set out above. The MAR abstraction borehole was switched off at the end of May 2021. Water levels in the southern compartment continued to drop and on 14 February 2022, the southern compartment production boreholes were also switched off due to poor recharge, even with above-average rainfall.

Afterward, the water level in another southern compartment borehole declined (data not shown). The southern compartment is extremely depleted and production boreholes are pulling water from the far side of the southern compartment with no influence from the northern compartment.

The findings from the Williston site indicate the urgent need to speed up additional MAR schemes, like the ones already in progress in the western area of the town, enabling a greater capture of rainfall and runoff, while diversifying sources of water supply. As part of this, the option to purposefully enhance recharge in the upper transfer scheme to the northeast needs to be considered.

10.5.4 Sutherland: Sustaining Natural Rainwater Harvesting and Recharge Systems

Site Information

This MAR scheme, which is located south of Sutherland in quaternary catchment D51A, is still being built (Fig. 10.13). The catchment has many non-perennial rivers, while Sutherland town has manmade stormwater channels meant to divert rainfall runoff from the higher-lying areas to avoid flooding. All the rainfall is being diverted around the town to the river or to the Sutherland Dam. The river alluvium in this area is often targeted for shallow boreholes for the water it holds (Esterhuyse et al. 2004). The new MAR development focuses on recharging the river alluvium via the fracture zone of a specific dolerite intrusion that is associated with the non-perennial river that runs past Sutherland (Fourie 2021b). The fractured Jurassic-age dolerite dykes of the Beaufort Group and the fractured Salpeterkop volcanic pipe are common groundwater targets in this area (Mulder 2021).

Sutherland is located in a transitional zone between summer and winter rainfall (Roffe et al. 2020) and receives late summer rainfall from March to May, with occasional winter rainfall and snow. The mean annual precipitation is 420 mm/yr (Esterhuyse et al. 2004). In this area, recharge mostly occurs during high rainfall and runoff events (>25 mm). MAR schemes can however facilitate recharge even with rainfall events of as low as 10 mm (Fourie 2021b).

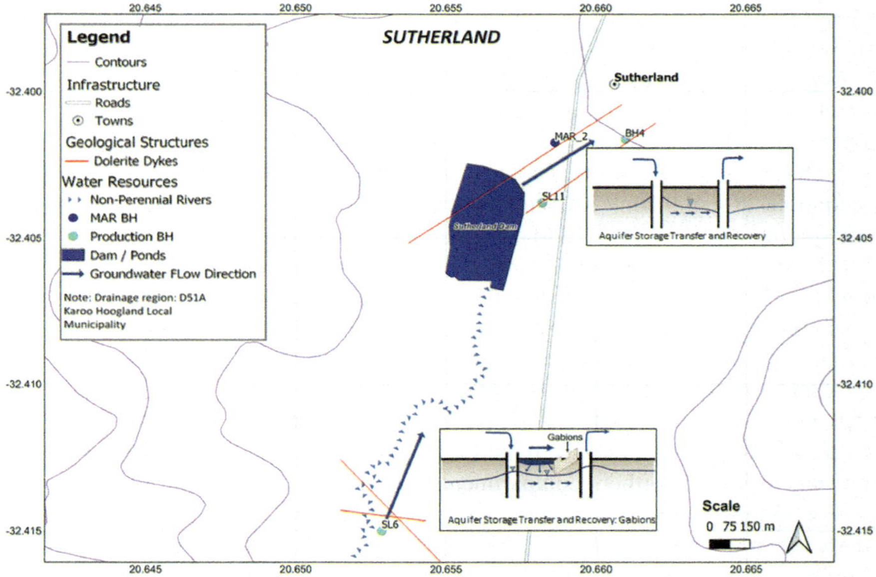

Fig. 10.13 Detailed locality map showing the MAR boreholes and production boreholes of the Sutherland MAR schemes

In 2021, the water level in all the original water supply boreholes of Sutherland's shallow and deep fractured aquifers dropped to below long-term average levels. Various methods can be used to enhance infiltration to ensure better and more regular groundwater recharge. At Sutherland, the proposed recharge improvement methods include trenches, infiltration ponds, and gabions that are placed perpendicular to the flow direction within the non-perennial river. To date, gabions have been installed at the SL6 production borehole located upstream of the Sutherland town dam. This is currently an active scheme. Further designs are planned to augment recharge upstream of borehole SL11 and a MAR scheme in the vicinity of BH4.

Water Levels and Isotopes

Figure 10.14a shows that the groundwater levels for production borehole BH4 in the MAR scheme just south of town declined from 2015 to 2022 but increased again after a flood in 2016. Thereafter the water level declined steadily until April 2021, at which time it ran dry and was taken out of production.

Water restrictions and very good rainfall events in 2021 helped with the recharge of the town dam which is feeding BH4. Currently, this borehole continues to recover while providing all the water to the town and being pumped at 3 L/s for 22 hr/d.

Production borehole SL6 is part of an in-stream infiltration scheme south of town. The gabions introduced slow down the flow of the intermittent river and increase the infiltration of water. SL6 was drilled together with SL11 in November 2021 and put into production in June 2022 after good rainfall. The logger showed a constant increase in water level, even during pumping (Fig. 10.14b).

Isotope samples of BH4 were taken in October 2021 and March 2022 (Fig. 10.15). The samples plot very close to each other with little variation, which possibly indicates the stability of the system over that period. In March 2022, the isotope samples show a little enrichment, possibly due to recharge from the dam water. The isotope sample from SL11 does not resemble either the dam or BH4 water and could be derived from a different aquifer system, though further evidence is needed.

Isotope samples of the rainfall event of 21–22 October 2021 are very similar to the data of the dam water for 23 October 2021 (Fig. 10.15). The dam water was collected at the entry point of the dam. The similar signature for the rainfall and the dam water could be because the stream and dam both contain water from the rainfall event.

Successes and Challenges

If this scheme operates successfully, it will be a low-cost, low-maintenance scheme with no required mechanical treatment of water (Nature-Based Solutions approach). This scheme could ensure water security for the town in the future. The success will depend on the adequate and rapid capture and recharge of rainwater through the two systems during relatively intense rainfall events. The future will shed more light on recharge options as continuous monitoring takes place.

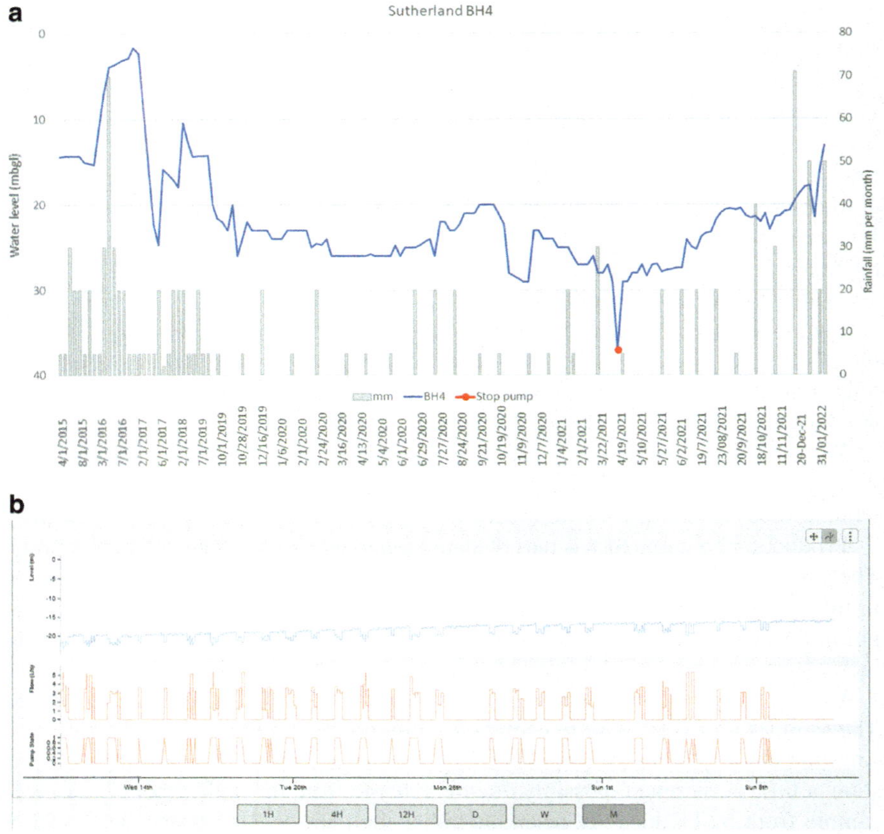

Fig. 10.14 **a** Monthly rainfall and groundwater levels for BH4 for the Sutherland scheme from 2015 to 2022. 'Pump stop' indicates the stop of the pump at borehole BH4 in April 2021. **b** Groundwater level for SL6 for the Sutherland scheme from Dec 2022

10.6 Socioeconomic Aspects of MAR

Overall, the communities in the Western Karoo are poor. The people mostly live from farming sheep and goats and rely on tourism where possible, while few other sources of income are being generated in these towns (Redelinghuys 2016). Water supply planning and water demand predictions were done for the Western Karoo towns, but longer-than normal droughts and increasing water demand from a growing population (though estimates are uncertain) have put pressure on the water resources beyond what was predicted (Tables 10.2 and 10.3). This has resulted in depleted aquifers and forced the implementation of water rationing and cuts to manage the water supply sustainably, as was the case in Carnarvon and Vanwyksvlei in December 2020 and Sutherland in April 2021 (Hohne and Fourie 2021a, b). In most cases, droughts of 2–3 years were included in groundwater planning models when pump

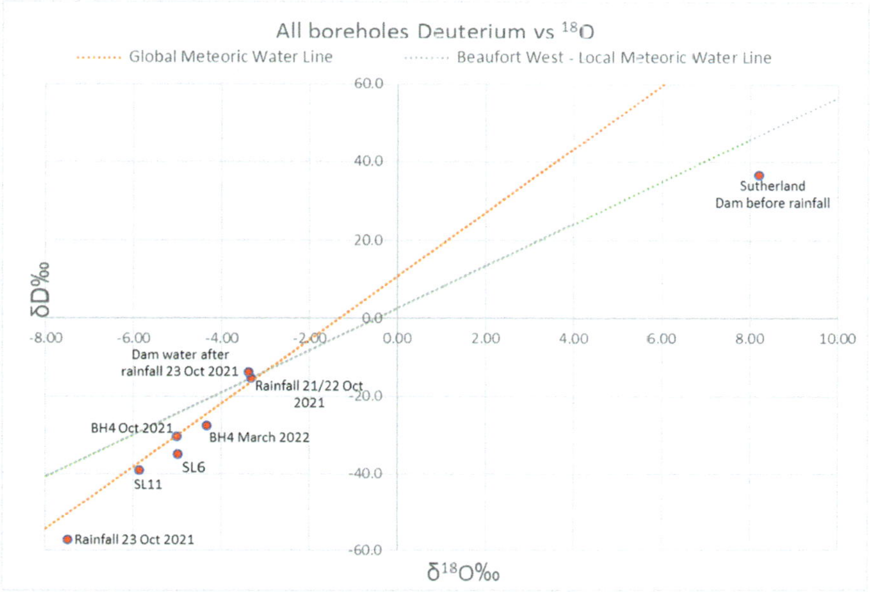

Fig. 10.15 Isotope data for the Sutherland MAR scheme for October 2021 and March 2022

testing was done (Van der Westhuizen et al. 2007; Esterhuyse 2014a, b), but from 2019, they have included a conservative no-recharge volume in the pump testing model to increase assurance of sustainable supply of boreholes (Esterhuyse 2019a, b). Renewable energy developments are also entering the rural Karoo, which brings with it much more working force people not planned for in planning models (Hohne and Fourie 2021a, b).

To make up for this deficit, towns need to rely on getting water by either expanding wellfields or in some cases developing new wellfields, with the risk of new water being brackish. Ultimately, the towns may need to rely on pipeline water, which is much more expensive than MAR schemes. An example is Vanwyksvlei, where it

Table 10.2 Population estimates

Source	All towns 2011			All towns 2015	Greenbook 2019
Year	2008	2025	2030	2035	2050
Carnarvon[1]	5,184	4,742	4,658	3,916	10,936
Vanwyksvlei[2]	2,311	2,867	2,992	2,263	3,367
Williston[3]	2,968	2,653	2,600	4,268	3,841
Sutherland[4]	1,821	1,898	1,947	No data	3,237

[1]DWA (2011a), DWS (2015a); [2]DWA (2011c), DWS (2015b); [3]DWA (2011d), DWS (2015c); [4]DWA (2011b)

Table 10.3 Water requirements

Town		2008	2011	2030	2035
Carnarvon[1]	ML/d	1,18	1,03	1,25	1,60
	Mm^3/yr	0,430	0,375	0,455	0,584
	L/p/d	227	156	268	180
Vanwyksvlei[2]	ML/d	0,309	0,186	0,477	0,285
	Mm^3/yr	0,113	0,068	0,174	0,104
	L/p/d	134	108	159	126
Williston[3]	ML/d	0,447	0,636	0,452	0,885
	Mm^3/yr	0,163	0,223	0,165	0,323
	L/p/d	151	189	174	207
Sutherland[4]	ML/d	0,274		0,367	
	Mm^3/yr	0,121		0,134	
	L/p/d	150		188	

[1]DWA (2011a), DWS (2015a); [2]DWA (2011c), DWS (2015b); [3]DWA (2011d), DWS (2015c); [4]DWA (2011b)

was decided to build a cheaper pipeline from the Saaipoort aquifer rather than one from Copperton along the Orange River. In 2012, the total comparative estimated project life cycle cost for the two pipelines was: Saaipoort to Vanwyksvlei: $2,9 million (ZAR 52 million) and from Copperton to Vanwyksvlei: $3.3 million (ZAR 60 million) (IRS 2012).

The economic cost of developing groundwater compared to surface water, indicate that groundwater should be further developed and properly recharged, to avoid depletion and salinity issues. In the Western Karoo, surface water sources evaporate fast due to the extreme heat. All the rivers in these areas, except the Orange River, are non-perennial. Supplying towns with surface water would therefore mean costly pipelines from the Orange River. Groundwater that is stored underground is not exposed to evaporation and can also often be supplied where it is needed, without using costly pipelines. Since climate change will lead to more droughts and more variable rainfall over the long term, MAR is an efficient way to enhance groundwater storage in the Western Karoo. MAR sites in specific towns can also provide water to neighbouring towns. However, since MAR relies on a surface or other water sources, it is important to co-manage surface and groundwater as well as other waters, like wastewater, to optimise water availability and water quality.

Table 10.2 shows the population of the four MAR sites as projected for 2050, while Table 10.3 shows the water requirements of these towns as projected for 2035. Table 10.4 indicates actual water use up to 2019.

From these tables, it is seen that population growth (Table 10.2) is driving an increase in water demand in all towns (Table 10.3). Sutherland experienced a large decline in water availability in 2021 (Table 10.4). The newly drilled boreholes now

Table 10.4 Actual water use, newly added water, and water quality

Town	Water used (m^3/d)				Newly added water (m^3/d)	Water quality[5]
Carnarvon[1]	2014	No data	No data	No data	2014 and 2019	
	302	No data	No data	No data	458	Class 2
Vanwyksvlei[2]	2007	No data	No data	No data	No data	
	203	No data	No data	No data	No data	Class 2
Williston[3]	2014	No data	No data	No data	2014 and 2019	
	527	No data	No data	No data	411	Class 1
Sutherland[4]	2004	2011	2020	2021	2022	
	792	1577	583	126	1810	Class 4

[1]Esterhuyse (2014a, b); [2]Van der Westhuizen et al. (2007), Potgieter (2014); [3]Esterhuyse (2014a, b); [4]Fourie and Hohne (2021), Mulder (2021); [5]See Table 10.5

Table 10.5 Water quality classes[1]

Class 0 Ideal water quality	Class 1 Good water quality	Class 2 Marginal water quality	Class 3 Poor water quality	Class 4 Unacceptable water quality

[1]DWAF (1999)

can provide enough water if the MAR system is properly managed. In 2014, new boreholes were drilled for Williston and Carnarvon, and in 2019, even more, boreholes were needed to satisfy water demand.

10.7 Cost-Effectiveness and Maintenance of the MAR Schemes

The cost of the MAR scheme in Saaipoort completed in 2015 was moderate at ZAR150,000 (USD9,036) (Esterhuyse 2015). The cost of the water supply of the Kareeberg Municipality, which is charged to the water user, ranges from ZAR1.70 to ZAR12.42 per cubic meter on a sliding scale. With the 12 cm increase in the aquifer water levels in 2019, a pending water crisis was averted. For three months, water could be supplied to the constituents at a cost of ZAR79, 560 to ZAR581,256.[6] The scheme, therefore, was paid off within the first rainfall event. Maintaining the scheme has limited costs as it is self-operational and to date, no clogging or bacteria were encountered.

[6] The cost range was calculated as follows: [3 months] x [an average volume of 520 m^3/d of water being pumped] x [the sliding scale municipal water supply cost (ranging from ZAR1.70 to ZAR12.42)] = ZAR79,560 to ZAR581,256.

The Smouskolk MAR scheme has proven to help recharge the Smouskolk aquifer, which provided most of the freshwater to Vanwyksvlei before the completion of the pipeline from Saaipport to Vanwyksvlei. At Smouskolk, three infiltration boreholes were drilled to a depth of 80 m to ensure recharge of shales. Based on the Quality of Domestic Water Supply Assessment Guide (DWAF 1999), the water quality at Smouskolk is classified as a Class 2 (Potgieter 2014). When the aquifer is recharged, no clogging occurs, and the water does therefore not need extensive treatment before recharge. Cleaning a borehole involves cleaning it by drilling and making use of chlorine to scrub and pump the borehole. The scheme can be kept operational at lower costs since the water that recharges via the infiltration boreholes that are located in the dam and river naturally migrates to the production boreholes, which is one of the environmental services of this type of MAR. The cost to develop this scheme was moderate, as the DWS drilling team drilled the boreholes, and a more expensive commercial drilling team was not used. The three infiltration boreholes were only drilled to 80 m deep, which kept drilling costs at a minimum. The required maintenance of these boreholes only amounts to cleaning the casing slots after a flood event.

The Williston transfer scheme is also to a large extent self-operational and only needs good management of pumping schedules. At the Bokvlakte scheme of Williston, two boreholes were drilled to a depth of 75 m, where water is abstracted from one compartment and injected into the other. The infiltration boreholes were drilled to depths of between 50 m (in 1983) and 150 m (in 2019). The infiltration scheme is low cost as it is only the pond and dam basins that need regular cleaning. No clogging has ever been reported in the Williston injection scheme, even after extreme drought periods.

Costs to develop the Sutherland MAR scheme have been moderate, but only one scheme (BH4 and the town dam) currently recharges the aquifer effectively. At Sutherland, one infiltration borehole was drilled, and five gabions were installed in the riverbed. Other costs involved scraping the bottom of the dry dams to ensure better infiltration. The water that is recharged to BH4 has low trace metal concentrations, which minimises the need for water treatment. In 2021, while the scheme was not recharging, bacteria were detected in BH4, probably deriving from the over-abstraction of the aquifers (Hohne and Fourie 2021a, b). When extreme droughts occur, or when aquifers are over-abstracted, bacterial growth can reduce the yields of the boreholes. However, as in the Smouskolk case, the bacteria disappeared when these boreholes were recharged again. At SL6, the gabions in the riverbed were dug out approximately once a year to ensure good recharge. To ensure good management of the schemes, boreholes were equipped with telemetry systems, in which critical pumping levels and groundwater levels for stopping pumping were set strictly to ensure enough reserve water in the aquifer to prevent over-abstraction of the aquifer system.

10.8 Institutional Aspects of the MAR Schemes

DWS views MAR schemes as one of the main tools to increase water supply and ensure water security for Western Karoo towns (DWS 2016a). The source water is envisaged to derive mostly from episodic rainwater that is intermittently available to recharge the aquifers. DWS regulates the MAR schemes under General Authorisations[7] and water use licenses that focus on Sect. 21 of the NWA, for (a), abstraction of water, (b) storing of water, (c) & (i) impeding and diverting water of rivers or channels, and (g) the intentional recharging of an aquifer with any water containing waste.

Despite its importance for water supply in water-scarce areas, MAR is still under-utilised in the Western Karoo, with only one major scheme (the Williston transfer scheme) and a few minor schemes (DWS 2016a), altogether amounting to an estimated 1.4 Mm^3/year of facilitated recharge. Within the National Groundwater Strategy (DWS 2016a), priority actions include promoting the implementation of MAR schemes. The National Water and Sanitation Master Plan (DWS 2018) envisioned that more artificial recharge schemes would be developed in the future. MAR schemes have been piloted at several sites across Southern Africa (Braune and Israel 2021), and are continuously being developed. Most schemes are managed by the local municipalities with some oversight by DWS.

However, it is important to bear in mind that groundwater is still, despite its unrivalled importance in arid and semi-arid areas, like the Western Karoo, perceived by many (if not most) municipal role players and communities in South Africa as an unreliable and difficult resource to manage (DWS 2016a). Most municipal managers prefer water supply from surface water sources—ideally with bulk potable water supplied by a water board.

The specific MAR schemes in the Western Karoo have been designed to be relatively simple and mostly self-operational so that municipalities do not have to spend large amounts of resources to maintain them. Iron bacteria or other clogging is also not a concern in these MAR systems. MAR schemes must however be monitored to ensure sustainable and good quality water abstraction. Monitoring of the Williston MAR system, for example, only focused on one compartment. Municipal officials, therefore, did not notice that the other compartment was being depleted. Monitoring must be comprehensive to ensure sustainable water supply from MAR systems. MAR systems in water-scarce areas must also be designed in such a way that even the smallest amount of rainfall can be captured into a MAR system to enhance sustainable water supply.

To make more efficient use of MAR systems, the NWA allows for 'the intentional recharging of an aquifer with any waste or water containing waste'. Recharge with wastewater should only be allowed if it can be proven that no alteration of the water quality within the groundwater system will take place. There are currently only two MAR schemes in South Africa—at Atlantis (just north of Cape Town) and Polokwane in the Limpopo Province—that are using treated wastewater to recharge

---

[7] General Authorisations do not require a water use license under the NWA.

groundwater (DWAF 2007). Also, two licenses were issued in the last few years to the Kolomela mine near Postmasburg (Northern Cape) and Elandsfontein mine near Saldanha (Western Cape) to abstract groundwater to dewater the aquifer for mining purposes and then recharging (through injecting) the abstracted water elsewhere (Fourie et al. 2016).

10.9 Conclusion

In water-scarce countries or regions, it is often a challenge to ensure a secure water supply to all users while meeting the growing requirements with an optimised mix of available water resources. The medium- and long-term available water resources include a mix of conventional and unconventional water sources, including increased groundwater use (including through MAR), desalination, and re-use of water such as wastewater from oxidation dams. MAR, in as much as additional water resources can be harvested, will become a strategic measure in ensuring sustainable and reliable water sources in the future.

MAR schemes are a viable long-term water supply option for the Western Karoo towns. Climate change will increase rainfall variability, and larger MAR schemes will have to be implemented to account for this variability. Interestingly, mostly unknown schemes in places like Loeriesfontein and Williston, where retention dams were constructed in the 1950s, still seem to help with the recharge of aquifers 60 years later. These historic schemes assist in recharging MAR systems during the rare occasions when rainfall does occur in these dry areas.

MAR schemes can be developed to be mostly self-operational, which makes them relatively cost-effective to implement. MAR schemes, especially the infiltration-based systems, apply nature-based processes, like infiltration and soil treatment, which means that the environmental impacts can be managed. The schemes discussed in this chapter ensured the recharge of four aquifers (Saaipoort, Smouskolk, Williston, and Sutherland). The water supply was stabilised, and a good deal of drought proofing was achieved, though issues do arise in some of the schemes during long-term droughts where the aquifers may get exhausted, and/or aquifer recovery may take extended time (Williston transfer and injection scheme). Ongoing water quality sampling has indicated that overall good quality is attained in retrieved MAR water, though further work is required (e.g. for the Sutherland scheme). One scheme (Williston) indicated bacterial growth during drought periods with little flushing from new recharge water. Little source water treatment needed before recharge ensured the low cost of schemes. In the Western Karoo, these localised MAR schemes are the best way to ensure that aquifers get recharged in priority areas to get water to the towns.

The following lessons are important for MAR schemes in water-scarce areas. A good understanding of the geology is required as this influences the effectiveness of recharge to the aquifers and recovery of groundwater. Good continuous data on rainfall, as well as monitoring of abstraction and groundwater levels and estimation of recharge of the aquifer, are needed. Isotope data can be useful, but a relevant sampling plan should be developed to ensure useful data are collected that may help in interpreting recharge processes.

Good monitoring data will ensure the proper management of not only the MAR schemes but also the associated aquifers. Monitoring data will indicate declines in water levels so that timeous action can be taken to protect the integrity of the MAR schemes. Cooperation is needed between the aquifer managers (DWS), the municipalities, and the communities. If over-abstraction takes place, short- or long-term water restrictions must be implemented. Here, clear communication between authorities and local stakeholders is crucial. Also, as illustrated in this chapter, MAR schemes benefit from incremental development/expansion as water demands increase and climate conditions change. It is important to have medium to long-term plans for how to address such expansions.

The notable, yet slow, growth of MAR at 0.2% in Southern Africa (Dillon et al. 2019) where the climate is variable and water stress significant, illustrates the importance of MAR schemes in this region. MAR becomes even more important when considering further climate change, especially MAR that re-uses treated wastewater.

References

Ackerman P (2022) Pers Comm. Department of Water and Sanitation

Avenant M, Watson M, Esterhuyse S, Seaman MT (2016) Potential impact of unconventional gas mining on surface water systems of the karoo. In: Hydraulic fracturing in the karoo: critical legal and environmental perspectives. Juta, Claremont, South Africa, pp 222–244

Braune EM, Israel S (2021) Managed Aquifer recharge: Southern Africa. The Groundwater Project. https://gw-project.org/books/managed-aquifer-recharge-southern-africa/

Carvalho Resende T, Longuevergne L, Gurdak JJ, Leblanc M, Favreau G, Ansems N, Van der Gun J, Gaye CB, Aureli A (2019) Assessment of the impacts of climate variability on total water storage across Africa: implications for groundwater resources management. Hydrogeol J 27(2):493–512

Chevallier LP, Goedhart ML, Woodford AC (2001) Influence of dolerite sill and ring complexes on the occurrence of groundwater in karoo fractured aquifers: a morpho-tectonic approach: report to the water research commission. Water Research Commission

Clark ID, Fritz P (2013) Environmental isotopes in hydrogeology. CRC Press

Cuthbert MO, Taylor RG, Favreau G, Todd MC, Shamsudduha M, Villholth KG, MacDonald AM, Scanlon BR, Kotchoni DV, Vouillamoz JM, Lawson FM (2019) Observed controls on resilience of groundwater to climate variability in sub-Saharan Africa. Nature 572(7768):230–234

DWA (Department of Water Affairs) (2010a) Strategy and guideline development for national groundwater planning requirements. A practical guide to using Artificial Groundwater Recharge, Water Banking. https://www.artificialrecharge.co.za/booklet/AR_Booklet_13Jan2011.pdf

DWA (2010b) The Atlantis Water Resource Management Scheme: 30 years of artificial groundwater recharge. http://www.artificialrecharge.co.za/casestudies/Atlantis_final_10August2010.pdf

DWA (2010c) Potential artificial recharge schemes: Planning for implementation. http://www.artificialrecharge.co.za/casestudies/Activity_No_36Report_26_RM_161210_low%20res.pdf

DWA (2010d) Artificial recharge: Recent initiatives in Southern Africa. http://the-eis.com/elibrary/sites/default/files/downloads/literature/Artificial%20Groundwater%20Recharge%20recent%20Initiatives%20SA%20DWA%202010.pdf

DWA (2010e) Groundwater strategy 2010. https://www.dws.gov.za/Groundwater/Documents/GSDocument%20FINAL%202010_MedRes.pdf
DWA (2011a) Reconciliation strategy for Carnarvon town in Kareeberg local municipality in the lower orange WMA 2011. Contract WP 9713 Development of reconciliation strategies for all towns in the central region. Pixley Ka Seme district municipality I the Northern Cape province
DWA (2011b) Reconciliation strategy for Sutherland town in Karoo Hoogland local municipality in the lower orange WMA 2011. Contract WP 9713 Development of reconciliation strategies for all towns in the central region. Namakwa district municipality I the Northern Cape province
DWA (2011c) Reconciliation strategy for Vanwyksvlei town in Kareeberg local municipality in the lower orange WMA 2011. Contract WP 9713 Development of reconciliation strategies for all towns in the central region. Pixley Ka Seme district municipality I the Northern Cape province
DWA (2011d) Reconciliation strategy for Williston town in Karoo Hoogland local municipality in the lower orange WMA 2011. Contract WP 9713 Development of reconciliation strategies for all towns in the central region. Namakwa district municipality I the Northern Cape province
DWAF (Department of Water Affairs and Forestry) (1999) Water quality classes: Quality of domestic water supplies Volume 1: Assessment Guide. https://www.wrc.org.za/wp-content/uploads/mdocs/TT-101-98.pdf.
DWAF (2004) License No B191/2/450/1/2
DWAF (2007) Artificial recharge strategy, version 1.3. In: Strategy development: a national approach to implement artificial recharge as part of water resource planning, Pretoria, South Africa. http://www.artificialrecharge.co.za/strategydocument/yearRStrategySectD.pdf
DWAF (2009) Strategy and guideline development for national groundwater planning requirements. Potential Artificial Recharge Areas in South Africa, PRSA000/00/11609/1-Activity 14 (AR04)
DWS (Department of Water and Sanitation) (2015a) Reconciliation strategy for Carnarvon town in Kareeberg local municipality in the lower orange WMA 2015. Contract WP 10589 Development of reconciliation strategies for all towns in the central region. Pixley Ka Seme district municipality in the Northern Cape province
DWS (2015b) Strategy for Vanwyksvlei town in Kareeberg local municipality in the lower orange WMA 2015. Contract WP 10589 Development of reconciliation strategies for all towns in the central region. Pixley Ka Seme district municipality in the Northern Cape province
DWS (2015c) Reconciliation strategy for Williston town in Karoo Hoogland local municipality in the lower orange WMA 2015. Contract WP 10589 Development of reconciliation strategies for all towns in the central region. Namakwa district municipality in the Northern Cape province
DWS (2016a) National groundwater strategy. https://www.dws.gov.za/Groundwater/Documents/NGS_Draft-Final_04012017.pdf
DWS (2016b) Reconciliation strategies—all towns. Directorate National Water Resource Planning, Department of Water and Sanitation, Pretoria, South Africa.
DWS (2018) National water and sanitation master plan volume 1: call to action. https://www.gov.za/sites/default/files/gcis_document/201911/national-water-and-sanitation-master-plandf.pdf
Dennis I, Dennis R (2012) Climate change vulnerability index for South African aquifers. Water SA 38(3):417–426
Dillon P, Stuyfzand P, Grischek T, Lluria M, Pyne RDG, Jain RC, Bear J, Schwarz J, Wang W, Fernandez E, Stefan C, Pettenati M, van der Gun J, Sprenger C, Massmann G, Scanlon BR, Xanke J, Jokela P, Zheng Y, Rossetto R, Shamrukh M, Pavelic P, Murray E, Ross A, Bonilla Valverde JP, Palma Nava A, Ansems N, Posavec K, Ha K, Martin R, Sapiano M (2019) Sixty years of global progress in managed aquifer recharge. Hydrogeol J 27(1):1–30. https://doi.org/10.1007/s10040-018-1841-z
Ebrahim GY, Lautze JF, Villholth KG (2020) Managed aquifer recharge in Africa: taking stock and looking forward. Water 12:238–245. https://doi.org/10.3390/w12071844
Edokpayi JN, Makungo R, Mathivha F, Rivers N, Volenzo T, Odiyo JO (2020) Influence of global climate change on water resources in South Africa: toward an adaptive management approach. In: Singh P, Milshina Y, Tian K, Gusain D, Bassin JP (eds) Water conservation and wastewater

treatment in BRICS nations—technologies, challenges, strategies and policies. Amsterdam, Elsevier, pp 83–115
Esterhuyse CJ (2014a) Results of the drilling and yield testing programmes conducted at Carnarvon, Northern Cape Province. SRK Project Number 472351 November 2014
Esterhuyse CJ (2014b) Williston hydrocensus and groundwater exploration report. Report number SRK 471367/Draft June 2014
Esterhuyse CJ (2015) Vanwyksvlei IRS-Enhanced Recharged Proposal for Saaipoort
Esterhuyse CJ (2019a) Williston drilling and yield testing report, Northern Cape Province Report prepared for Neil Lyners and Associates (RF) (Pty) Ltd. Report Number SRK 483634_Draft
Esterhuyse CJ (2019b) Carnarvon Water Borehole Drilling and Yield testing report, Northern Cape Province. Report number SRK 538207-Final
Esterhuyse CJ, Fortuin M, Visser D (2004) A preliminary assessment of groundwater resources and production boreholes at Sutherland. Report No 342224
Fourie F (2021a) A changing environment focusing on drought. In: 17th Biennial groundwater conference. Menlyn Maine/ Online Pretoria, Gauteng. https://gwd.org.za/abstracts?combine=fourie&keys=&field_gwd_year_presented_value=&items_per_page=10
Fourie F (2021b) Aquifer recharge enhancement schemes in Sutherland Report No: 14/14/3/5/3/3/2
Fourie F, Hohne D (2021) Emergency drought intervention report for Sutherland Draft V3 Report No:14/14/3/5/3/3/2
Fourie F, Vermaak N, Hohne D, Murray R (2016) Authorising MAR projects within the context of South Africa's National Water Act. In: Proceedings of the 9th international symposium on managed aquifer recharge (ISMAR9). Mexico City, Mexico
Hohne D, Esterhuyse C, Fourie F, Gericke H, Esterhuyse S (2021) Enhancing groundwater recharge in the main Karoo, South Africa during periods of drought through managed aquifer recharge. J Afr Earth Sc 176:104007
Hohne D (2016) Preliminary report Sutherland Hydro-census. GH4234
Hohne D, Fourie F (2021) Assuring supply during a changing environment focussing on drought. In: 17th Biennial groundwater conference. Menlyn Maine/Online Pretoria, Gauteng. https://gwd.org.za/abstracts?combine=hohne&keys=&field_gwd_year_presented_value=.&items_per_page=10
IRS (Implementation Readiness Study) (2012) Kareeberg bulk water supply. Scoping Option Analysis 14 September 2012
Johnson MR, Anhauesser CR, Thomas RJ (2006) The geology of South Africa. Geological Society of South Africa
Lehohla P (2016) The state of basic service delivery in South Africa: In: Depth analysis of the Community Survey 2016 data. https://www.statssa.gov.za/publications/Report%2003-01-22/Report%2003-01-222016.pdf
Louw D (2022) Pers Comm. Karoo Hoogland Municipality
Mulder D (2021) Groundwater investigation at Sutherland in the Northern Cape, for the Towns emergency bulk water supply. Geoss Report No: 2021/12-12
Murray EC (2002) The feasibility of artificial recharge to the Windhoek aquifer. Unpublished Doctor of Philosophy dissertation thesis, University of Free State, South Africa
Murray R (2007) Bitou municipality groundwater management and artificial recharge feasibility study. http://www.artificialrecharge.co.za/casestudies/plettenberg.pdf
Murray EC (2016) An overview of managed aquifer recharge in Southern Africa. International Association of Hydrogeologists Newsletter, https://recharge.iah.org/files/2018/02/Southern-Africa-Murray_MAR-in-Southern-Africa-2016.pdf
Murray K (2021) Water Use License Application Geohydrology Assessment: Calvinia, Hantam Municipality, Northern Cape. Geoss report No 2021/07–25
Murray EC, Tredoux G (1998) Artificial recharge: a technology for sustainable water resource development. Water Research Commission Report, number 842/1/98, http://www.wrc.org.za/wp-content/uploads/mdocs/842-1-98.pdf

Murray EC, Tredoux G (2002) Pilot artificial recharge scheme: testing sustainable water resource development in fractured aquifers. Water Research Commission Report, number 967/1/02, http://www.wrc.org.za/wp-content/uploads/mdocs/967-1-02.pdf
Murray R, Harris J (2010) Water banking—a practical guide to using artificial groundwater recharge. https://www.artificialrecharge.co.za/booklet/AR_Booklet_13Jan2011.pdf
Murray R, Ravenscroft P, Munch Z (2009) Potential artificial recharge areas in South Africa. https://www.artificialrecharge.co.za/potential/AR_areasREPORT.pdf
Mvandaba V, Mwenge Kahinda J, Nzuza P, Hobbs P (2019) Green Book—the impact of climate change on groundwater availability. Technical report, Pretoria: CSIR
Nel M (2017) Groundwater: the myths, the truths and the basics. Water Research commistion SP108/17. https://www.wrc.org.za/wp-content/uploads/mdocs/Groundwater%20book_web.pdf
Neumann ER, Svensen H, Galerne CY, Planke S (2011) Multistage evolution of dolerites in the Karoo large igneous province, Central South Africa. J Petrol 52(5):959–984
Nkosi M, Mathivha FI, Odiyo JO (2021) Impact of land management on water resources, a South African context. Sustainability 13(2):701
Potgieter D (2014) National groundwater quality monitoring project lower orange water management area 1996–2013. GH Report: GH4174 https://www.dws.gov.za/ghreport/Home/View.aspx?Id=7115
RSA (1998) National Water Act. https://www.gov.za/sites/default/files/gcis_document/201409/a36-98.pdf
Redelinghuys N (2016) Effects on communities: the social fabric, local livelihoods and the social psyche. In: Hydraulic fracturing in the Karoo: critical legal and environmental perspectives. Glazewski and Esterhuyse. JUTA, Cape Town
RSA (Republic of South Africa) (1971) Water Research Act. https://www.gov.za/sites/default/files/gcis_document/201504/act-34-1971.pdf
Roffe SJ, Fitchett JM, Curtis CJ (2020) Investigating changes in rainfall seasonality across South Africa: 1987–2016. Int J Climatol 41:E2031–E2050
Schulze RE (2005) Climate change and water resources in southern Africa. Water Research Commission report No. 1430/1/5. http://www.wrc.org.za/wp-content/uploads/mdocs/14302.pdf
Seward P (1983) The groundwater potential of the Williston area. Department of Water Affairs and Forestry. Division of geohydrology. Technical geohydrology report 3288
Seward P (1986) Motivation to study the geohydrological characteristics of the dolerite basin North-East of Williston. Department of Water Affairs and Forestry. Division of geohydrology. Technical geohydrology report 3470
Shepard D (2019) Global warming: severe consequences for Africa: new report projects greater temperature increases. Africa Renewal 32(3):34–34
Tredoux G, Murray EC, Cave L (2002) Infiltration systems and other recharge systems in Southern Africa. In: Tuinhof A, Heederik (eds.)Management of aquifer recharge and subsurface storage. Netherlands National Committee of the International Association of Hydrogeologists [NNC-IAH], publication number 4, https://www.hydrology.nl/images/docs/iah/publications/4_Management_of_Aquifer_Recharge_and_Subsurface_Storage.pdf
Van der Westhuizen CR, Visser D, Esterhuyse C (2007) Report on the feasibility of various short term options to improve the potable water supply to Vanwyksvlei. SRK Report No: 374919 (Draft V1)
Van Wyk E (2013) Correlations between rainwater and groundwater geochemistry signatures with reference to episodic rainfall events in semi-arid environments, South Africa. In: Abiye T (ed.) The use of isotope hydrology to characterize and assess water resources in South Africa. Water Research Commission report TT570/13, pp 102–110
Visser D, Fortiun M, Esterhuyse CJ (2004) Final report on assessment of groundwater resources around Vanwyksvlei March 2004 Report No 319056 Final
Vogel C, van Zyl K (2016) Drought: In search of sustainable solutions to a persistent, 'wicked' problem in South Africa. In: Climate change adaptation strategies–an upstream-downstream perspective. Springer, Cham, pp 195–211

Woodford A, Rosewarne P, Girman J (2005) How much groundwater does South Africa have? In: SANCIAHS conference. Midrand

Xi X (2014) A review of water isotopes in atmospheric general circulation models: Recent advances and future prospects. Int J Atmos Sci V2014:250920

Chapter 11
Artificial Recharge of Groundwater in Tunisia: A Long and Fruitful Experience

Faten Jarraya-Horriche and Habib Chaieb

Abstract Groundwater Artificial Recharge (AR) is considered as an alternative to safeguard groundwater against their overexploitation and quality degradation. This practice which has demonstrated positive results in many aquifers around the world could be helpful for water management mainly in arid and semi-arid areas such as Tunisia. Climate change impact makes the situation even worse; it is expected that groundwater recharge will be reduced and groundwater salinization will be increased due to seawater intrusion. In Tunisia, AR experiments started since the 70thand it has been considered among the national strategy actions for water development since the 90th. Several techniques of AR have been used such as the infiltration of water throughout the unsaturated zone in basins or water course beds, or recharge wells. However, the AR amount which is dependent on the surface water availability is variable in space and time. Thus, the AR effectiveness is generally not achieved compared to the expectations. The total volume of water used for AR in Tunisia, was equal to 1026×10^6 m^3 during 1992–2019. The annual volume used for the year 2019 was around 58×10^6 m^3 which has been practiced for 30 aquifers, approximately. In order to quantify the AR impact, monitoring networks have been set up all around the sites to follow-up the groundwater level and water quality variability. A significant positive impact has been recorded in many cases, which has encouraged the decision makers to continue promoting the AR practices. The long experience of AR practices in Tunisia should be considered as instructive and could be further developed in order to achieve the maximum benefits. It will be interesting to put more emphasis on the assessment of their impacts in order to improve their effectiveness. Research actions using modeling tools coupled with new technologies, should be further developed in this context.

F. Jarraya-Horriche (✉)
Georesources Laboratory, Centre for Water Research and Technologies, Borj Cedria Echopark, BP273 Soliman, Tunisia
e-mail: faten.horriche@topnet.tn

H. Chaieb
Free Water Resources Expert, Ex Chief of the Hydraulic and Rural Equipment Division in CRDA of Ben Arous, Ben Arous, Tunisia

D. Saha et al. (eds.), *Managed Groundwater Recharge and Rainwater Harvesting*, Water Resources Development and Management,
https://doi.org/10.1007/978-981-99-8757-3_10

Keywords Groundwater · Artificial recharge · Overexploitation · Monitoring · Tunisia

11.1 Introduction

Groundwater represents the most important freshwater on the earth with an amount of 11×10^6 km^3 representing almost 25% of the total freshwater whereas the rest is in the form of ice that are more or less unusable by humans (Kotwicki 2009). In fact, accessible freshwater represents only 1% of the global volume of water in the earth. Despite the technical and economic investments that are necessary for groundwater access unlike surface water, groundwater represents the most important water resource in arid and semi-arid regions (Margat and Gun 2013; Salar et al. 2018; Hamed et al. 2018; Agoubi 2021). In these regions, surface waters are limited and are often dependent on climatic conditions that often endured by erratic precipitations (Zammouri and Feki 2005; Al-Assa'd and Abdulla 2010; Hashemi et al. 2015; Bird et al. 2016; Sellami et al. 2016; Hamed et al. 2018; Ibn Ali et al. 2019; Agoubi 2021). Thus, it is necessary to undertake national and regional studies to safeguard the current freshwater resources in these regions which are subject to climate change impact (Ibn Ali et al. 2017; Hussain et al. 2019). Particular attention should also be considered to develop and manage water resources ensuring their sustainability in the face of their scarcity and growing needs (Quintana Ashwell et al. 2018; Singh et al. 2019; Hamed et al. 2022).

During the last decades, groundwater registered an overexploitation and a deterioration of its quality (Asano 1985; Al-Assa'd and Abdulla 2010) which is due to the increase of freshwater demand as a result of economic development mainly for irrigation purposes which are the higher consumer of groundwater with an average of about 70% (Margat and Gun 2013; FAO 2011; Mokarram et al. 2021; Aloui et al. 2022). In such conditions, the groundwater Artificial Recharge (AR) could help the restoration of the groundwater equilibrium partially or totally and the development of non-conventional water resources. AR consists in storing water in underground reservoirs when it is available by using several techniques to promote infiltration of water (Asano 1985; Bouwer 2002; Hamed et al. 2022). We can notice different types of AR such as direct infiltration throughout the unsaturated zone, indirect infiltration by enhancing the recharge of rainfall in the bed of water courses, injection of water in wells, or increasing the storage capacity of groundwater reservoir by underground dams. To achieve the optimal performance of AR, several crucial matters should be considered before the implementation of such a project. We can cite those in relation with the geomorphological and hydrogeological parameters, the hydrodynamic and quality of the groundwater, and the origin of the water planned to be used as AR (Brahim et al. 2017; Sallwey et al. 2019; Jarraya-Horriche and Benabdallah 2020; Jarraya-Horriche et al. 2022, Khan et al. 2020; Mokarram et al. 2021; Mahmoudi et al. 2021). Surface waters which are considered conventional water resources are often used for AR during excessive rainfall periods. However, Treated Wastewater

(TWW), an unconventional water resource that is available throughout the year, could be used in a sustainable way if it represent a suitable quality for recharge.

Groundwater AR is becoming an indispensable alternative to surmount the problem of overexploitation. This technique is practiced in many regions over the world and it has demonstrated good results on groundwater level and quality (Bouri and Ben Dhia 2010; Abdalla and Al-Rawahi 2013; Ibn Ali et al. 2019; Dillon et al. 2019). It represents an appropriate measure to support the groundwater management at local and basin levels and to protect groundwater against the impact of climate change (Rupérez-Moreno et al. 2017; Dillon et al. 2019). Although the assessment of the climate change impact on groundwater is complicated compared to its impact on surface water, it is obvious that harmful effects, direct or indirect, are confirmed in several aquifers (Hamed et al. 2018). Even more, along the coastal areas where population are often concentrated (Hussain et al. 2019; Agoubi 2021), the overexploitation leading to groundwater level depletion, causes a deterioration of groundwater quality due to seawater intrusion (Bear et al. 1999; Kouzana et al. 2009; Ben Hamouda et al. 2011; Abdalla and Al-Rawahi 2013; Hussain et al. 2019; Agoubi 2021; Abdelkarim et al. 2022).

The quantification of the AR impact on groundwater is a challenging issue as it depends on several inputs related to physical systems and the variability of natural and artificial inflows and outflows which are often poorly assessed in space and time (Singh et al. 2019; Jarraya-Horriche et al. 2020). Several methods are used to assess the impact of AR. We can cite the hydrological water balance which is used to establish the global inflow and outflow during a specified period, the monitoring of groundwater levels and quality, the hydrochemistry and isotopic analyses, and the numerical modeling (Shanafield and Cook 2014; Hamed et al. 2014; Jarraya-Horriche and Bogacki 2017; Ibn Ali et al. 2017; Zammouri et al. 2018; Melati et al. 2019; Jarraya-Horriche et al. 2020).

The biggest problem of AR in porous media is the clogging which results in the reduction of the rates of infiltration (Bouwer 2002; Barry et al. 2017; Chu et al. 2019). This phenomenon is due to several parameters related to physical, biological, and chemical processes. The suspended solid in recharged waters remains the most relevant one mainly in the case of using rainfall water. Several techniques of filtration are thus often considered to avoid this pattern like water pretreatment, the use of sands or membranes, periodic disinfection, or regular pumping in recharged wells (Bouwer 2002; Barry et al. 2017).

In this chapter, focus is given to the case study from Tunisia and related issues are highlighted. Tunisia is characterized by variable economic (agricultural, industrial, and touristic) activities which are high consumers of water. Population is more concentrated along the coastal regions where economic activities are the most present. Since the 90th, national strategies were set up to manage water resources and to develop guidelines for recognition of further groundwater resources and to increase groundwater recharge by direct or indirect methods. These actions should ensure the availability of water resources to meet the increase in freshwater demand and to deal with the impact of climate change (DGRE 1990; MARH and GIZ 2007; MALE 2015). Thus, the AR of groundwater was set up among the national strategies and several sites were identified to practice the AR which has been expanded according to the requirements and the availability of water.

11.2 Experimentations of AR in Tunisia

AR of groundwater was first initiated in Tunisia in the 70th (Chaieb 1995a, 2014a; Bouri and Ben Dhia 2010) to overcome a high decline in groundwater levels in some areas due to groundwater overexploitation. The first experimentations of AR were practiced in the Mateur aquifer in north Tunisia, which was heavily used for drinking water supply (Chaieb 1995a). A volume of 4,000 m^3 was first injected over 50 h in a shallow well dug in the Campanian limestone of Ras El Ain. Experimentation of AR in this aquifer was also done in the Khelij water course bed by constructing a basin of 100 m length and 5 m width. A volume of about 65,000 m^3 pumped from the Joumine water course bed, has been injected in the basin for 65 days. This operation allowed the assessment of the infiltration capacity to an average of 10 L/s. In 1984 and 1988, the AR was relaunched following the occurrence of the emergency from the limestone structure of the Joumine dam (Rassas 1988). The water flow which was initially around 320 L/s has significantly decreased to 70 L/s. A monitoring network of wells allowed the assessment of the AR impact on groundwater, and a rise in groundwater levels was measured between 0.35 and 2.6 m/year (Chaieb 2014a).

The AR in the Teboulba coastal aquifer, in the center of Tunisia, started in 1971 with the use of surface water from the Nebhana dam, following the impact of the uncontrolled groundwater exploitation (Chaieb 1995a; Jarraya-Horriche 1995; Bouri and Ben Dhia 2010). This exploitation has resulted in a decline in groundwater level and degradation of the groundwater quality due to the intrusion of saline water from the sea and the nearby sebkha of Moknine (a salty water lake). In 1971, before the practice of the AR, the groundwater level was equal to 30 m below the seawater level in the heavily exploited regions (Jarraya-Horriche 1995; Bouri and Ben Dhia 2010). The annual recharged volumes ranged between 20,000 m^3 and 400,000 m^3 during the period 1971–2000 with an interruption between 1979 and 1991 because of the lack of water in the dam. Water was initially injected in four wells with a volume of 186,000 m^3 to assess the infiltration capacity (Chaieb 1995a). After that, a total volume of 1.5×10^6 m^3 was injected in several wells (reaching 17 wells) during the period 1972–1978. It was noted that clogging due to suspended matters has resulted in a reduction in water infiltration rates in the wells. Despite this limitation, the groundwater recovery was quite noticeable in the previous depression areas and it has reached values of less than 5 m below the seawater level (Chaieb 1995a; Bouri and Ben Dhia 2010). In addition, groundwater salinity registered a significant reduction (i.e., quality improvements) in the zones in the vicinity of the injection wells. During the period 1992–2008, the total volume of recharged water was about 4.5×10^6 m^3 (DGRE 2014). However, in 2009; the water of Nebhana dam was allocated to drinking water supply and irrigation and the AR was then stopped.

Fig. 11.1 AR in the underflow oued siliana using dikes (Chaieb 2014a)

The AR was also practiced in the aquifer of Underflow Oued Siliana in 1972 where six artificial dikes (1 m high and 120 m width) were constructed in the upstream area to enhance the recharge (Chaieb 1995a) (Fig. 11.1). Differential hydrometric stations were constructed in the Siliana water course to assess the infiltration flow. The groundwater level registered a rise varying between 1.0 and 2.4 m/year (Chaieb 2014a). The second AR campaign was done in 1973 during which infiltration capacity has increased to 60 L/s following the systematic cleaning of the water course bed after each AR event.

The first experimentations of AR in the Cap Bon (Northeastern Tunisia) were started in 1975 in Menzel Bouzelfa-Beni Khalled (Grombalia aquifer) using water from the Bezigh dam, which was injected in wells and infiltrated from reservoir bottom (Chaieb 1995a). A monitoring network of 12 piezometers was set up to follow the impact of the AR on the groundwater level. The increase in groundwater level near the AR areas varied between 0.5 and 6 m/year (Chaieb 2014a). The AR was stopped because of the clogging problem of the basin's infiltration area due to suspended matters and the collapse of some wells, despite the positive impact of the AR. In 1984, the AR was relaunched in the Sidi Said water course bed (south of Menzel Bouzelfa) by injecting water into several wells leading to a rise of groundwater level varying between 2 and 3 m (Chaieb 1995a). Following these AR experimentations, it has been deduced a recharge capacity in these regions, ranging between 1 and 4 L/s as recharge in the wells and between 1 and 2 m^3/day as infiltration in the basins.

The AR of the Kairouan aquifer was first experimented with in 1988, following the decrease of natural recharge in the Zeroud and Merguellil water course beds

Fig. 11.2 AR in Zeroud and Merguellil water course beds (Kairouan aquifer) (Chaieb 2014a)

due to the construction of the Sidi Saad and El Houareb dams in 1982 and 1988, respectively (Jarraya-Horriche 2004; Nazoumou and Besbes 2005) (Fig. 11.2). The water was spread in the Zeroud water course bed from the Sidi Saad dam in 1988 to assess the infiltration capacity (Chaieb 1995b, 2014a; Nazoumou and Besbes 2005). Between 1988 and 1996, six AR events were done (5 in Zeroud and 1 in Merguellil water course beds) with a total recharged volume of about 90×10^6 m^3 (Nazoumou and Besbes 2005). More than 1,000 surface water flow measurements were done to assess the amount of infiltrated water (Saadaoui et al. 1989; Bouzaiane et al. 1993; Chaieb 1995b). During the period 1997–2007, the volume of surface water released from Sidi Saad dam was about 20×10^6 m^3 (Salem et al. 2012). A significant rise of groundwater level reaching 5 m, was registered after a recharge of 13×10^6 m^3 by the water release in the Zeroud water course bed for about five months (May–Sep. 2005) (Salem et al. 2012).

Several techniques were used during all these AR experimentations such as wells, natural/constructed basins, and beds of the water courses (Chaieb 2014b; Aloui et al. 2022). It was concluded that infiltration from spreading water was the best technique as it could recharge up to 80% of the released water (Chaieb 2014a). These AR cases were instructive and a recovery of groundwater level was registered which encouraged the decision makers to consider these practices among the important axis of water management in Tunisia in order to sustain the groundwater resources and optimize the water management.

After assessing the AR several studies, it has been confirmed the effectiveness of the use of TWW for AR when conventional freshwater is getting short (Bouwer 2002; Cazurra 2008; Jarraya-Horriche and Benabdallah 2020; Valhondo et al. 2020). In Tunisia, the first experience of AR with TWW was started in Dec. 1985 in Oued Souhil (Nabeul–Hammamet aquifer) (Fig. 11.3a) to protect the groundwater resources against seawater intrusion and to enhance the reuse of large amounts of TWW initially discharged into the sea (Kallali et al. 2007; Chaieb 2014a, b). This project consists of 4 infiltration basins with an area of about 350 m^2 and a depth of about 1.6 m, each. The TWW was reused from the Wastewater Treatment Plant (WWTP) of Beni Khiar SE4 (Fig. 11.3b). The control of the AR impact was ensured by 18 piezometers. The first experiment was done over 69 days using a volume of about 58,000 m^3. The infiltration capacity of the basins has been assessed to 1.5 m^3/

Fig. 11.3 **a** AR in Oued Souhil (Nabeul–Hammamet aquifer) (Chaieb 2014a); **b** WWTP of Beni Khiar SE4 (Nabeul)

m^2/day (Chaieb 1995a). This AR has been dropped since 2012 due to quality issues of the TWW and the aging of the adduction network.

The use of TWW in irrigation in several areas has demonstrated a beneficial impact on the groundwater. Despite the objective of these practices is not to recharge the groundwater, it is not obvious to consider them as AR sites but these practices should be taken into consideration to assess their impact on groundwater. In Tunisia, the impact of these practices was first observed in 1965 following the irrigation of the Soukra public area with TWW from the Charguia WWTP (Chaieb 2014b).

11.3 Historical Practices of AR

Groundwater AR was therefore launched in Tunisia as a strategic axis in 1992, in seven sites using the excess of surface water of about 24×10^6 m^3/year during rainy periods (DGRE 2013). The recharged aquifers were located mainly in both the north of Tunisia such as Ras Jabel, Mornag, Nabeul–Hammamet, and Grombalia, and in the center of Tunisia such as Nadhour and Teboulba aquifers. Groundwater monitoring networks were set up to assess the impact of AR on groundwater level and quality. These practices were then subsequently multiplied and were expanded throughout the Tunisian territory (Fig. 11.4a). Furthermore, in the framework of the TWW valorization strategy, two experiences (Oued Souhil and Korba-elmida) have been practiced by using TWW for AR due to the shortage of surface water resources. Some other projects are planned in different sites in the framework of the national strategy of TWW valorization.

During the period 1992–2019, the total recharged volume was about 1026×10^6 m^3 (DGRE 2019) which represents three times the annual groundwater resources of the recharged aquifers (350×10^6 m^3/year), and about the half of the total groundwater resources in Tunisia (2200×10^6 m^3/year). The annual recharged volume varies according to the availability of water sources. In fact, it reached the maximum value, a little more than 60×10^6 m^3/year, during the rainy years, like 1995–1996

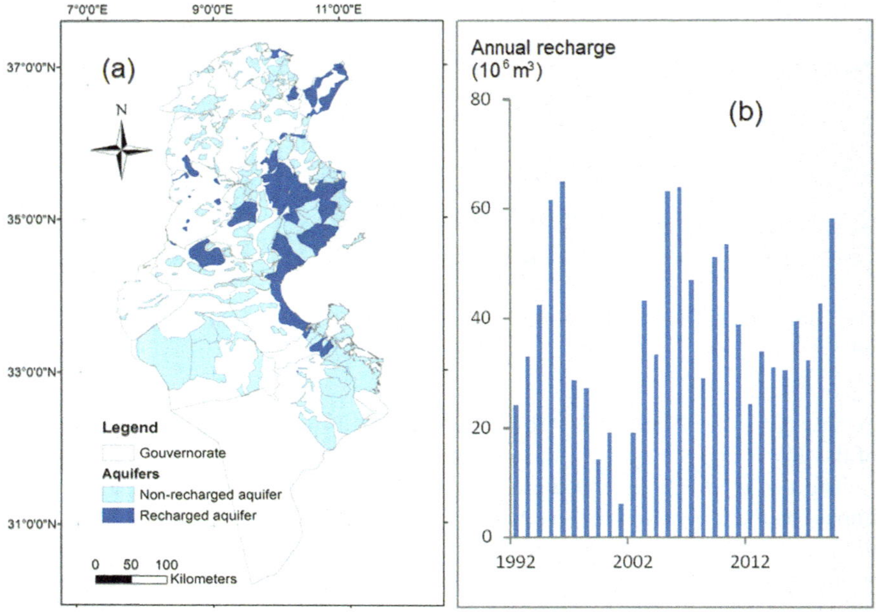

Fig. 11.4 Artificial recharge of groundwater in Tunisia: **a** Recharged aquifers; **b** Variation of AR annual volumes

and 2005–2006 (Fig. 11.4b). However, during some years, the AR has been very low due to drought, such as in 1999–2001 during which the annual recharged volume decreased to 20×10^6 m^3/year. A total number of 44 aquifers have benefited from AR during the period 1992–2019. The annual number varies according to the planned program at the regional scale and mainly to the availability of water to be used for AR. In fact, it varied between 7 in the beginning (1992) and 34 in 2019 with an average of 16 sites/year (Fig. 11.5). However, high numbers don't reflect usually the availability of water resources such as in 2008 during which the recharged volume registered a decrease compared to the three years before despite the number of recharged aquifers has risen. This is mainly due to the shortage of surface water availability in 2008 as a result of rainfall decrease. In fact, the efficiency of AR could be assessed by comparing the recharged amount to the number of recharged aquifers per year. Thus, we registered a high value of 7.7×10^6 m^3 per aquifer during 1995 corresponding to the period of AR strengthening according to the national strategy of water resources. The low value corresponds to 2001, during which only a volume of 0.6×10^6 m^3 was used per aquifer. This year has indeed been a dry one mainly in the center and north of Tunisia and only ten aquifers have been recharged with a total volume of 6.2 $\times 10^6$ m^3. One can see the example of Kairouan aquifer which is the most recharged one in Tunisia (Fig. 11.6). An average of 2.5×10^6 m^3 per aquifer has been used for AR during the period 1992–2019.

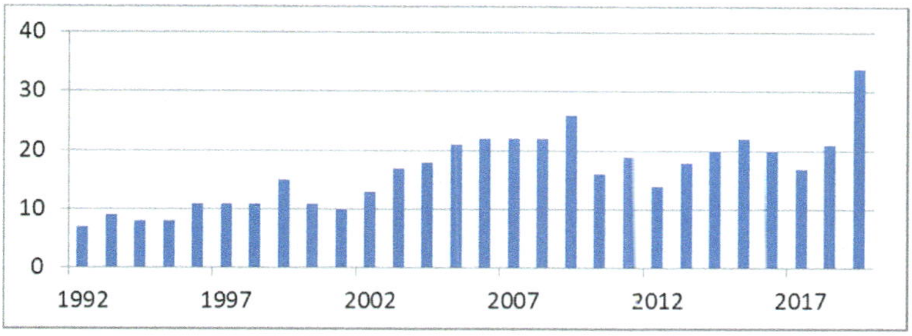

Fig. 11.5. Variation of annual recharged aquifer numbers

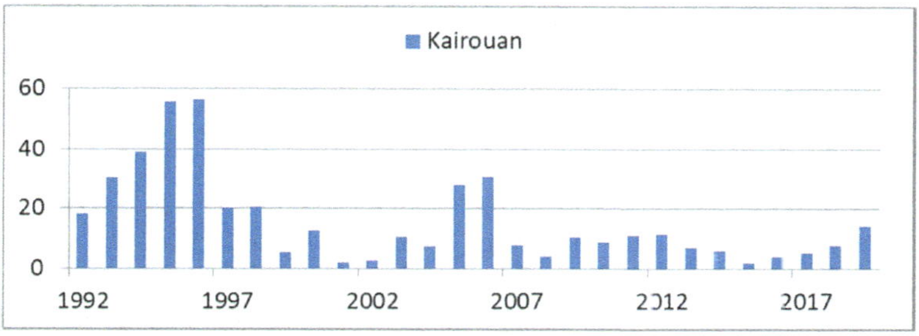

Fig. 11.6. Variation of recharged volumes in Kairouan aquifer

The impact of AR was significant for several aquifers since we registered an important rise in groundwater level and an improvement of groundwater quality. This impact is variable depending on the recharged amount and the technical approach used for the AR. In fact, many approaches were used in Tunisia and each of them has resulted in more or less impact on groundwater, but surely a positive impact.

11.3.1 Artificial Recharge Using Wells

The AR experiences in shallow and deep wells were done in several areas like the aquifers of the Cap Bon (Grombalia and Eastern Coastal), Ras Jebel, Guenniche, Teboulba, and Bekalta. In the Cap Bon, some wells were first used for AR experimentation. Injection rates were around 3,000 m^3/month in Menzel Bouzelfa (Grombalia aquifer) and 4,500 m^3/month in Lebna (Eastern coastal aquifer) (Chaieb 1995a). In Ras Jebel aquifer where salinization of groundwater has been recorded as a result of marine intrusion, AR has been tested in 10 wells located in the Beni Ata region (Chaieb 1995a; Lachaal et al. 2022). The recharged water was brought from the hill dams of Beni Ata, Chouk El Felfel, and Chaabat Eddoud. The first experimentations in wells (Fig. 11.7) started in April–May 1993 with an injection rate of about

Fig. 11.7 Artificial recharge in well **a** Ras Jebel; **b** Oued El Khairat; **c** Zeuss-Koutine; **d** Chenchou

20 L/s. A water level rise of 1 m was registered in the nearby monitoring wells, following this experimentation. AR in wells were also done in Oued El Khairat aquifer (Ketata et al. 2014; Zammouri et al. 2018), Bekalta, Zeuss-Koutine and Chenchou (Chaieb 2014a) (Fig. 11.7).

11.3.2 Artificial Recharge into Basins and Quarries

The AR in quarries is done in several aquifers such as the Mornag aquifer (north Tunisia) (Fig. 11.8a, b) by discharging water in the abandoned quarry outcrops of Oligocene sandstone using water from Medjerdah-Cap Bon channel (Chaieb 2014a). In the Bekalta aquifer, (center Tunisia) and Ras Jebel aquifer (Fig. 11.8c) (north Tunisia), quarries of sand were used for AR which started in 1993.

Fig. 11.8 **a** AR in Mornag aquifer in Oligocene outcrops; **b** Medjerdah-Cap Bon Channel; **c** Sidi Guebbari quarry (Ras Jebel aquifer)

11.3.3 Artificial Recharge Through Water Course Beds

Infiltration in water course beds and cultivated soils is practiced in many regions in Tunisia, as presented in the following cases. In the aquifer of Mateur (north Tunisia) in the Khelij water course bed using water release from the Joumine dam. In the Nadhour-Saouef aquifer (Center Tunisia), the AR is done by water release from Essahel and EL Ogla dams to enhance the infiltration of water. During the period 1990–2000, 6 AR events were done by water release from El Ogla dam and 5 other events from Essahel dam with a total discharged volume between 50 and 43 L/s (Zammouri and Feki 2005). The flow rate was measured in 7 hydrometric stations and the groundwater monitoring was performed by Essahel well. The recharge efficiency was assessed at 96% with an infiltration rate varying between 390 and 680 mm/day (Zammouri and Feki 2005). The AR of the El Fahs aquifer, located in the same region, started in May 1996 using water release from the El Abaya dam. The monitoring network was composed of 13 shallow wells and four piezometers. A rise of about 2.5 m/year was registered in the piezometer of Oued Tahouna. During the period 2002–2015, the annual released water flow in the Oued El Khairat (Center Tunisia), varied between 1.2×10^6 m^3 and $10.7 \times 10^6 m^3$. According to the calibrated period modeling, the infiltration rates were computed to values ranging between 40 and 80%

of the released water flow (Zammouri et al. 2018).In the Chougafia aquifer (Centre Tunisia), the recharge was done by water release in the Oued El Fej with a flow rate of about 100 L/s. A monitoring network composed of 24 shallow wells, allowed the control of the AR impact. The impact varied between 0.4 and 1 m rise in groundwater level per year.

11.3.4 Artificial Recharge and Irrigation with TWW

Following the positive impact of AR with TWW in Oued Souhil, another project was thereafter launched in the site of Korba-Elmida (Fig. 11.9) in Dec. 2008, to create a barrier against the seawater intrusion and to protect the groundwater in upstream areas used mainly for irrigation purposes (Chaieb 2014b; Jarraya-Horriche and Benabdallah 2020; Jarraya-Horriche et al. 2020). This site is composed of three rectangular basins with an area of 1,500 m^2 each which were dragged to a depth of 1.5 m (Fig. 11.9). The TWW is brought from the WWTP of Korba located about 300 m in southeast of the AR site and at 1.5 km far from the sea. The daily recharged volume into the basins was planned to be 1500 m^3, however, the average value measured during 2009–2014 was about 850 m^3 due to the reduced duration of the recharge (Jarraya-Horriche et al. 2020).

Irrigation with TWW is also considered among the AR sites since it has an impact on groundwater. This impact is followed by monitoring networks around the irrigated areas with TWW such as those of El Hajeb-Sidi Abid started in 1989 (Fig. 11.10) and following the AR with TWW in the beds of Zrata and Ayoum water courses resulting in an increase of groundwater levels of Boumerdes aquifer in 2003 and Mahdia-Ksour Essef aquifer in 1994 (Neubert and Benabdallah 2003; Ben Brahim-Neji et al. 2014; Chaieb 2014b).

Fig. 11.9. AR with TWW in Korba-Elmida site (North Eastern aquifer of Cap Bon)

Fig. 11.10 Irrigation with TWW in Hajeb-Sidi Abid aquifer

11.3.5 Artificial Recharge with Hydraulic Structures

Another technique of AR has been noted following the construction of structures to preserve the water and soil and to control runoff and flooding. Although the objective of these structures is not to recharge the aquifer, we registered a significant impact on the groundwater levels in the vicinity of these structures, following rainfall events. We can cite some areas like the case of El Ferch, Metameur, Zeuss-Koutine, Gafsa Nord, in the south of Tunisia; and the El Fekka water course bed (Fig. 11.11a), Bouhafna and Zermedine-Beni Hassen in the center, and EL Fahs and Oued Siliana in the north (Chaieb 1995a, 2014a).

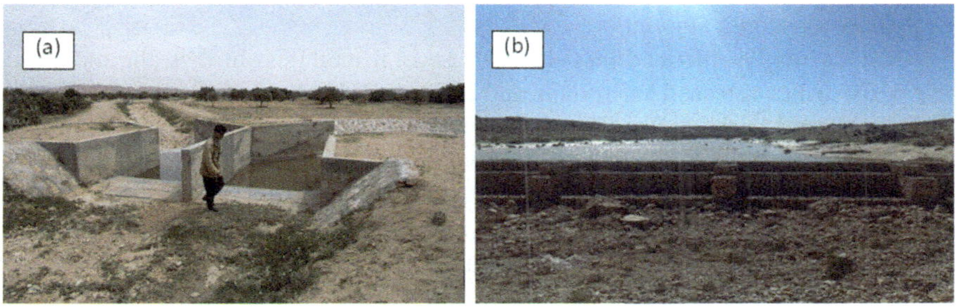

Fig. 11.11 **a** El Fekka water course bed **b** AR in Zeuss-Koutine aquifer

Fig. 11.12 Ongoing construction of the OumLagsab underground dam

11.3.6 Artificial Recharge Through Underground Dams

The only experience of this mechanism in Tunisia was the underground dam downstream of Oum Lagsab aquifer in the center of Tunisia. It has increased the storage capacity of the groundwater reservoir and has raised significantly the groundwater levels (DGRE 2005) (Fig. 11.12).

11.4 Current Status

According to the last status of AR in Tunisia (DGRE 2019), the annual recharged volume was around 58 × 10^6 m^3 which registered a rise of 15 × $10^6 m^3$ compared to the value used in 2018. Approximately, 30 aquifers located in the overall of Tunisia, have benefited from this AR. The used water was taken from dams (27 × 10^6 m^3), Medjerdah-Cap Bon channel (3 × 10^6 m^3) and TWW plants (2 × 10^6 m^3) (Fig. 11.13). We estimated an average volume of 15 × 10^6 m^3 and 12 × 10^6 m^3 used for AR, following the use of water and soil conservation structures and natural emergence recharge, respectively. During 2019, the AR contribution to the natural recharge was estimated to an average of 47% (DGRE 2019) which represents a valuable additional amount.

AR is currently practiced in Tunisia by several methods. In fact, the most used one is the water released from dam and infiltrated from the bed of water courses or from natural depressions. This most practiced method is due to the facility of recharging mainly when an excess of water is available in a nearby dam reservoir and also because of the large area and the high infiltration capacity as they represent the natural areas for rainfall infiltration. We notice also that the aquifer of Kairouan in the center of Tunisia is the most artificially recharged one which is done by water release from the Sidi Saad and EL Houareb dams, through the water course beds of Zeroud and Merguellil, respectively. The total recharged volume represents almost 25% of the whole volume used in 2019. AR in wells and managed/ constructed basins is quite low due to the small areas subject to water infiltration. Authorities

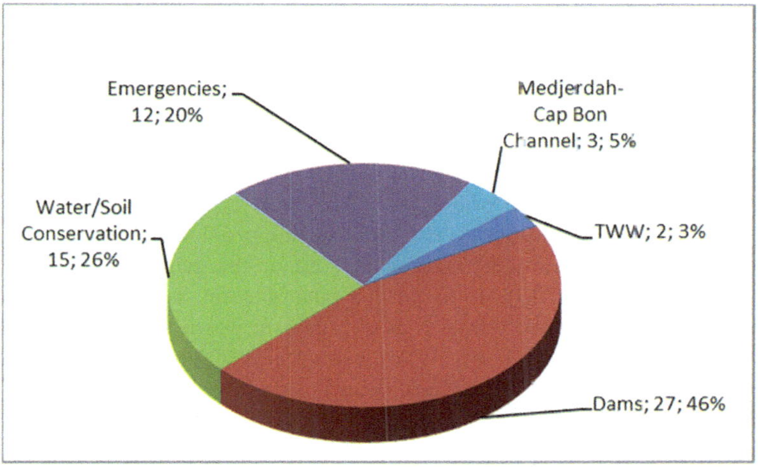

Fig. 11.13 Distribution of AR volumes ($\times 10^6 m^3/y$) according to the water sources

also used structures for the conservation of water and soils that are generally used mainly to promote the water infiltration as an irrigation purpose but also to benefit for the groundwater recharge. The use of TWW for AR is still low in Tunisia compared to the other techniques.

11.5 Lessons Learned from AR in Tunisia

The long experience of the AR practices in Tunisia has allowed for deepening the capacity of their management and the controlling of their impacts on groundwater. In fact, several experimentations were tested using different technical approaches such as injection in wells or infiltration through basins and water release in water course beds. Also, the water used for AR was brought from several sources such as dam reservoirs, channels, and TWW. These experimentations have allowed the assessment of the infiltration/recharge capacity and the setup of a strategy plan for AR. Monitoring networks were often used to assess the impact of the AR on groundwater, whether on groundwater level or quality.

Although this long experience of AR in Tunisia, the assessment of their impact is subject to uncertainty due to the lack of accurate data in some cases. In fact, the dynamic and the quality of the groundwater, which depend on several variables and parameters make it difficult to assess the AR impact. Among these variables, we can cite groundwater exploitation which is often poorly evaluated due to the shortage of real field control of a huge number of monitoring wells (Kerrou et al. 2006). The natural recharge of aquifer is another variable that should be known before the practices of AR in order to evaluate accurately the impact of AR regardless of the rainfall infiltration. Although the recharge assessment was done for some aquifers (Salem et al. 2012; Ghouili et al. 2017; Jarraya-Horriche et al. 2018; Jarraya-Horriche et al. 2022), it should be generalized to better evaluate the AR impact on groundwater. Another problem is related to the application of the AR program which is usually not implemented according to the initially planned. This is usually due to the availability of water mainly during dry seasons or drought periods; supply for drinking and irrigation has always been the priority compared to AR.

As for all AR projects in the world, the clogging problems always handicap the AR continuity due to mainly the suspended solids in the treated wastewater (Barry et al. 2017; Chu et al. 2019). To avoid this phenomenon, several technical works should be frequently carried out such as cleaning the basin and pumping in wells. If these works are not done correctly, we cannot prevent the clogging phenomena and we can risk irreversible damage to the AR structure. This case was registered in several sites in Tunisia such as Teboulba, Grombalia, and Korba aquifers (Chaieb 1995a; Bouri and Ben Dhia 2010; Jarraya-Horriche et al. 2020).

The assessment of the AR impacts using modeling that allows the prediction of the reliability of such projects is a challenging approach. The modeling could be used to simulate the AR as a hypothetical scenario before the setup of the AR project and to help the managers in the planning phase of the projects, like the works done by Fillion (1996), Jarraya-Horriche and Bogacki (2017), Jarraya-Horriche and Benabdallah (2020). It can also be used after real practices of AR by calibrating the model referring to real measurements in the monitoring networks such as the works undertaken by Nazoumou and Besbes (2005), Zammouri and Feki (2005), Zammouri et al. (2018). Thus, the modeling simulations would be more accurate and the results should be helpful for the decision makers to manage the AR projects and assess their reliability.

Despite the limitation of tools and sufficient funds, the experiences of AR undertaken in Tunisia are instructive and have allowed the development of supplementary non-conventional water resources where the shortage of conventional water resources could not meet the increasing water needs. The use of TWW for AR is not enough developed in Tunisia, as only two projects are now operational and particular procedures should be considered in such situations to avoid contamination of groundwater. However, these practices are planned to be promoted in the framework of the national strategy to enhance the valorization of the huge volume of TWW which is now discharged into the bed of water courses or deposed into the sea, except the amount reused for irrigation.

11.6 Conclusion

The practice of groundwater AR in Tunisia has been a successful experience and has contributed to the sustainability of groundwater resources. It has helped to reduce the impact of groundwater overexploitation and quality degradation in many aquifers. These results have encouraged decision makers to expand AR practices throughout the country. However, additional efforts need to be made in order to increase the efficiency of the AR and to achieve the expected results. Indeed, adequate spatiotemporal monitoring of several variables and parameters is needed to minimize uncertainty in assessing the impact of AR. These include injected volumes, pumping rates, natural recharge, clogging, etc. Modeling tools should be most useful when input data and field recognition are most accurate, mainly for calibration. Thus, the accuracy of model prediction would be better and decision makers could take it for the further management.

References

Abdalla OAE, Al-Rawahi AS (2013) Groundwater recharge dams in arid areas as tools for aquifer replenishment and mitigating seawater intrusion: example of AlKhod, Oman. Environ Earth Sci 69:1951–1962

Abdelkarim B, Telahigue F, Agoubi B (2022) Assessing and delineation of groundwater recharge areas in coastal arid area southern Tunisia. Groundw Sustain Dev 18:100760

Agoubi B (2021) A review: saltwater intrusion in North Africa's coastal areas—current state and future challenges. Environ Sci Pollut Res 28:17029–17043

Al-Assa'd TA, Abdulla FA (2010) Artificial groundwater recharge to a semi-arid basin: case study of Mujib aquifer, Jordan. Environ Earth Sci 60:845–859

Aloui D, Chekirbane A, Stefan C, Schlick R, Msaddek MH, Mlayah A (2022) Use of a GIS-multi-criteria decision analysis and web-based decision support tools for mapping and sharing managed aquifer recharge feasibility in Enfidha plain, NE of Tunisia. Arab J Geosci 15:658

Asano T (1985) Artificial recharge of groundwater. Butterworth, Boston

Barry K, Vanderzalm J, Miotlinski K, Dillon P (2017) Assessing the impact of recycled water quality and clogging on infiltration rates at a pioneering Soil Aquifer Treatment (SAT) Site in Alice Springs, Northern Territory (NT). Australia. Water 9:179

Bear J, Cheng AH-D, Sorek S, Ouazar D, Herrera I (1999) Seawater intrusion in coastal aquifers—concepts, methods and practices, theory and applications of transport in porous media. Springer Netherlands, Dordrecht

Bel Hadj Salem S, Chkir N, Zouari K, Cognard-Plancq AL, Valles V, Marc V (2012) Natural and artificial recharge investigation in the Zéroud Basin, Central Tunisia: impact of Sidi Saad Dam storage. Environ Earth Sci 66:1099–1110

Ben Brahim H, Horriche FJ, Slama F (2017) Multicriteria decision aid for choosing groundwater artificial recharge sites with treated wastewater: Case study of Mornag aquifer. In: Impacts of global changes on western mediterranean aquifers. presented at the congress on groundwater and global change in the western mediterranean, international association of hydrogeologists. Spanisch Chapter, Granada, Spain, p 6

Ben Brahim-Neji H, Ruiz-Villaverde A, González-Gómez F (2014) Decision aid supports for evaluating agricultural water reuse practices in Tunisia: the Cebala perimeter. Agric Water Manag 143:113–121

Ben Hamouda MF, Tarhouni J, Leduc C, Zouari K (2011) Understanding the origin of salinization of the Plio-quaternary eastern coastal aquifer of Cap Bon (Tunisia) using geochemical and isotope investigations. Environ Earth Sci 63:889–901

Bird DN, Benabdallah S, Gouda N, Hummel F, Koeberl J, La Jeunesse I, Meyer S, Prettenthaler F, Soddu A, Woess-Gallasch S (2016) Modelling climate change impacts on and adaptation strategies for agriculture in Sardinia and Tunisia using AquaCrop and value-at-risk. Sci Total Environ 543:1019–1027

Bouri S, Ben Dhia H (2010) A thirty-year artificial recharge experiment in a coastal aquifer in an arid zone: The Teboulba aquifer system (Tunisian Sahel). CR Geosci 342:60–74

Bouwer H (2002) Artificial recharge of groundwater: hydrogeology and engineering. Hydrogeol J 10:121–142

Bouzaiane S, Chaieb H, Maalel F (1993) Recharge des nappes de Kairouan par les lâchures du barrage de El Haouareb sur l'oued Merguellil, campagne du 19 avril au 8 mai 1993. Direction Générale des Ressources en Eau, Tunisie

Cazurra T (2008) Water reuse of south Barcelona's wastewater reclamation plant. Desalination 218:43–51

Chaieb H (2014a) Tunisian experience in artificial recharge. In: Report of intensive course on: "Aquifer Recharge", TEMPUS project "Water Education and Training in Maghreb." ENIT, Hammamet, Tunisia

Chaieb H (2014b) Tunisian Experience in Artificial Recharge using Treated Waste Water. Presentation. In: Regional conference on "Sustainable Integrated Wastewater Treatment and Reuse". Egypt

Chaieb H (1995a) L'expérience tunisienne en matière de recharge artificielle des nappes. Direction Générale des Ressources en Eau, Tunisie

Chaieb H (1995b) Note sur la recharge de la nappe de Kairouan par les lâchers du barrage de Sidi Saâd, campagne du 19 juin au 19 juillet 1995. Direction Générale des Ressources en Eau, Tunisie

Chu T, Yang Y, Lu Y, Du X, Ye X (2019) Clogging process by suspended solids during groundwater artificial recharge: evidence from lab simulations and numerical modelling. Hydrol Process 33:3226–3235

DGRE (2019) Annuaire de recharge artificielle des nappes en Tunisie (Rapport interne). Direction Générale des Ressources en Eau, Tunisie

DGRE (2014) Annuaire de recharge artificielle des nappes en Tunisie (Rapport interne). Direction Générale des Ressources en Eau, Tunisie

DGRE (2013) Annuaire de recharge artificielle des nappes en Tunisie (Rapport interne). Direction Générale des Ressources en Eau, Tunisie

DGRE (2005) Annuaire de recharge artificielle des nappes en Tunisie (Rapport interne). Direction Générale des Ressources en Eau, Tunisie

DGRE (1990) Stratégie pour le développement des ressources en eau de la Tunisie au cours de la décennie 1991–2000 (Rapport National). Direction Générale des Ressources en Eau, Tunis, Tunisie

Dillon P, Stuyfzand P, Grischek T, Lluria M, Pyne RDG, Jain RC, Bear J, Schwarz J, Wang W, Fernandez E, Stefan C, Pettenati M, van der Gun J, Sprenger C, Massmann G, Scanlon BR, Xanke J, Jokela P, Zheng Y, Rossetto R, Shamrukh M, Pavelic P, Murray E, Ross A, BonillaValverde JP, Palma Nava A, Ansems N, Posavec K, Ha K, Martin R, Sapiano M (2019) Sixty years of global progress in managed aquifer recharge. Hydrogeol J 27:1–30

FAO (2011) The state of the world's land and water resources for food and agriculture. FAO, Rome and Earthscan, London

Fillion E (1996) Modélisation de la nappe de Téboulba (Tunisie): Simulation des phases de recharge artificielle; Etude des intrusions salines depuis le sebkra et la mer. BRGM, France

Ghouili N, Horriche FJ, Zammouri M, Benabdallah S, Farhat B (2017) Coupling WetSpass and MODFLOW for groundwater recharge assessment: case study of the Takelsa multilayer aquifer, northeastern Tunisia. Geosci J 21:791–805

Hamed Y, Ahmadi R, Demdoum A, Bouri S, Gargouri I, Ben Dhia H, Al-Gamal S, Laouar R, Choura A (2014) Use of geochemical, isotopic, and age tracer data to develop models of groundwater flow: a case study of Gafsa mining basin-Southern Tunisia. J Afr Earth Sc 100:418–436

Hamed Y, Hadji R, Ncibi K, Hamad A, Ben Sâad A, Melki A, Khelifi F, Mokadem N, Mustafa E (2022) Modelling of potential groundwater artificial recharge in the transboundary Algero-Tunisian Basin (Tebessa-Gafsa): the application of stable isotopes and hydroinformatics tools. Irrig Drain 71:137–156

Hamed Y, Hadji R, Redhaounia B, Zighmi K, Bâali F, El Gayar A (2018) Climate impact on surface and groundwater in North Africa: a global synthesis of findings and recommendations. Euro-Mediterr J Environ Integr 3:25

Hashemi H, Berndtsson R, Persson M (2015) Artificial recharge by floodwater spreading estimated by water balances and groundwater modelling in arid Iran. Hydrol Sci J 60:336–350

Hussain MS, Abd-Elhamid HF, Javadi AA, Sherif MM (2019) Management of seawater intrusion in coastal aquifers: a review. Water 11:2467

Ibn Ali Z, Gharbi AA, Kallel A, Ghzel LL, Zairi M (2019) Efficiency of the groundwater artificial recharge from dam water release in arid area. IJW 13:12–25

Ibn Ali Z, Triki I, Lajili-Ghezal L, Zairi M (2017) A method to estimate aquifer artificial recharge from a hill dam in Tunisia. J Arid Land 9:244–255

Jarraya Horriche F, Benabdallah S (2020) Assessing aquifer water level and salinity for a managed artificial recharge site using reclaimed water. Water 12:341

Jarraya Horriche F, Bogacki W (2017) Modelling of groundwater artificial recharge in a heavily exploited coastal aquifer: The case of Mornag (Tunisia). In: Abstract Book. Presented at the 1st atlas georessources international congress AGIC2017. Hammamet, Tunisia, pp 133

Jarraya Horriche F, Mgaidi N, Ghouili N, Zammouri M (2018). Groundwater recharge assessment using wetspass: case study of the Sidi Marzoug-Sbiba Aquifer (Tunisia). In: Calvache ML, Duque C, Pulido-Velazquez D (eds) Groundwater and global change in the western mediterranean area, environmental earth sciences. Springer International Publishing, Cham, pp 161–167

Jarraya-Horriche F (2023) Caractérisation et évaluation de l'aquifère côtier prioritaire Ras Jebel et des écosystèmes associés. Composante CP 2.1, MedProgramme Gestion des aquifères côtièrs et des écosystèmes associés. UNESCO

Jarraya-Horriche F (2004) Contribution à l'analyse et à la rationalisation des réseaux piézométriques (Thèse de Doctorat). Ecole Nationale d'Ingénieurs de Tunis, Tunisie

Jarraya-Horriche F (1995) Élaboration d'un modèle de transport de masse par la méthode des trajectoires des particules (DEA). Faculté des Sciences de Tunis, Tunisie

Jarraya-Horriche F, Ben Brahim H, Slama F (2022) The compromise programming method to choose artificial recharge sites in Mornag aquifer (Tunisia). Arab J Geosci 15:1418

Jarraya-Horriche F, Benabdallah S, Ayadi M (2020) Groundwater monitoring for assessing artificial recharge in the Mediterranean coastal aquifer of Korba (Northeastern Tunisia). Environ Monit Assess 192:442

Kallali H, Anane M, Jellali S, Tarhouni J (2007) GIS-based multi-criteria analysis for potential wastewater aquifer recharge sites. Desalination 215:111–119

Kerrou J, Lecca G, Renard P, Tarhouni J (2006) Uncertainty in exploitation rates and numerical modeling of its impact on seawater intrusion in the Korba aquifer (TUNISIA). In: Proceedings of the 1st SWIM-SWICA joint saltwater intrusion conference. Cagliari-Chia Laguna, Italy, pp 27–34

Ketata M, Gueddari M, Bouhlila R (2014) Hydrodynamic and salinity evolution of groundwaters during artificial recharge within semi-arid coastal aquifers: a case study of El Khairat aquifer system in Enfidha (Tunisian Sahel). J Afr Earth Sc 97:224–229

Khan A, Govil H, Taloor AK, Kumar G (2020) Identification of artificial groundwater recharge sites in parts of Yamuna River basin India based on Remote Sensing and Geographical Information System. Groundw Sustain Dev 11:100415

Kotwicki V (2009) Water balance of Earth. Hydrol Sci J 54:829–840

Kouzana L, Mammou AB, Felfoul MS (2009) Seawater intrusion and associated processes: case of the Korba aquifer (Cap-Bon, Tunisia). CR Geosci 341:21–35

Lachaal F, Chargui S, Jebalia N, Ayari K, Triki L, Gabtni H (2022) Adapting groundwater artificial recharge to global and climate change in water-stressed coastal region: the case of Ras Jebel aquifer (North Tunisia). Arab J Geosci 15:1202

Mahmoudi M, Aydi A, Ibrahim H (2021) Site selection for artificial recharge with treated wastewater with the integration of multi-criteria evaluation and ELECTRE III. Environ Sci Pollut Res 28:46748–46763
MALE (2015) Convention cadre des Nations Unies sur les changements climatiques. Contribution prévue déterminée au niveau national. Ministère des Affaires Locales et de l'Environnement, Tunisie
Margat J, van der Gun J (2013) Groundwater around the World. CRC Press, Taylor & Francis Group, New York, A Geographic Synopsis
MARH, GIZ (2007) Stratégie nationale d'adaptation de l'agriculture tunisienne et des écosystèmes aux changements climatiques. (Rapport National). Ministère de l'Agriculture et des Ressources Hydrauliques, Tunis, Tunisie
Melati MD, Fan FM, Athayde GB (2019) Groundwater recharge study based on hydrological data and hydrological modelling in a South American volcanic aquifer. CR Geosci 351:441–450
Mokarram M, Negahban S, Abdolali A, Ghasemi MM (2021) Using GIS-based order weight average (OWA) methods to predict suitable locations for the artificial recharge of groundwater. Environ Earth Sci 80:428
Nazoumou Y, Besbes M (2005) Simulation de la recharge artificielle de nappe en oued par un modèle à réservoirs. Rev Sci Eau 13:379–404
Neubert S, Benabdallah S (2003) La réutilisation des eaux usées traitées en Tunisie. Deutsches Institut für Entwicklungspolitik, Bonn
Quintana Ashwell NE, Peterson JM, Hendricks NP (2018) Optimal groundwater management under climate change and technical progress. Resour Energy Econ 51:67–83
Rassas S (1988) Recharge artificielle de la nappe de Mateur à partir de l'ancien lit de l'oued Joumine. Faculté des Sciences de Tunis, Tunisie
Rupérez-Moreno C, Pérez-Sánchez J, Senent-Aparicio J, Flores-Asenjo P, Paz-Aparicio C (2017) Cost-benefit analysis of the managed aquifer recharge system for irrigation under climate change conditions in Southern Spain. Water 9:343
Saadaoui M, Jeridi M, Abbasi M (1989) Recharge de la nappe de Kairouan par les lâchures du barrage de Sidi Saâd. Direction Générale des Ressources en Eau, Tunisie
Salar SG, Othman AA, Hasan SE (2018) Identification of suitable sites for groundwater recharge in Awaspi watershed using GIS and remote sensing techniques. Environ Earth Sci 77:701
Sallwey J, Bonilla Valverde JP, Vásquez López F, Junghanns R, Stefan C (2019) Suitability maps for managed aquifer recharge: a review of multi-criteria decision analysis studies. Environ Rev 27:138–150
Sellami H, Benabdallah S, La Jeunesse I, Vanclooster M (2016) Climate models and hydrological parameter uncertainties in climate change impacts on monthly runoff and daily flow duration curve of a Mediterranean catchment. Hydrol Sci J 61:1415–1429
Shanafield M, Cook PG (2014) Transmission losses, infiltration and groundwater recharge through ephemeral and intermittent streambeds: a review of applied methods. J Hydrol 511:518–529
Singh A, Panda SN, Uzokwe VNE, Krause P (2019) An assessment of groundwater recharge estimation techniques for sustainable resource management. Groundw Sustain Dev 9:100218
Valhondo C, Martínez-Landa L, Carrera J, Díaz-Cruz SM, Amalfitano S, Levantesi C (2020) Six artificial recharge pilot replicates to gain insight into water quality enhancement processes. Chemosphere 240:124826
Zammouri M, Brini N, Jarraya Horriche F (2018) Assessment of artificial recharge efficiency against groundwater stress in the El Khairat Aquifer. In: Calvache ML, Duque C, Pulido-Velazquez D (eds) Groundwater and global change in the western mediterranean area, environmental earth sciences. Environmental earth sciences, pp 305–312
Zammouri M, Feki H (2005) Managing releases from small upland reservoirs for downstream recharge in semi-arid basins (Northeast of Tunisia). J Hydrol 314:125–138

Chapter 12
Artificial Groundwater Recharge in Santa Marta and Bogotá, Colombia

C. Carlos E. Molano

Abstract This paper describes two cases of artificial recharge in Colombia; the first one located in the coastal aquifer of Santa Marta in Northern Colombia, where a recharge structure was built on the Manzanares River; the second located in the capital district of Bogotá, wherein Aquifer Storage and Recovery (ASR) pilot test was implemented using a single well for both recharge and pumping. In 1993, the Santa Marta artificial recharge project began, the objective of which was to store part of the high flows of the Manzanares River in the unconfined alluvial aquifer to augment the city's water supply. A recharge structure was built on the riverbed in 1994, operated efficiently for 10 years, and was then removed in 2004 because water quality conditions deteriorated, the city expanded, and therefore river flow decreased. The ASR test site, with a single injection and abstraction well and an observation well, both up to about 300 m deep, was established in 2002 in a Cretaceous sandstone aquifer located in a mountainous area of Bogotá, near a water treatment plant, enabling the use of clean water which should avoid well clogging during injection. Since it is an area with complex structural geology with several local and regional faults and confined conditions, the water level in the injection well was very close to the surface, and the observation well showed artesian conditions. This implies that positive pressure on the injected water was required for the ASR to function. The tests indicated good potential for ASR in the Bogotá region, especially in areas of aquifer depletion.

Keywords Artificial recharge · Santa Marta alluvial aquifer · Bogotá Cretaceous aquifer · Aquifer storage and recovery

---

C. Carlos E. Molano (✉)
Department of Civil and Environmental Engineering, Los Andes University, Cra. 1 Este #19A-40, Edificio Mario Laserna, Piso 6, Bogotá, Colombia
e-mail: cmolano@uniandes.edu.co

D. Saha et al. (eds.), *Managed Groundwater Recharge and Rainwater Harvesting*, Water Resources Development and Management,
https://doi.org/10.1007/978-981-99-8757-3_11

12.1 Introduction

Managed aquifer recharge is not prevalent in Columbia, and relatively little review work on existing schemes exists. In this chapter, two cases of artificial recharge are covered. They are located in different climatic regions and hydrogeological settings. In Santa Marta, a scheme has been implemented in an unconfined alluvial coastal aquifer, recharged by a river, in a warm region where climate change is accentuated. The second case, in the Bogotá region is located in the mountainous inland area called Cordillera Oriental with a cold climate, and the ASR pilot test is in a confined fractured consolidated rock setting. Both projects can be adapted and scaled to other regions of Colombia. In addition, there is an artificial recharge project in the Morroa multilayer aquifer, located in Sucre state (northern Colombia), in consolidated tertiary rocks (Navarro 2020), which however, is not further covered in this chapter, as the experience is still relatively limited.

12.2 Artificial Groundwater Recharge in the Santa Marta Coastal Aquifer

12.2.1 Location

Santa Marta is a touristic city located on the Caribbean coast (Fig. 12.1). When the project began in 1993, its population was 360,000. By 2021, it had increased to 534,000.

12.2.2 Hydrogeological Setting

The hydrogeological framework corresponds to a phreatic alluvial aquifer consisting mainly of sand and gravel, delimited by a mountainous hard rock morphology at depth (up to 50 m) and to the inland side by igneous and metamorphic rocks (Fig. 12.2). The main source of aquifer recharge is the Manzanares River. The mean river flow is 1.2 m^3/s, which previously discharged into the sea. The main objective of the scheme was to augment water supply to the city.

12.2.3 Monitoring Program

PVC piezometers were installed at various depths in the riverbed to measure the pressure head (Fig. 12.3), which, along with the hydraulic conductivity, was applied to calculate the combined natural and artificial recharge from the river, as well as any

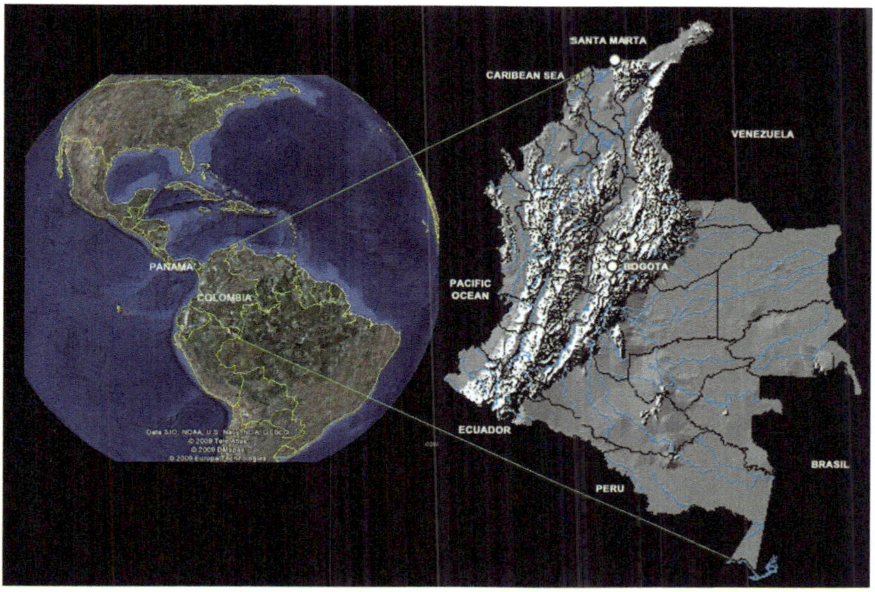

Fig. 12.1 Location of artificial recharge projects in Santa Marta and Bogotá, Columbia

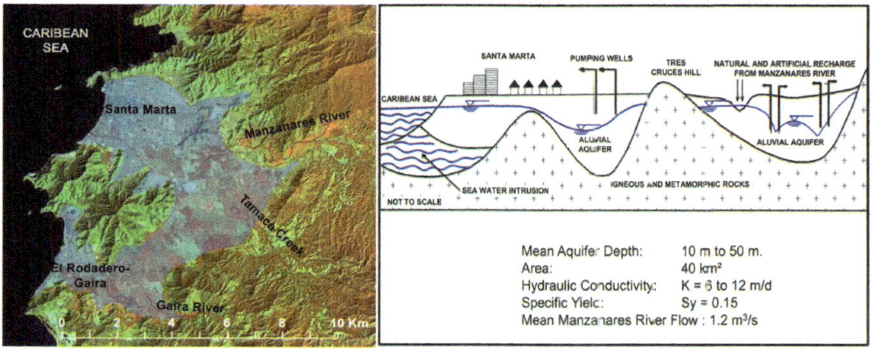

Fig. 12.2 Map of the Santa Marta alluvial aquifer (blue-purple area), Santa Marta city, rivers flowing into the coastal lowland; and schematic hydrogeological cross section relevant to the MAR case

discharge from the aquifer to the river. In the dry season, and under pre-development conditions, the predominant flow direction is from the aquifer to the river, while in the rainy season, it is from the river to the aquifer.

For the monitoring program, 32 piezometers were installed, most of them located near the artificial recharge site.

Fig. 12.3 Installation of piezometers to measure hydraulic head and recharge in the riverbed

Fig. 12.4 In-situ infiltration tests prior to construction of the recharge structure in the riverbed

In-situ riverbed recharge tests were carried out initially to estimate recharge. To enhance water retention and recharge, sandbags were placed across the river transect, increasing the water level in the river (Fig. 12.4). Piezometers and observation wells were used to calculate recharge. The children playing in the river indicates the original good quality conditions of the water used for recharge.

12.2.4 Groundwater Modeling

As part of the scoping and design phase, a steady-state 2D numerical groundwater flow model was applied via a spreadsheet. Output for groundwater level contours in

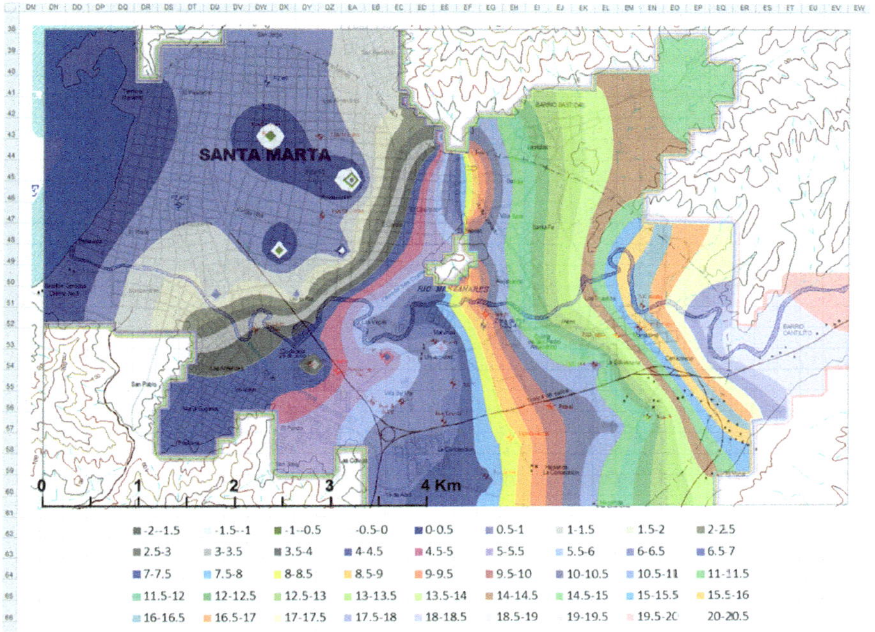

Fig. 12.5 Groundwater level contours (1992) determined by a spreadsheet model

1992 is shown in Fig. 12.5. Subsequently, the 3D Modflow© and Feflow© models were applied. Figure 12.6 shows the 3D finite element grid used in the FeFlow model, and Fig. 12.6 the simulated intrusion of seawater at steady state conditions (Molano and Corrales 2009), principally caused by the "La Carcel" pumping well that operated from 1989 to 2003. The model performance, in terms of water flow and salinity, was calibrated against observation and pumping wells. The impact of seawater intrusion is a large problem very close to the shoreline, where the abstraction from several pumping wells was stopped. Although the salinity in the model adequately represented that observed, the managed aquifer recharge scheme was not applied to manage seawater intrusion, since the geometry of the basement protects the aquifer around the city against the intrusion to some extent (see Fig. 12.2).

12.2.5 Recharge Structure Design

Figure 12.7 shows the conceptual design of the recharge system for the Manzanares River in Santa Marta (Molano et al. 1994). The design was based on the Santa Ana River Recharge scheme in Orange County, California, USA. In the California case, inflatable rubber dams are used to enhance water retention and recharge. In the case of the Manzanares River, sandbags were used between concrete dams to enhance

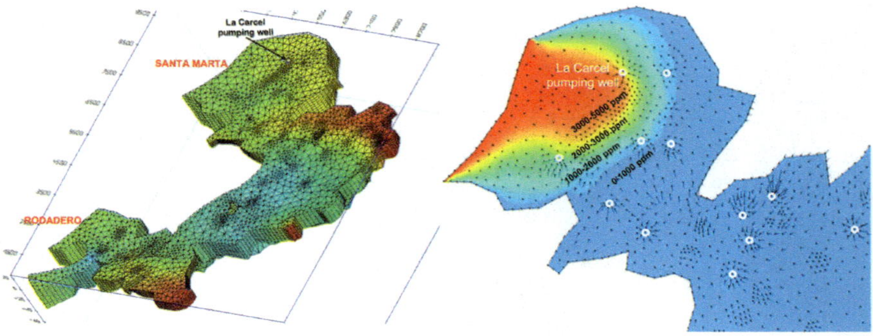

Fig. 12.6 3D finite element grid model; and model output of seawater intrusion

recharge. When peak flows occur, the sandbags tend to break, which works to evacuate excessive flood waters; on the other hand, the shape design of the concrete structure can evacuate extreme flows. Pumping wells were built next to and downstream of the MAR site along the Manzanares River to take advantage of artificial recharge. This water was used for domestic water supply.

Figure 12.7 also shows: the lab hydraulic model to test the reliability of the structure (Saldarriaga et al. 1995), especially to evacuate flows during extreme events;

Fig. 12.7 Conceptual design; lab model; construction; and operation of the recharge structure

the final recharge structure at the end of its construction; and the recharge system in operation. The artificial recharge system increased the groundwater supply from 250 L/s to 550 L/s (Molano and Corrales 2009).

12.2.6 Results and Conclusions of the Santa Marta Scheme

1. The Santa Marta Artificial Recharge System increased the mean groundwater supply to Santa Marta city from 250 L/s in 1993 to 550 L/s in 1995. In non-rainy season the mean groundwater supply was 45% of the total water demand of the city, while in rainy season the groundwater supply was 35% of the total water demand (Molano and Corrales 2009).
2. Conjunctive use of surface water and groundwater, including artificial recharge, is the way to manage water supply for Santa Marta. Results from relatively few years of operation indicate that management of the Santa Marta water supply after the implementation of artificial recharge works most optimally with the following source distributions: in the dry season: 60% groundwater and 40% surface water, and in the rainy season 40% groundwater and 60% surface water.
3. Manzanares River quality conditions deteriorated over time, the population of Santa Marta city increased, and the city expanded into upstream catchments. Also, the runoff from rain to the river diminished, partly due to climate change (Cifuentes and Molano 2016). For these reasons, the river recharge structure was removed in 2004.
4. Santa Marta currently suffers from a lack of sufficient water supply. The alternative to satisfy the city's demand for water is still the joint use of surface and groundwater, including artificial recharge, but with the improvement of the environmental conditions of the Manzanares River and its upper catchments and possibly treatment of river water prior to recharge, which however significantly increases the cost. Possibly, inter-basin water transfer will be necessary but could also include MAR components for better and more climate-resilient water storage.

12.3 Aquifer Storage and Recovery Pilot Test in Bogotá

12.3.1 Introduction

The Cretaceous sandstone aquifer of the Sabana de Bogotá (Bogotá plain), known as the Guadalupe Aquifer, is a confined aquifer. Apart from its outcrops in high mountainous areas, where the aquifer is unconfined, it is generally located in the flat Bogotá plain area between 100 and 1000 m depth. This aquifer is suffering from depletion in several areas where water levels have dropped up to 150 m compared to pre-development times. Although its use for the city of Bogotá is low, it is also

Fig. 12.8 ASR well, before installation of the submergible pump; ASR well, in the back, a tank built to store water for injection; and ASR well after adaptation of the wellhead for pump installation

used for water supply in some municipalities and for agriculture outside the city, which can be very water intensive. Several institutions, including the water supply company for Bogotá (EAAB-Empresa de Acueducto y Alcantarillado de Bogotá), have been interested in its artificial recharge to enhance water security of the area, and to counteract aquifer depletion. The approach suggested was through injection wells, taking advantage of excess water during heavy rains. An existing water treatment plant in the area was used to treat water, enabling the injection of clean water, free of sediments, thereby avoiding clogging of the test well.

12.3.2 Recharge and Observation Wells

In the beginning of 2002, an 8-inch diameter ASR dual well (for injection and pumping) was installed to 300 m depth in the Eastern Bogotá Hills (Fig. 12.8), where the Cretaceous sandstone is the predominant rock (JICA and EAAB 2003). In addition, a 4 inch observation well was installed to 240 m depth 120 m away from the ASR well (Fig. 12.9). The Cretaceous sandstone is the aquifer targeted for the ASR well, which is recharged by water from the Fucha River after treatment in the water treatment plant. The test was carried out during the month of November 2002.

12.3.3 Hydrogeological Conditions

The structural geology is complex with local and regional faults, including the regional Bogotá fault, which places cretaceous permeable rocks into contact with impervious tertiary rocks. This plays an important role in head conditions since water (and pressure) "builds up" against the fault, without a natural outlet. This is reflected in both the observation well (Fig. 12.9), showing artesian conditions (free

Fig. 12.9 **a** Free-flowing observation well **b** Observation well with a pipe to measure the pressure, at the back uphill, the water tank for injection, and the ASR well

flowing well) and the injection well showing a static water level close to, but below, the ground level.

12.3.4 Test Results

The results of a sample of the cyclical ASR tests in Nov 2002 are presented in Fig. 12.10. The pressure head applied for injection was approximately 8.5 m for a mean injection rate of 10 L/s (864 m^3/d). The injection and pumping were cyclical because the innate head in the ASR test well is close to the surface (8 m below the ground). The recharge was made through the injection tank (Fig. 12.8) that has a head of 0.5 m above the ground. This can be seen as the immediate head jump of 8.5 m after pumping (Fig. 12.10). An operational injection well is expected to operate continuously depending on the pressure head in the aquifer and the availability of suitable water for recharge. Although there is no operational injection well in Bogotá plain presently, the results indicate that MAR schemes for ASR wells are feasible and can be applied in the flat area, where the Cretaceous aquifer is depleted, confined, and located between 100 and 1000 m depth.

12.3.5 Results and Conclusions of the Bogotá Scheme

1. The cyclical tests (pumping and injection) showed that an Aquifer Storage and Recovery (ASR) scheme works well for the Bogotá Cretaceous confined aquifer

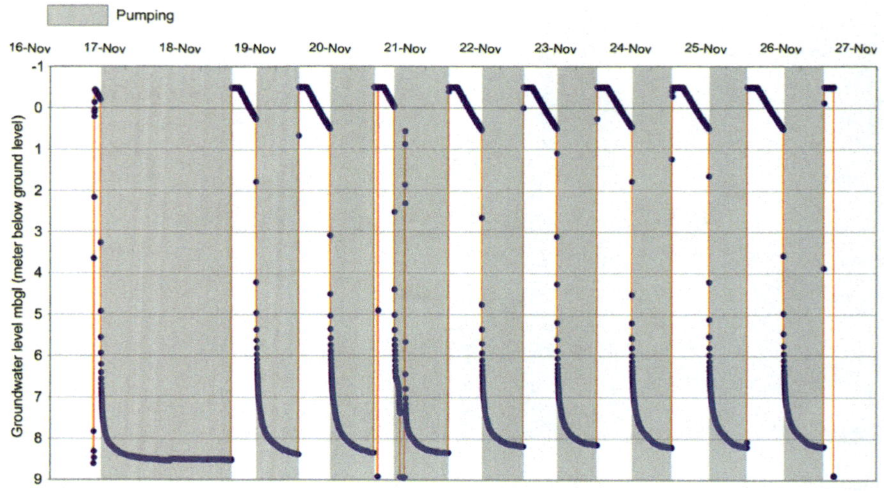

Fig. 12.10 CyclicalASR test results from Nov 2002

using water from the nearby Fucha River and that it can be an alternative to ordinary artificial recharge.

2. Even the tests with a relatively low-pressure head (8.5 m) showed a good artificial recharge flow rate (864 m^3/d). It is expected that in depleted areas, a pressure head of up to 150 m can be applied, substantially increasing the recharge flow rate.
3. Additional tests are required in depleted aquifer areas. Water treatment is expected before well injection, so that water quality is suitable for domestic, agricultural, or industrial use, and to prevent clogging of the well.

References

Cifuentes G, Molano CE (2016) Groundwater modelling of the Coastal Aquifer of Santa Marta, Colombia, under different hydrological and pumping scenarios, including seawater intrusion, artificial recharge, and inter-basin transfer. In: Proceedings of the 9th international symposium on managed aquifer recharge, ISMAR 9. México

Japan International Cooperation Agency (JICA) and BogotáAqueduct and Sewage Company (Empresa de Acueducto y Alcantarillado de Bogotá, EAAB) (2003) Sustainable Groundwater Development for Bogotá Plain, Colombia. Final Report. EAAB, Bogotá

Molano CE, Bonilla F, Mejía J, Rodríguez C (1994) Artificial recharge of the Santa Marta Aquifer, Colombia. In: Artificial recharge of ground Water II, ASCE, pp 446–454

Molano CE, Corrales J (2009) Artificial recharge, groundwater modeling, and management of the coastal aquifer of Santa Marta, Colombia. NGWA Groundwater for the Americas Conference. Panamá

Navarro JL (2020) Monitoring of artificial recharge pilot works in the Morroa aquifer, Sucre, Colombia. Degree project. In Spanish. Earth Sciences Dept., EAFIT University, Medellin, Columbia

Saldarriaga JG, Obregon N, Amaya JA (1995) Hydraulic design of the damming structures for the artificial recharge system of the Santa Marta aquifer. In: Spanish Civil and Environmental Engineering Dept., Los Andes University, Bogotá, Columbia

Chapter 13
Groundwater Artificial Recharge in Jordan—Case Studies and Potential Areas

Elias Salameh and Ghaida Abdallat

Abstract In this article, managed aquifer recharge (MAR) projects and high-potential areas for groundwater recharge in Jordan are presented, discussed, and assessed, and is based on the availability of recharged water, aquifers, and the potential purpose of the recharged water use. Therefore, general information about implemented recharge schemes and the high-potential areas for MAR in Jordan such as the Jordan Valley, Wadi Araba, Jafr, and Azraq areas were analyzed. The main findings of this study show five high-potential sites of MAR along the northern part of the Jordan Valley (Yarmouk River, Wadi Ziglab, Suleikhat, Wadi Shueib, and Hisban (32° 37′ 58″ N, 35° 45′ 55″ E to 31° 51′ 29″ N, 35° 44′ 07″ E)), which are built of alluvial fan deposits. There are high-potential sites that are located along Wadi Araba in the southern part of the Jordan Rift Valley (30° 40′ 12″ N, 35° 35′ 50″ E) to (29° 40′ 24″ N, 35° 06′ 28″ E). In addition, Jafr Depression (30° 22′ 20″ N, 36° 11′ 23″ E) shows high-potential MAR sites. Furthermore, very high-potential sites exit along Sirhan Depression main side wadis (31° 00′ 41″ N, 36° 59′ 30″ E). Wadi Rajil northeast of Azraq (31° 54′ 24″ N, 37° 10′ 07″ E) presents another high-potential MAR site. Finally, South of Aqaba, the Durra area (29° 24′ 04″ N, 35° 03′ 15″ E) shows very high potential for MAR in the alluvial fans. However, a minimum level of studies for any artificial recharge project must be carried out before starting and executing any MAR scheme. Such studies include defining the aquifer parameters, the water quality evolution of the recharged, and the exiting groundwater and their interactions. In addition, it is of utmost importance to understand the fate of the recharged water before starting with recharge schemes. Managed aquifer recharge can alleviate the impacts of climate change by making use of unused water, and in the case of Jordan, it can alleviate the implications of depleting groundwater resources.

E. Salameh
Department of Geology, University of Jordan, P.O. Box 11942, Amman, Jordan
e-mail: salameli@ju.edu.jo

G. Abdallat (✉)
Economic and Social Council of Jordan, Amman, Jordan
e-mail: ghidaaabdallat@yahoo.com

D. Saha et al. (eds.), *Managed Groundwater Recharge and Rainwater Harvesting*, Water Resources Development and Management,
https://doi.org/10.1007/978-981-99-8757-3_12

Keywords Artificial recharge · Groundwater · Alluvial deposits · Jordan Valley

13.1 Introduction

Jordan lies in the arid to semi-arid climatic zone with about 3–5% of its area, which is classified as of Mediterranean climate (JDoM, open files 2016). The total amount of natural surface and groundwater resources of the country sums up to around 900 MCM/yr (Salameh et al. 2018a). For a population of more than 10 million, this means a per capita share in these resources of less than 90 cubic meters per year for all use purposes of municipal, industry, and irrigation (MWI 2016a). That renders water to be a limiting factor for any development and turns each cubic meter of water countable so that the country cannot tolerate the loss of any water quantity. In addition, due to a growing population, improving living standards, and urban/ industrial developments, water requirements are projected to increase significantly.

Jordan has developed its precious and available water resources in a very good way. These development efforts include: the construction of dams to collect flood, management of base flow and treated wastewater, building of irrigation canals, and drilling of groundwater wells all over the country including the reuse of treated sewage effluent for irrigation purposes.

Indeed, there are aspects that remain to be developed for maximizing the water resources such as flood flows of small wadis, downstream floods of dammed wadis, desert wadis, and treated wastewater at certain times of the year when irrigation water is not needed. Rainfall harvesting is still one of the potential windows for water resources development in Jordan.

Such waters, although in international understanding, are very small in their amounts (a few million cubic meters per year) but for a country with very scarce water resources, they are of utmost importance, especially because they can be made available in remote areas to serve the local population. Therefore, their storage and utilization are essential for the country.

Ancient storage facilities in surface and underground reservoirs are well known from historic sites in Jordan where water pools, cisterns, and underground water reservoirs are found distributed all over the country, without which the existence and development of life have not been possible (AbdelKhaleq and Alhaj Ahmed 2007). Of course, scales are now different and the per capita water requirements increased dramatically compared to ancient water demands (Salameh et al. 2018a).

This article sheds light on successful and less successful water collection and recharge schemes in Jordan. It tries to avoid failure of such projects by assessing and defining the conditions for successful projects. With the hope that the experience gained in Jordan can serve Jordan and other countries with similar climatic conditions and water stresses.

The potential areas for groundwater recharge according to the availability of excess water, even when this availability is only seasonal, potential aquifers and need for water are shown in Fig. 13.1. Table 13.1 lists the general areas that can be recharged with their sites and types of aquifers (Salameh and Abdallat 2020).

As a matter of fact, there are limited numbers of articles and published work that can be found in the literature about groundwater recharge in Jordan. However,

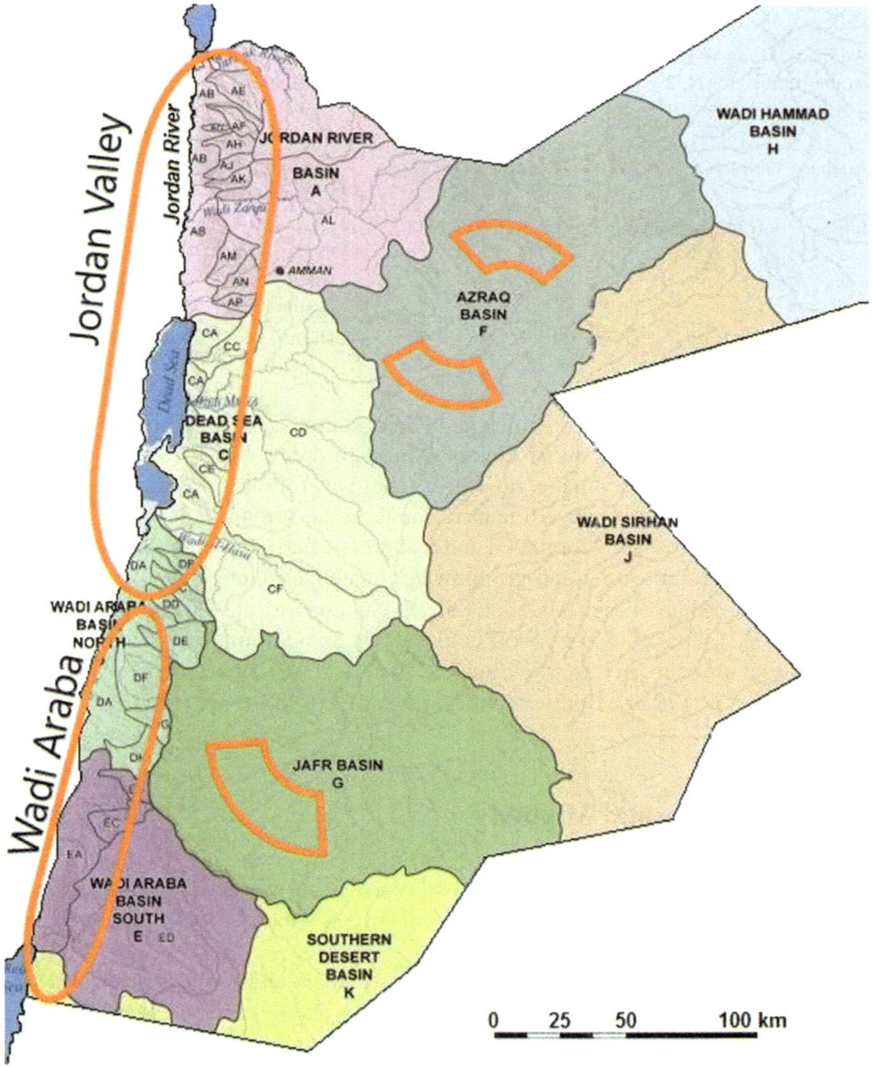

Fig. 13.1 Potential areas for managed aquifer recharge by considering the availability of recharge water, aquifers, and the need for water. The figure shows the Lower Jordan Valley and Wadi Araba with their numerous side wadis in addition to the potential highland areas of Azraq and Jafr Basins (modified after Salameh and Abdallat 2020)

Table 13.1 Areas with high practical potentiality for managed aquifer recharge (after Salameh and Abdallat 2020)

Area	Site Name and Coordinates	Potential for MAR	Comments
Lower Jordan Valley	Yarmouk River, Wadi Ziglab, Suleikhat Wadi Shueib and Hisban (N: 32° 37′ 58″, E: 35° 45′ 55″ to N: 31° 51′ 29″, E: 35° 44′ 07″)	Very high	Alluvial fans
Wadi Araba	Along Wadi Araba (N: 30° 40′ 12″, E: 35° 35′ 50″) to (N: 29° 40′ 24″, E: 35° 06′ 28″)	High: restricted by recharge water availability	Alluvial fans
Aqaba	Durrah (N: 29° 24′ 04″, E: 35° 03′ 15″)	Very high	Alluvial fans and wadi deposits
Jafr	Jafr Depression (N: 30° 22′ 20″, E: 36° 11′ 23″)	High: restricted by recharge water availability	Alluvial fans and Tertiary aquifers
Azraq	Wadi Rajil (N: 31° 54′ 24″, E: 37° 10′ 07″)	High	Alluvial fans

recently and after the 1980s, Salameh and others started a study on artificial groundwater recharge in the Muwaqqar area in central Jordan on three weirs constructed for water harvesting (Salameh et al. 1991). That was followed by another study on the potential for groundwater artificial recharge in the Jordan Valley area (Salameh 2001). After that, research continued and a number of articles were published about studies on water harvesting and groundwater artificial recharge such as Eggleston et al. (2021), Salameh and Abdallat (2020), Salameh et al. (2019a, b), Laat and Nonner (2012), Steinel (2012), Wolf et al. (2007). In addition, use was made of the international literature in this article such as Asano (2016), Heviánková et al. (2016), Chawdhury et al. (2010), Bouwer (2002), EXACT (1998), BGR (1977).

13.2 Materials and Methods

This article summarizes the previous work carried out in relation to groundwater artificial recharge, especially the most recent works, which have been published by different authors.

It is a review of previous work providing assessment and some critical reviewing of their success or failure. For that purpose, all available previous work published or reported on, M.Sc. and Ph.D. theses, consulting studies, and other formats have been used to compose this article. Some additional work is included in this article concerning some figures and tables and the general situation of groundwater artificial recharge in the country, which may serve other countries in preparing and implementing groundwater artificial recharge projects.

13.3 Results and Discussion

13.3.1 Conditions and Methods of Groundwater Artificial Recharge

The need for and the conditions of successfully managed aquifer recharge projects and their relevance to the prevailing circumstances in Jordan, are summarized (Salameh et al. 2019a, b).

13.3.1.1 MAR Can Be a Very Useful Technique in the Following Cases

1. Storage for available water in times and its need in others
2. Create temporary storage of water (riverbank infiltration)
3. Alleviate saltwater intrusion
4. Partial treatment of impaired water quality
5. Improving the quality of groundwater by recharging it with water of better quality
6. Relieve flooding effects
7. Injection of water into oil and gas fields to enhance their productivity.

13.3.1.2 The Pre-conditions for Implementing MAR Projects Are

1. Need for the recharged water to achieve one or more of the goals stated above
2. Availability of water for recharge
3. Suitable aquifers for storage
4. Depending on the purpose of recharge, the extraction potentials of the recharged water must be there
5. Necessity and potentials for the pre-treatment of the recharge water
6. Suitable topographic set-up to implement the appropriate recharge technique
7. Conveyance facilities for water to reach recharge sites
8. Potential for the construction of recharge structures
9. Depending on the purpose of recharge, water quality of rechargeable aquifers, quality of vadose zone, and quality of recharge water must be appropriate
10. Construction, maintenance, and repair capabilities and financial affordability.

For the case of Jordan, all the above points, 1–10 must be fulfilled in order to go ahead with any MAR scheme.

13.3.1.3 Methods of Groundwater Artificial Recharge

Groundwater artificial recharge can be achieved through a variety of methods and engineered work which are mainly:

- Recharge dams and weirs,
- Leaking dams and weirs,
- Infiltration ponds or basins,
- Infiltration wells,
- Injection wells,
- Bank filtration,
- Land application or excess irrigation,
- Trenches and galleries.

In Jordan, MAR has become a necessity to make utmost use of all available and limited water resources, especially, the resources that are still lost due to a lack of available storage facilities and which the country badly needs due to its severe water shortages and the expected negative impacts of climate change. Such interventions will certainly not reverse climate change, but they can alleviate the immediate negative impacts on water supplies.

13.3.1.4 Advantages of Underground Storage Compared with Surface Storage

In Jordan, with its arid and semi-arid climate, underground water storage is preferable to surface water storage because of the following:

1. It minimizes the evaporation losses compared to surface water storage in dam's reservoirs, weirs, and other facilities not protected from evaporation.
2. Protection of the stored water underground from direct pollution and pollution accidents that surface water storage is exposed to.
3. Prevention of eutrophication processes that inflict water stored in surface reservoirs.
4. A relatively cheaper way of storing water compared to surface water dams.
5. Improvement of natural groundwater quality in many areas.
6. According to the used technology of recharge, MAR can treat some biologically impaired recharge water qualities.

13.3.1.5 Conditions Governing the Allocation of Water for Recharge

- The water used in MAR must have no other use and recharging it into aquifers will save it from being lost. That is unless MAR methods are used to improve the water quality by recharging it for the purpose of treatment or to be mixed with the existing better-quality groundwater.
- Storing surface runoff water of upstream water courses underground without adequate knowledge of its natural fate might be absurd because that water might, on its way to its downstream destination, infiltrate and naturally recharge the aquifers in the downstream areas.

- Water has many environmental services and therefore, if recharging it in the underground proves to be detrimental to those services, then MAR is not the right choice.

In accordance with these conditions, sites in Jordan for MAR have been sorted out with the following results about potential areas for MAR:

1. In the area of the Jordan Rift Valley (Jordan Valley, Dead Sea, and Wadi Araba area), many small undammed wadis discharge flood and base flow water, especially during the rainy season, where that water ends in the Dead Sea or the Red Sea, where it mixes with these seas salty water without being utilized.
2. In addition, in those areas where dams exist, flood and some base flow waters find their way to the Dead Sea without being used.
3. In the Durra area, south of Aqaba City, wadis originating in the high mountains discharge their floodwaters into the Gulf of Aqaba without being used for any purpose. Here, big alluvial fans are found, which can store hundreds of thousands of cubic meters of water.
4. In the Plateau of Jordan, two local base levels exist, namely Jafr and Azraq Depressions with catchment areas of 12.200 and 11.588 km^2. The floodwater of the Jafr catchment that is collected in the playa, becomes salty and evaporates without finding any use. In addition, the floodwater of the Azraq catchment that is collected in the, by now, salty playa, dissolves salts, infiltrates into the groundwater, and builds a recharge mouths, and adds to the salinity of the existing groundwater. Collecting Jafr and Azraq floodwaters before reaching the playas and recharging it in potential areas will certainly add this fresh water to the water resources of the country and will certainly alleviate their present negative impacts on the fresh groundwater resources of these areas
5. Sirhan and Hammad catchments are very large catchments with many dry wadis and playas receiving hundreds of thousands of cubic meters of floodwater, which either collect in the salty playas and evaporate or flow unused outside the country. Such water can be captured and used by applying suitable MAR.

13.3.2 Assessment of Environmental Setting of the Study Areas

13.3.2.1 Jordan Valley Study Area

The eastern side of the Jordan Rift Valley is a major promising study area for MAR (Fig. 13.2). This is due to floodwater originating from precipitation over the eastern mountains of the rift valley drains to that valley through numerous wadi courses, many of which cannot be dammed due to conditions governing each of them (Eggleston et al. 2021; Salameh and Abdallat 2020). In addition, downstream areas of some dammed wadis receive appreciative amounts of floodwater with no storage facilities. Furthermore, treated wastewater with no available storage facilities and no reuse, in

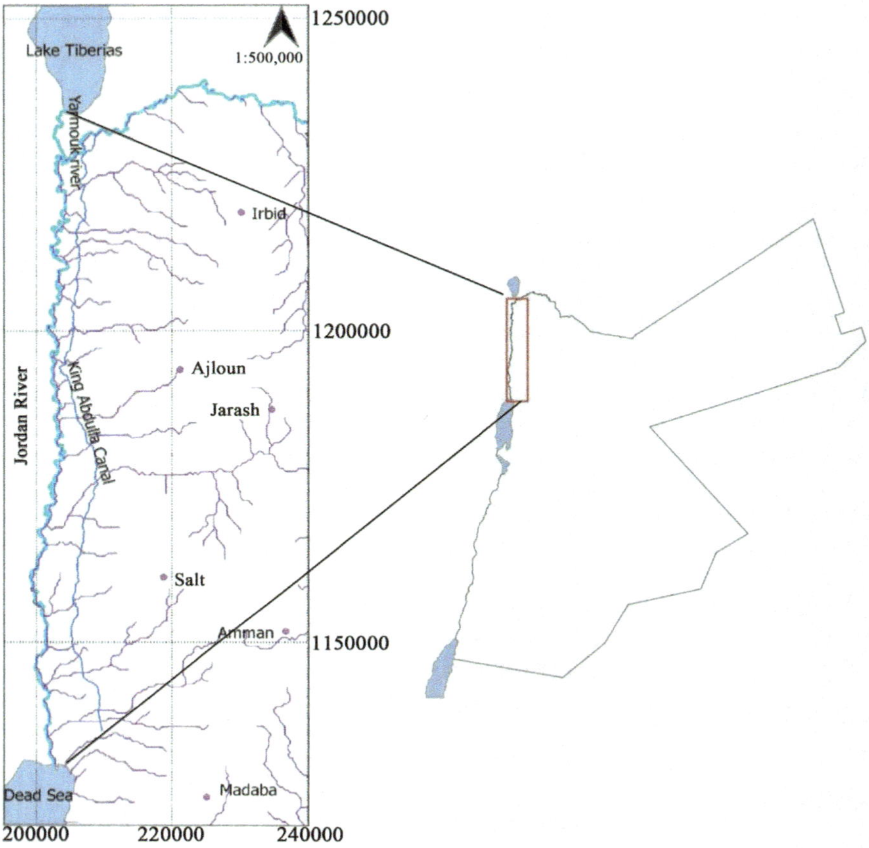

Fig. 13.2 Location map of the Jordan rift valley

certain times, e.g., during and after rain events, can be stored in the underground and extracted in time of need. The suitability for MAR is given here by the presence of alluvial fans resulting from the geologic development of the rift valley.

13.3.2.2 Jordan Rift Valley Area

The Jordan Rift Valley is a floodplain extending from Lake Tiberias in the north to the Red Sea in the south with a length of around 400 km and a width of 4–30 km as shown in Fig. 13.2. To the north of the Dead Sea, in the middle of that floodplain, the Lower Jordan River has during the Holocene, cut its meandering course into the Dead Sea. The eastern and western wadis drain the shoulder mountains, which built at their entrances into the rift valley extended alluvial fans covering other Pleistocene clastic sediments and some lacustrine sediment in the middle of the valley. Between

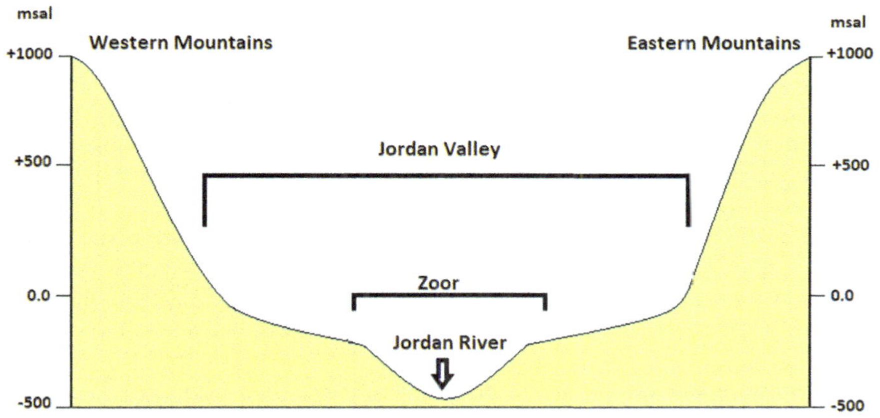

Fig. 13.3 General topographic E-W cross-section through the Jordan valley

the Dead Sea and the Gulf of Aqaba (Wadi Araba), the rift valley is geologically composed of alluvial fans deposited by wadis coming from its eastern and western sides and inter-fingering in the middle of the wadi. Wadi Araba partly drains to the Dead Sea and partly to the Red Sea with a watershed in its middle part at an elevation of 230 m above mean sea level (amsl) as shown in Fig. 13.3.

13.3.2.3 The Lower Jordan Valley

The Lower Jordan Valley forms a floodplain deposited by the Jordan River and tributary wadis with a length of 100 km extending from Lake Tiberias in the north at an elevation of −210 m amsl to the Dead Sea in the south at an elevation of about −432 m amsl (Pierson 2021). The width ranges from 30 km north of the Dead Sea narrowing to 4 km in the area lying 30 km south of Lake Tiberias and widening again to 10 km just south of that lake.

The middle part of the Lower Jordan Valley forms a topographic depression of 10–30 m below the Jordan Valley (JV) floodplain level and up to 500 m in width, locally called the Zoor area. This topographic depression has been cut into a flat floodplain by the Jordan River during the Holocene (Fig. 13.3).

The JV is bordered on both sides by high, steep mountains with differences in elevations between the valley floor and the surrounding mountains of 1000–1500 m. The main features of the JV area are the relatively flat terraces on both sides of the Jordan River, which constitutes the bed of the valley (Fig. 13.3). The east–west slope of the Jordan Valley ranges from 3 to 5/1000, which is twice that of the north–south slope that of around 2%.

13.3.2.4 Climate of Jordan Valley

The elevation of the Jordan Valley Depression as mentioned above is at −200 to −400 m below mean sea level (amsl), which entails a unique climate, very different from the climate of its eastern and western shoulder mountains. The Jordan Valley area is characterized by its hot arid climate with temperatures of 8–12 °C higher than the surrounding mountains and a mean maximum annual temperature of 31.5 °C and a mean maximum monthly temperature in summer of 38.8 °C (June). The highest ever-registered temperature is 52 °C. In winter, temperatures during the night may drop to a few degrees centigrade or even reach the freezing point. In the northern part of the valley, the average temperature is around 1.8 °C less than that of the southern part.

The relative humidity is low in the southern part with a long-term daily mean of 64% in the coldest period of the year (Jan. and Feb.) going down to 27% in the hottest summer months at midday. In the northern part, the humidity ranges from 30% during hot summer days to 70% in the cold winter.

Rain falls in the winter months, October to May, and ranges in average from 400 mm/yr in the Lake Tiberias area to 100 mm/yr at the northern shore of the Dead Sea (Fig. 13.4). In dry years, the amount of precipitation over the valley is half of its long-term average, and in rain-rich years, it almost doubles. The potential evaporation is very high, ranging from 2400 mm/yr at the shores of the Dead Sea to about 2300 mm/yr in the Lake Tiberias area.

13.3.2.5 Geology of Jordan Valley

The surface of the JV floor is mainly built of alluvial sediments consisting of gravel, sand, shale marl, and clay deposited by wadis and coming from the east areas such as Yarmouk, Yabis, Zarqa, Shueib, and many others. These unconsolidated sediments form alluvial fans of up to 100 m in thickness at their apexes in the mountain foothills decreasing to a few meters at their entrance into the Zoor area or into the Jordan River. The narrow shoulders of the Zoor consist of the older Lisan Marls, which precipitated from the Lisan Lake, the ancestor of the present Dead Sea during the Late Pleistocene (Fig. 13.5).

The foothills of the mountains consist of rocks ranging in age from Triassic to recent and are covered by the more recent JV deposits. Figure 13.6a, b show general N–S and E–W geologic cross-section to illustrate the relative setup of older geological formations to the more recent JV deposits along the Jordan Transform Fault. Table 13.2 contains the relevant information about the escarpment foothills formations; rocks, thicknesses, and hydraulic properties.

The Quaternary sediments deposited in the Jordan Valley are subjected here to study their potential for MAR. They were deposited in the Jordan Valley Depression after the last regression of the Tethys and the creation of the new base level for surface water in that depression. The sediments of relevance to the present study are of Lower Pleistocene and younger ages, and are subdivided into the following formations:

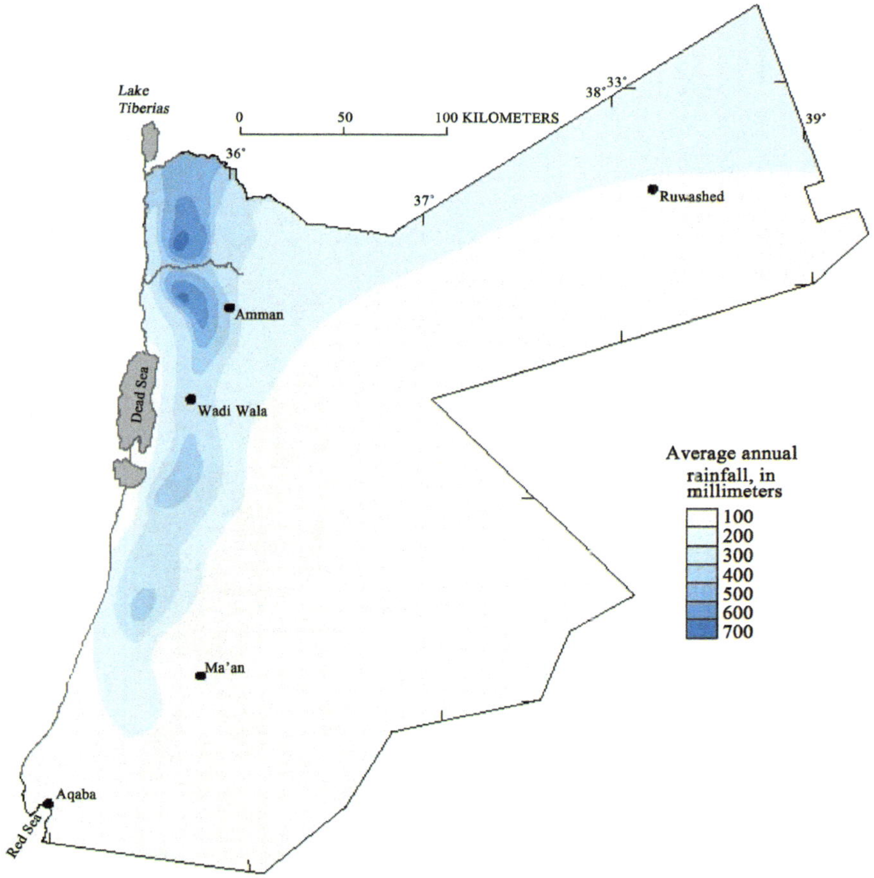

Fig. 13.4 Rainfall distribution in Jordan in mm (after EXACT 1998)

13.3.2.6 Jordan Valley 1 (JV1, Ghor El Kattar, Shagure, and Baqura Conglomerates

This formation was deposited in the Jordan Valley floor, accompanying the faulting and the formation of the Jordan Valley depression. It consists of well-cemented conglomerates with low porosities and permeabilities. Ghor el Kattar sediments with a thickness of about 350 m cover bedrock of different ages at the foothills of the Jordan Valley (Table 13.2).

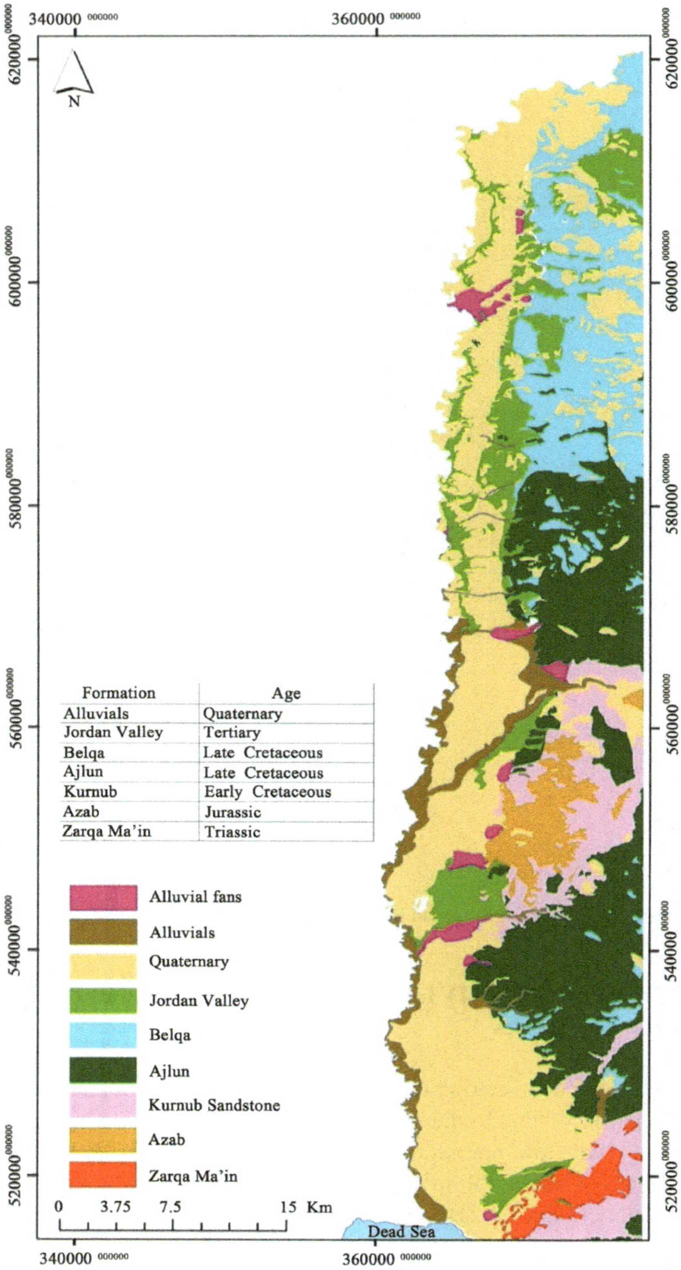

Fig. 13.5 Geological map of the eastern lower Jordan valley and its eastern foothills and highlands (Based on the natural resources authority geological maps of Jordan, NRA open files)

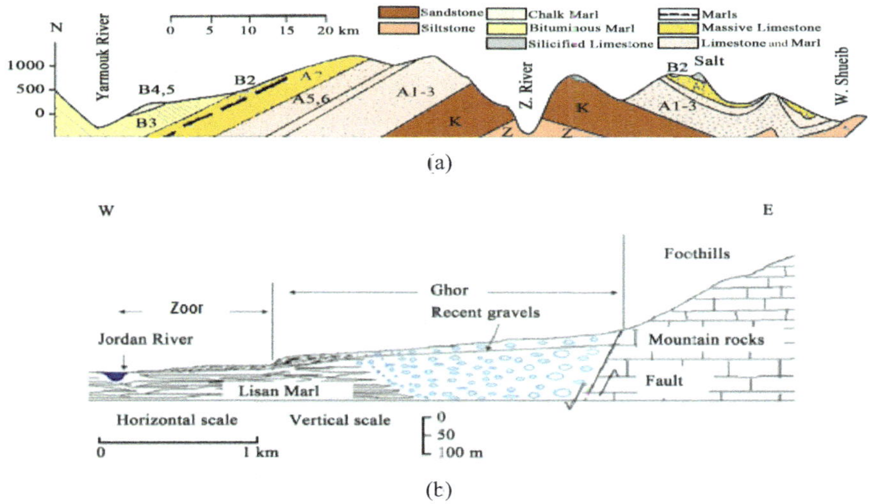

Fig. 13.6 **a** North–South generalized geologic cross-section along the eastern foothills of the Lower Jordan Valley, parallel to the Jordan Transform Fault extending from the Dead Sea in the south to the Yarmouk River in the north. It shows the main structural features of the eastern highlands and the geologic interfaces with the JV area (Z; Triassic and Jurassic, K; Lower Cretaceous, A1-3; lower part of Upper Cretaceous, A1-7, B1, B2, and lower part of B3; upper part of Upper Cretaceous and part of B3, B4, and B5; Tertiary) **b** East–West generalized geologic cross-section from the eastern foothills of the Lower Jordan Valley to the Jordan River showing the relationship of the mountain rocks, in general, to the recent sediments of the Jordan Valley across the Jordan Transform Fault

Table 13.2 Formations in the Jordan valley area (data after Bender 1968, 1975)

Formation name	Thickness (m)
Alluvial fans and recent deposits	0–100
Lisan	40–few hundreds
Samra	35
Kufranja gravels	100
Abu Habil series	100
Ghor El Kattar and Shagur Conglomerates	~350 m

13.3.2.7 Jordan Valley 2 (JV2, Abu Habil, Kufranja and Samra Formations)

Abu Habil Formation consists of clastic sediments with a total thickness of some 100 m. It overlies the JV1 with an angular unconformity and is related to another tectonic activity along the Jordan Rift Valley. Kufranja Conglomerates overlie Abu Habil Gravels in places and have a thickness of about 100 m. Samra Formation consists of fine to coarse-grained classics covering areas along the course of the

Jordan River. The groundwater in these formations is generally saline, except at the foothills of the mountains that are fed by lateral flows from the mountain aquifers

13.3.2.8 Jordan Valley 3 (JV3, Lisan Formation)

Lisan Formation consists of very thin-bedded marls, a few millimeters in thickness with a total thickness of the formation ranges from 40 m to more than a few hundred meters. It was deposited in Lisan Lake, which occupied the JV area from Lake Tiberias in the north to about 45 km to the south of the southern shore of the Dead Sea from about 80,000 yr ago until the end of the Pleistocene (Bender 1968, 1975). From the area of the entrance of the Zarqa River into the Jordan River to the Dead Sea, the Lisan Marls are gypsiferous, and salty, and therefore, its groundwater is saline (around 50 g/l) as measured in Karameh dam wells (GIBB report 1994). In the areas between the mountain foothills and the outcrops of the Lisan Marls, there are old alluvial fans containing fresh water inter-finger with the Lisan Marls.

13.3.2.9 Recent Alluvial Fans

The recent alluvial deposits of Late Pleistocene and Holocene ages cover all older Quaternary deposits and have been deposited by the different rivers and wadis at their entrances into the Jordan Valley area. These alluvial fans consist of gravels at their apexes and silt and fine sands at their toes near the Jordan River, which is reflected in the high porosity and permeability at the apexes decreasing toward the toes, which is in turn reflected in lower groundwater velocities toward the Jordan River. The thickness of these alluvial fans, as encountered by drilling and geoelectric sounding, may reach 100 m at the apexes thinning out toward the toes. Slope debris, which originates from mass wasting at the escarpment foothills inter-finger with and overlaps these fans.

13.3.3 Water Resources

13.3.3.1 Surface Water

The Jordan River's eastern tributaries including the Yarmouk River used to contribute to the Jordan River flow an average amount of flood and base flow of 543 MCM/yr, until the 1970s (Fig. 13.7). The surface water contributions of the different side wadis to the JV area are given in Table 13.3 together with the average discharge of inter-catchments. In the meantime, the main wadis have been developed by dam construction. Table 13.4 shows the sites, capacities, and year of construction of these dams. The average discharges from catchments located downstream of the dams are listed in Table 13.5.

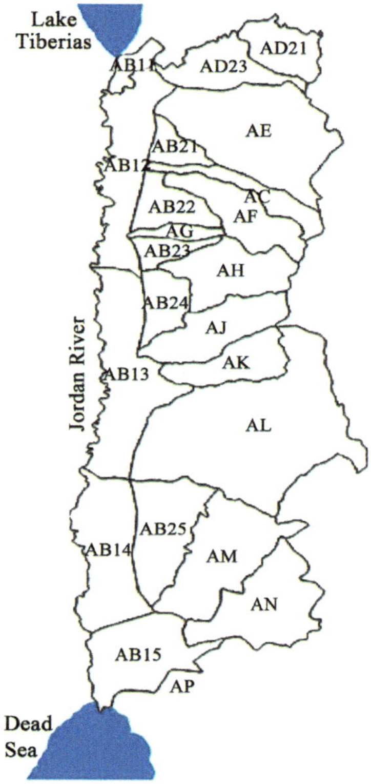

Catchment and Inter-catchments Name	Code	Area [Km²]
Yarmouk triangle	AB11	35
Yarmouk – Yabis	AB12	107
Yabis – Zarqa	AB13	197
Zarqa – Shueib	AB14	140
Shueib – Dead Sea	AB15	127
Wadi Al-Arab – Ziglab	AB21	98
Ziglab – Jurm	AB22	49
Jurm – Yabis	AB23	27
Yabis - Kufranja	AB24	73
Zarqa – Shueib	AB25	141
Teiba	AC	87
Yarmouk Jordan	AD21	126
Yarmouk Jordan	AD23	140
Wadi Al-Arab	AE	167
Ziglab	AF	106
Jurm	AG	22
Yabis	AH	124
Kufranja	AJ	111
Rajib	AK	82
Shueib	AM	178
Kafrain	AN	209
Hisban	Ap	82

Fig. 13.7 Surface water catchments of the eastern side of the lower Jordan valley area, their names, codes, and catchment areas

Flood flows of inter-catchments and of dam's downstream areas flow until now unused into the Jordan River and the Dead Sea, which makes their water available for use in groundwater recharge plans.

13.3.3.2 Groundwater

As an ultimate base level for surface and groundwater, the Jordan Valley bottom contains groundwater of different sources such as lateral groundwater flows from a variety of mountain aquifers containing fresh to brackish groundwater, surface water of side wadis, recharge water from precipitation, treated wastewater, irrigation return flows and others. Therefore, the different parts of the Jordan Valley area contain groundwater with diverse groundwater qualities. But at the mountain foothills, the alluvial fans contain in general fresh groundwater recharged by flood flows. Such alluvial deposits can still accommodate large amounts of recharge water.

Table 13.3 Average pre-development discharges of side wadis and inter-catchments into the eastern Jordan valley area (MCM/yr)

Catchment area	Flood flow	Base flow	Total
Yarmouk (at Adasiya)	182	218	400
El-Arab	6.48	24.90	31.38
Ziqlab	2.2	8.3	10.5
Yabis	1.63	6.2	7.83
El-Jurm	0.23	11.5	11.73
Rajib	1.31	3.0	4.31
Kufranja	1.02	5.8	6.82
Shueib	1.77	8.0	9.77
Kafrain	1.35	12.0	13.35
Hisban	0,34	6.3	6.64
Zerka	46.52	48.3	94.82
Inter-catchment areas			
Arab and Ziqlab (AB21)	5	1	6
Ziqlab and Jurm (AB22)	3	1.5	4.8
Jurm and Yabis (AB23)	1.5	0.0	1.5
Yabis and Kufranja (AB24)	0.1	1.3	1.4
Kufranja and Rajib (AB)	0.2	1.2	1.4
Rajib and Zerka (AB10)	0.1	0.3	0.4
Zerka and Shueib (AB25)	3.5	0.0	3.8
Shueib and Kafrain	2	0.4	2.4
Kafrain and Hisban (AB26)	0.0	1.3	1.3
Hisban and Udheimi (AP2)	0.34	0.3	0.64

Table 13.4 Dams on side wadis of the Jordan river (Water authority of Jordan)

Dam	Capacity (MCM)	Year of construction
Wahda on the Yarmouk	100	2007
El Arab	20	1989
Ziqlab	4.3	1966
Kufranja	8.0	2016
King Talal Dam (Zerka River)	89	1977 Raised in 1989
Shueib	2.5	1968
Kafrain	7.5	1968 Raised in 1996
Yabis	7	Under study
Karamah (in the Jordan Valley)	55	1995

Table 13.5 Average discharges of catchments lying downstream of existing dams constructed on side wadis east of the Jordan valley

Catchment areas in the downstream of	Flow in (MCM/yr)		
	Flood	Base	Total
Wadi Arab Dam	Negligible	0.0	0.0
Ziqlab Dam	0.5	0.0	0.5
King Talal Dam	0.7	0.0	0.7
Shueib Dam	Negligible	0.0	0.0
Kafrain Dam	Negligible	0.0	0.0

The decreasing gradients from the mountain foothills toward the Jordan River allowed the rivers and wadis to deposit their coarse-grained sediment loads at the foothills of the JV and their fine-grained sediments, such as fine silt and clay, close to the Jordan River. This has resulted in rapidly decreasing permeability of the sediments toward the Jordan River course. Therefore, the groundwater velocities in these alluvial sediments decrease rapidly in a westerly direction.

The banks of the Jordan River are built Lisan Marls, which in the southern half of the southern part of Lower Jordan Valley contain salty groundwater because these Marls contain dissolution salts (gypsum and halides). Therefore, the groundwater, which flows toward the Jordan River, becomes gradually salty the closer it moves toward the Jordan River course. The alluvial fans contain fresh water in their apex areas that becomes saltier in their toe areas.

The geological, hydrological, and geophysical surveys carried out to study the potentials of the different parts of the eastern Lower Jordan Valley area for MAR, allow us to define these potential areas from the standpoints of the presence of suitable aquifers, their capacities to accommodate recharge water, existing groundwater quality, availability of recharge water, extraction potential of the recharged water and its potential uses (Al-Amoush et al. 2012; Fehdi 2000; El-Hassan 1999; Hawi 1990; El Naqa 1989). These areas are illustrated in Fig. 13.8.

The quantities of water that can be recharged in the different potential areas were calculated using storage coefficient measurements obtained from field and pumping tests, extension of rechargeable aquifers, groundwater levels and the allowed water level rises so as not to cause water logging, which may threaten existing land use. However, this potential area for MAR is controlled by feasibility studies for real-time implementations.

As an example, the Suleikhat area (Yabis and Kufranja (AB24), Inter-catchment) was found a suitable site for MAR due to the availability of a large excavated gravel pit of around 140.000 m^2 and 15 m in depth (Fig. 13.9). Moreover, it lies very close to the fresh water (King Abdullah Canal). The groundwater table lies at a depth of 20–30 m below ground; therefore, it has a high capability of storing 5–10 MCM of recharged water with easy abstraction (hydraulic conductivity of gravel: 5.10^{-3} m/s). The salinity of the groundwater in the up-gradient area is 1000–1200 μS/cm.

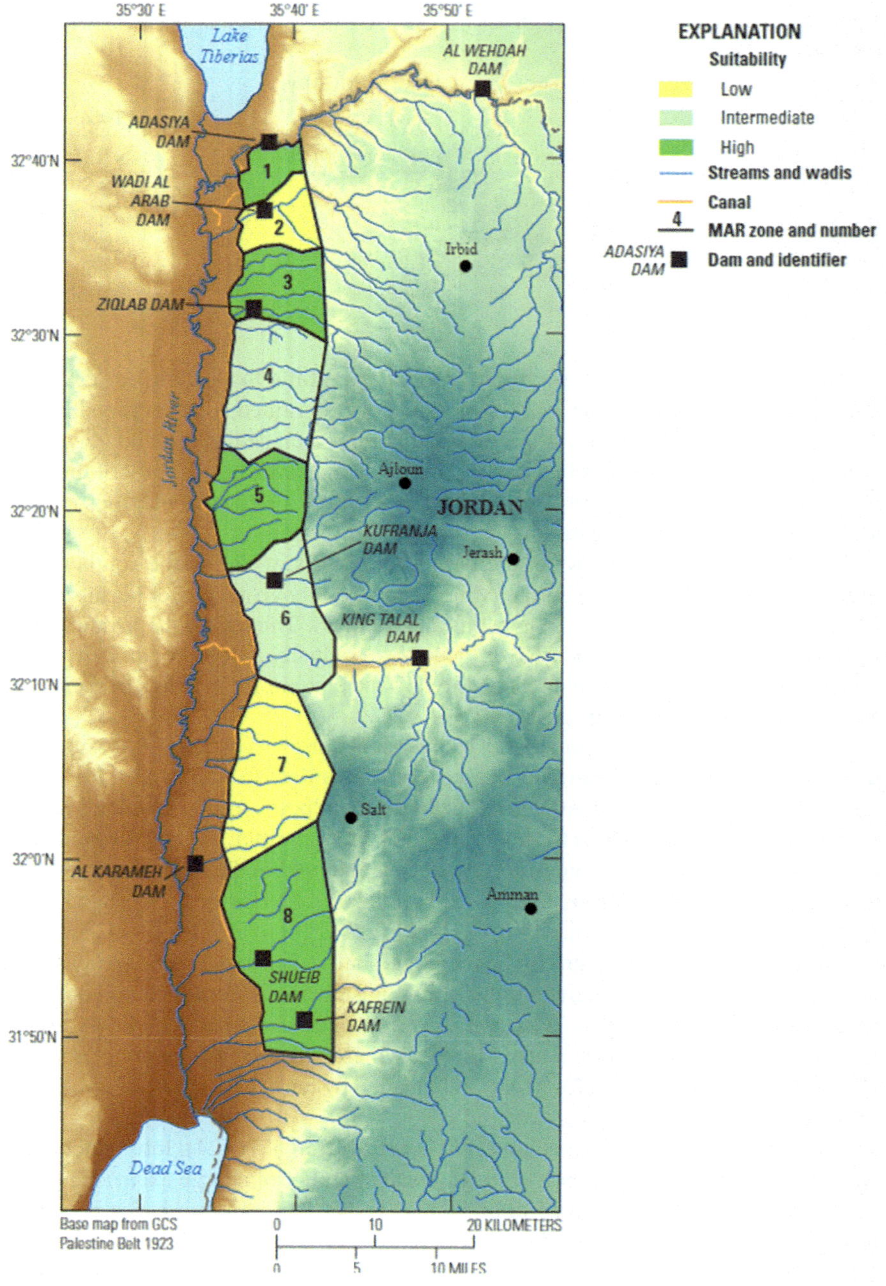

Fig. 13.8 Potential sites for MAR on the eastern side of the Lower Jordan Valley

Fig. 13.9 Suleikhat gravel pit as a MAR potential site (5) in the Yabis and Kufranja (AB24) Inter-catchment area of the Jordan valley

Potential recharge areas in the Jordan Valley from a hydrogeological point of view are given in Fig. 13.8. However, these areas do not tell anything about the potential uses of the stored water for a certain purpose considering the groundwater quality found in these areas.

Starting in the north in the potential recharge area between Yarmouk River and Wadi Arab, the groundwater in the area is fresh to brackish and is used at present for irrigation. The potential water sources, which can be used for recharge are fresh floodwater from the Yarmouk River and treated wastewater discharged from the Wadi Arab wastewater treatment plant. From a water salinity point of view, both sources will slightly deteriorate when upon recharge they mix with the higher salinity groundwater of 1600–4000 μS/cm compared with the Yarmouk River water salinity of around 1000 μS/cm and treated wastewater salinity of around 1200 μS/cm (MWI, open files 2016a). From a biological point of view, both recharge water sources will improve upon recharge, because their contents on microorganisms and organic matter will be decreased due to natural attenuation. The capacity of the aquifer to be recharged is limited because the depth to groundwater in the area is less than 10 m. It will not be appropriate to recharge this aquifer for drinking purposes, but some recharge activities can be practiced to improve the water quality of the aquifer and to enhance the quantity and quality of the groundwater used for irrigation.

The area between the wadi Teibeh and the wadi Ziglab is a high-potential area for groundwater recharge for later use in domestic purposes. The potential water sources are flood discharges of small inter-catchments between the wadi Arab and the wadi Ziglab, and downstream catchments of the wadi Arab and the Ziglab dams. The water quality of the floods is good and can serve the purpose of groundwater recharge for later use in domestic purposes. The potential storage is good and the area can accommodate several millions of cubic meters of water.

The area between the wadi Ziglab and the wadi Yabis is of intermediate potential for groundwater recharge. The recent sediments contain, in general, high percentages of marl and clay, and therefore, for any groundwater recharge project, a detailed study has to be carried out.

In the area extending from the wadi Yabis to Kreima, huge alluvial fans are found with an N-S extension of about 12 and an E–W of about 5 km. This alluvial deposit contains fresh water in its eastern parts but, in its most western parts, starts to acquire higher salinities of up to 2000 μS/cm. The water sources for recharge are the excess water of King Abdullah Canal and the floods of the area's eastern wadis. The recharged water could later be extracted for domestic water supply. In this area, a big excavated gravel pit exists with an N-S extension of about 700 m and an E–W of about 200–300 m. The depth of groundwater is about 20 m. Here, several millions of cubic meters of water can be recharged and later to be extracted along the western margins of the alluvial fan where grain sizes fine up and permeability decreases, and before the water comes in contact with the Lisan Formation, which may contain brackish water.

The area extending from Kreima to Ma'adi is composed of alluvial deposits and sands with marl. It is an intermediate aquifer with fresh to brackish water quality but with a high percentage of irrigation return flows originating from irrigation using King Talal dam water, which, consists, of more than 60% of treated wastewater mixed with 40% flood and base flow water. Hence, the groundwater in the area is of impaired quality with high salinity in places and high pollutant contents and hence cannot be used as a source of drinking water. Groundwater recharge in this area does not seem to be a good option, because irrigation return flows are recharging the semi-aquifer. In addition, since the groundwater is connected to the surface water of the Zarqa River, which mainly discharges treated wastewater, the groundwater in the area is naturally replenished by the river's water, which makes recharge schemes unnecessary.

The area extending from Ma'adi to Karama is composed of Triassic, Jurassic siltstones and limestones containing residues of evaporates, and Lower Cretaceous sandstone receiving water by upward leakages, the groundwater in the area is brackish. In addition, a salt dome in the Damya area, Thahret El Grain, contributes to the groundwater salinity. Therefore, recharging the groundwater deprives the objectives of groundwater recharge, especially because it will cause deterioration of the recharged water quality.

The area extending from Karama to Suweima (The Dead Sea) is composed of sandstones, siltstones, and limestones of Triassic and Lower Cretaceous ages containing brackish water, may be residues of salty water of Dead Sea ancestral Lakes, especially the Lisan Lake or a result of Triassic evaporate residuals dissolution. These deposits inter-finger with the Lisan Formation, which underlies the western parts of the area. The area receives appreciable amounts of side wadis flood and base flows in addition to treated wastewater from Salt and Fuheis treatment plants. The groundwater quality, in most parts of the area, is impaired with a relative high salinity of around 2200 μS/cm and high percentage of irrigation return flows. Groundwater artificial recharge of the area's aquifers is recommended, but only for later use.

13.3.4 Wadi Araba Side Wadis

Wadi Araba side wadis from north to south (Fig. 13.10) present further potential sites for MAR. These sites have several factors that are required for successful MAR as presented in the following sections (Eggleston et al. 2021; Salameh and Abdallat 2020).

13.3.4.1 Water Resources

Table 13.7 shows the surface water resources along Wadi Araba. Fifa side wadi (DB) has the highest water availability in the area, about 11.16 MCM/yr. Surface water catchments of the eastern side of the Wadi Araba study area (Table 13.6), their names, codes, and catchment areas are shown in Fig. 13.11 (Table 13.7).

13.3.4.2 Geology

Alluvial deposits of Pleistocene age cover older rocks of mainly Precambrian, Cambrian, and Ordovician ages of the eastern side of Wadi Araba. These alluvial

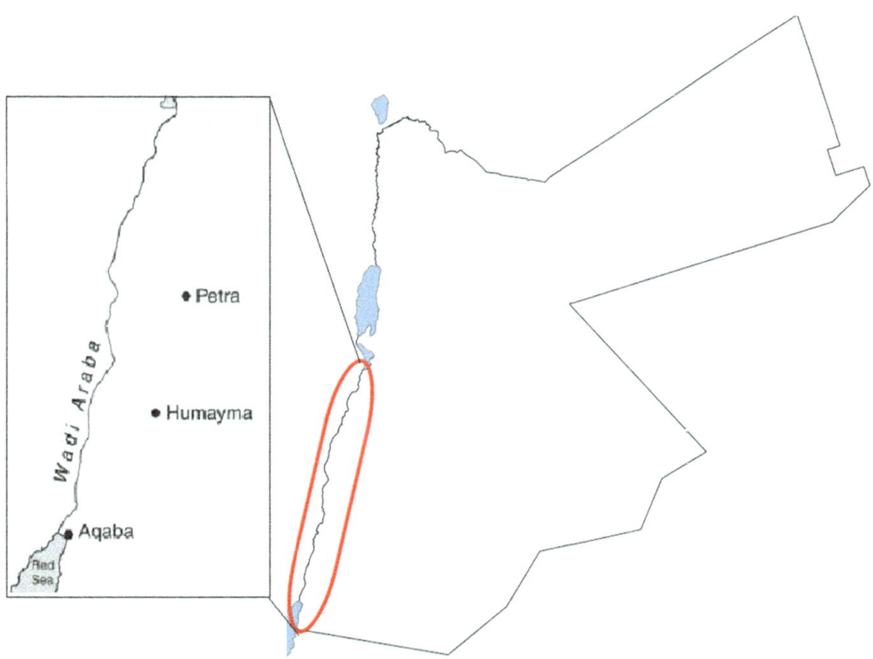

Fig. 13.10 Wadi Araba study area location

Table 13.6 Summary of findings of the potentials of MAR along the eastern side of the Lower Jordan Valley

Area	Aquifer type	Accommodation capacity of the aquifer	Groundwater level below ground surface	Source of recharge water	Recharge method	Existing groundwater quality	Benefits of MAR
Yarmouk to Wadi Arab	Alluvial fans	2–3 million	5–10 m	Yarmouk floods or Treated wastewater	Weirs or pools	Brackish	Improving groundwater salinity for irrigation
Teibeh	No suitable aquifer						
Waqqas to Ziglab	Alluvial deposits	A few millions	20–30 m	Floodwater of side wadis	Weirs or pools	Drinking	Drinking water
Ziglab to Yabis	Alluvials with clays	Not determined	10–25	Floodwater of side wadis	Weirs or pools	Irrigation quality	Irrigation
Yabis to Kufranja (Suleikhat)	Alluvium: Gravel, sand and silt	15–25 MCM	20–30 m	Excess water in King Abdullah Canal	Pool with a capacity of 2.4 MCM is there	Drinking grade	Drinking water
Kureima to Ma'adi	Alluvial fan	Irrigation-return flows, deeply incised wadis, and deteriorated groundwater quality do not allow MAR					
Ma'adi to Karama (Kibid)	Alluvials, Upper Cretaceous and Triassic sand stones	The groundwater in this area is saline, hence aquifer recharge is useless					
Karama to Sweima	Alluvial deposits overlying sandstones	Few tens of millions	20–35	Flood flow of side wadis and treated wastewater	Weirs and pools	Highly impaired by irrigation return flows	Irrigation water

Table 13.7 Wadi Araba side wadis from north to south

Name	Catchment area (km^2)	Precipitation (mm/yr) (MCM/yr)		Flows MCM/yr flood base total		
Fifa (DB)	161	206	33.2	1.16	10.0	11.16
Khuneizira (DC)	183	234	35.6	1.18	3.0	4.18
Dahal (DD)	96	197	18.9	0.30	0.04	0.34
Ghauba (DA25)	100	153	13.5	014	0.0	0.14
Fidan (DE)	280	235	65.8	1.32	4.1	5.42
Buweirda (DF)	518	156	81.1	2.76	0.8	3.56
Musa (DG)	165	142	23.5	0.09	0.3	0.39
Huwar (DH)	229	144	32.9	0.29	0.5	0.79
Darba (EA25)	116	-	19.4	0.29	0.0	0.29
Abu Burqa (EB)	136	136	18.5	0.22	0.12	0.34
Rakiya (EC)	182	128	23.3	0.09	0.2	0.29
Yutum (ED)	4390	112	69.8	1.45	0.0	1.45

(km: kilometer; mm: millimeters; yr: year; MCM: millioncubic meters)

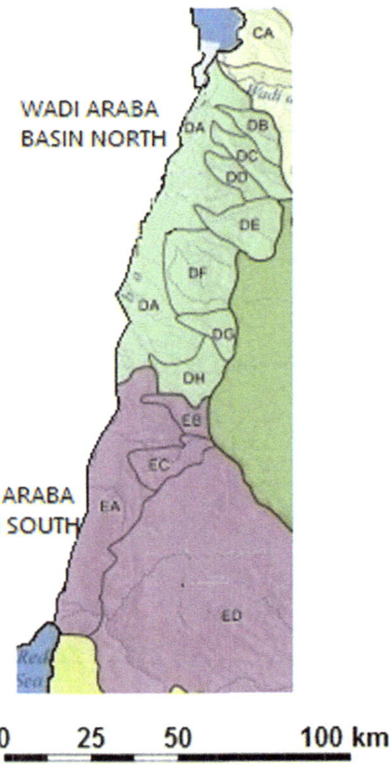

Fig. 13.11 Surface water catchments of the eastern side of the Wadi Araba study area, their names, codes, and catchment areas

fan deposits consist of gravels at their apexes and silt and fine sands at their toes (Fig. 13.12).

Some of the alluvial fans are large (a few kilometers by a few kilometers), contain fresh water, and have the capacity to accommodate additional recharge water.

In the following points, two examples are presented and discussed in short:

Rahma alluvial fan

This alluvial fan is located in the Middle of Wadi Araba between coordinates 920–930 N and 160–167 E and covers an area of about 25 km^2 at an elevation of 200 mamsl in the east and 110 m amsl at the road from Aqaba to Amman in the west with a slope of about 3%.

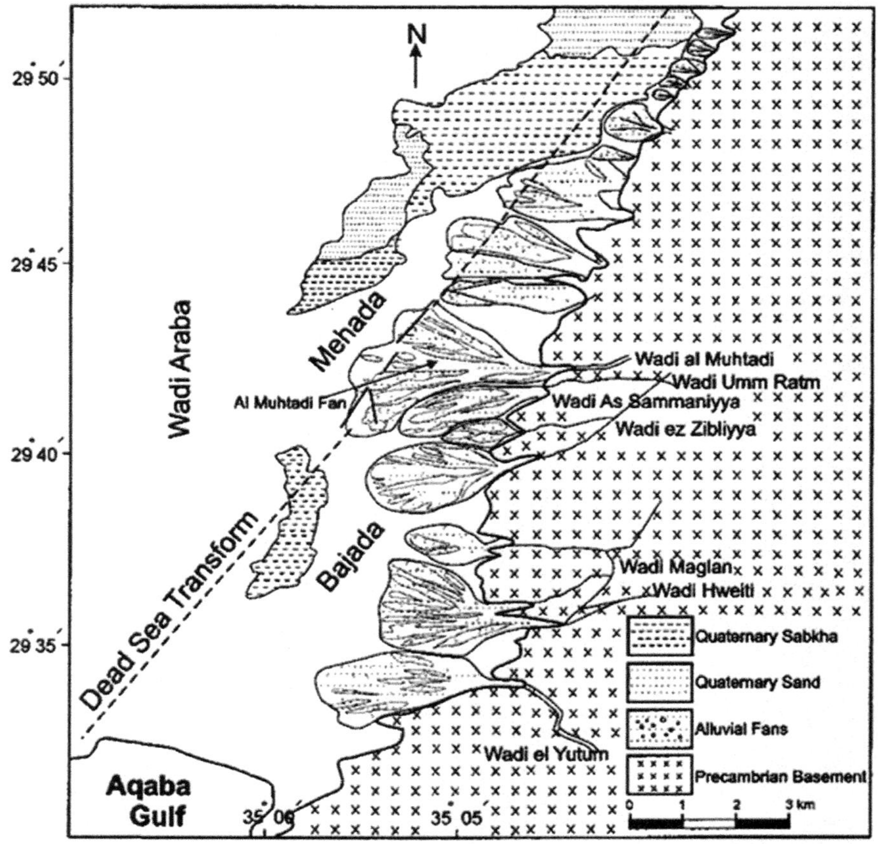

Fig. 13.12 Potential MAR sites along Wadi Araba. Alluvial fan deposits as a MAR potential site in Wadi Araba area, Wadi Buweirda (N 30° 35′, E 35° 25′)

At the mountain slope in the east (200 mamsl), the alluvial fan inter-fingers with slope debris consisting of boulders and coarse-grained sediments whereas, the alluvial fan consists of gravels and coarse-grained sand.

The groundwater level at the apex of the alluvial fan in the east lies at around 80 mamsl or 50 m below ground surface and at about 30 mamsl or 20–30 m below ground surface in the west. The groundwater flow is generally east–west oriented toward the center of Wadi Araba.

The groundwater salinity in the drilled wells ranges from about 500 μS/cm in the east to about 1100 μS/cm in the west, which is good fresh water. The drainage porosity of the alluvial fan is around 15% and its permeability is around 9.0×10^{-4} m/s.

Wadi Malaga Wadi Finan Area

This area extends from Humrat Finan in the west to Jabal Mubarak and Wadi Dana entrance into Wadi Araba in the east (Coordinates E: 186.5–194.0) and from the head of Wadi Malaga in the south to Wadi Finan in the north (Coordinates N: 001.0–004.5) with a total area of about 28 km^2. In the east, the topographic elevation is around 250 mamsl, and in the west, around 100 mamsl. Generally, it is a flat area incised by the shallow wadis of Malaga Feinan and Fidan.

The area is composed of the alluvial fans deposited along Wadi Dana and its Southern tributary wadis; Mubarak and Malaga. The thickness of the alluvial fan in the west is >200 m and in the east is >150 m, consisting of gravel, sand, and some silt, in addition to clay lenses. Underlying the alluvial fan, rocks of different geologic formations are found including Upper Cretaceous calcareous rocks, and Quaternary clastic rocks. The groundwater salinity ranges from 600 to 1300 μS/cm. The drainage porosity of the alluvial fan is around 15% and its permeability is around 2.5×10^{-3} m/s.

13.3.5 Aqaba Area

In this area, three sites are found suitable for groundwater artificial recharge. There, three areas are located: (a) in the southernmost part of Wadi Araba, just north of Aqaba City, (b) in the Durrah area, just north of the Saudi Arabian border, and (c) along Wadi Yutum, which originates in Disi area and flows southwards to the Gulf in Aqaba City (Salameh and Abdallat 2020). These three sites are shown in Figs. 13.13 and 13.14.

13.3.5.1 North of Aqaba Area

The area north of Aqaba City is covered by clastic sediments deposited from the eastern side wadis of Wadi Araba and contains brackish water with a salinity of 1500–2500 μS/cm. The alluvial deposits are very thick hundreds of meters with a

Fig. 13.13 Alluvial fan deposits as a MAR potential site in the Aqaba shore area

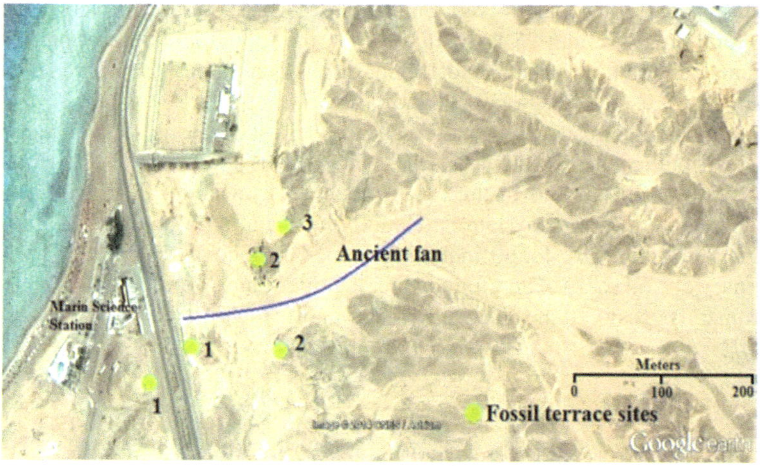

Fig. 13.14 Alluvial fan in Aqaba (Wadi Durrah)

porosity of 15–20% and groundwater levels at depths of 20–50 m. Excess water for recharge is provided by the treated wastewater of Aqaba city. The treated wastewater is collected in several pools and it infiltrates to join the brackish groundwater and reduces herewith its salinity. The infiltrating treated wastewater undergoes also natural attenuation processes during infiltration passage and underground flow and its quality becomes improved from a biological point of view. The natural groundwater current is oriented southward to the Gulf of Aqaba, which will allow drilling wells south of the recharge area to extract the recharged water and reuse it in irrigation.

Wadi Yutum

Wadi Yutum is a deep-cut wadi into the granitic Basement Complex of Aqaba in the area extending from Disi in the north and to the Gulf of Aqaba in the south, with a total length of around 40 km. It drains areas in Disi that are covered by sandstones, sands, and other areas on both sides of the wadi that are covered by granitic rocks. The width of the wadi at its granitic bottom ranges from around 100 m to around 500 m. Alluvial sediments of gravels, sands, and some silt build the aquifer with a total thickness of about 130 m at the top of the Wadi and decreasing to about 40 m at the entrance of the Wadi into the alluvial fan in Aqaba.

The groundwater level is below the ground surface and ranges from 18 to 50 m indicating the potential for groundwater recharge. The permeability (hydraulic conductivity) ranges from 1.0×10^{-4} m/s to 7.8×10^{-3} m/s. The salinity of the groundwater ranges from 840 to 1100 μS/cm, with almost total absence of pollution parameters. The flood flow of the wadi, which can be used for recharge is about 1.4 MCM/yr.

Durrah area

This area is composed of a wadi, 300–400 m wide near the shore of Aqaba close to the border with Saudi Arabia, narrowing up in an easterly direction to less than 50 m. The wadi bottom and the surrounding hills are composed of alluvial sediments of around 100 m in thickness at the wadi bottom. The groundwater in this alluvial aquifer builds an interface of fresh/salt water with seawater. The catchment area of the wadi measures around 460 km^2 sloping very strongly from more than 1000 mamsl to sea level with a distance of 10–15 km. Impermeable granitic rocks cover most of this catchment area. The rare rain events are committed as storm rain producing huge short flash floods discharged into the Gulf of Aqaba without being utilized. Hence, it is important to make use of these floodwaters. Hydraulic weirs, along this wadi course, seem to be the viable solution to collect the floodwater for recharge purposes and to alleviate the hazards induced by the flooding events, which year after year cause damage to infrastructure and the landscape. The flood flow amount that can be used for recharge is about 1.5 MCM/yr.

13.3.6 Jafr and Azraq Enclosed Basins

Jafr and Azraq basins are considered potential sites for MAR. These enclosed basins receive a high amount of floodwater. Jafr Depression lies some 200 km east of the Jordan Rift Valley (Fig. 13.15). At times, during the Pleistocene, this basin held one of the largest freshwater lakes in Southwest Asia. However, currently, its water has become salty through high evaporation rate and dryness. Thus, MAR processes can be implemented up-gradient of the saline water zone as shown in Fig. 13.16, within the red circle. Small dams or weirs can stop freshwater before it reaches the saline zone and provide more time for infiltration. Figure 13.16 shows Azraq enclosed basin has several high-potential sites for MAR where, the collected fresh water becomes salty due to the high evaporation rate in the area (Table 13.8).

Fig. 13.15 Jafr depression in red circle

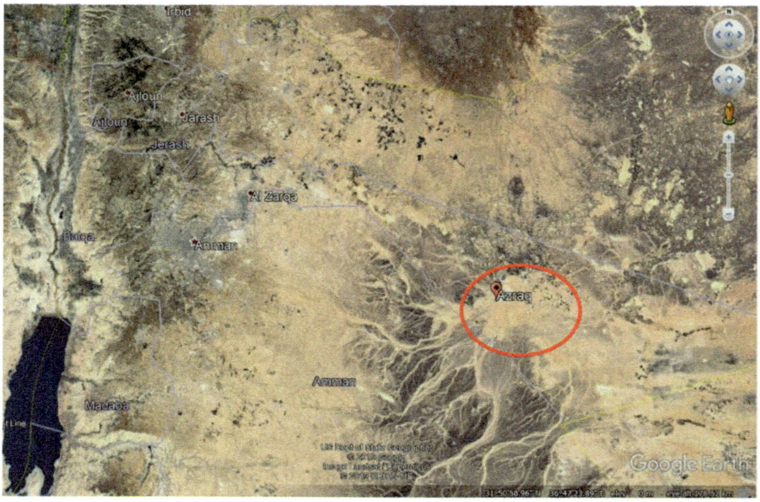

Fig. 13.16 Azraq enclosed basin

Table 13.8 Hydrogeologic parameters in the Jafr area

Aquifer thickness	80–100 m
Saturated thickness	40–60 m
Depth to Water	30–40 m
Salinity	1000–8000 μS/cm
Total pumping	35 MCM/yr
Recharge	27 MCM/yr
Porosity 7%	
Permeability 6×10^{-5} m/s^{-1} $\times 10^{-4}$ m/s	
Vadose zone Chert and limestone	
MAR Capacity 50 MCM	

13.3.7 Other Sites and Projects

In different parts of Jordan, studies have been carried out on potential locations for MAR (Al-Kharabsheh 1995; Al-Alami 1996; Al-Alami 1999; Abu-Taleb 2003; Al-Amoush et al. 2012; De Laat and Nonner, 2012; Salameh et al. 2018b). However, MAR projects have only been implemented at some of these sites. These sites are presented and evaluated in the following.

13.3.7.1 Wadi Madonnah and Wadi Butum

This area has been selected after a reconnaissance study carried out by the Ministry of Water and Irrigation supported by USAID to locate potential sites for MAR in the highlands of Jordan (de Laat and Nonner 2012). The catchment area of the wadi Madonnah is 37 km^2, with an average long-term precipitation of 169 mm/year (MWI, open files 2018). Three weirs were constructed along the wadi Madonnah located in the Badia area, South East (SE) of Amman with a height of 3–6 m in-channel to collect floodwater for MAR purposes by discharging the water for infiltration in the downstream area.

The capacities of the weirs ranged from around 2000–66,400 m^3. The evaluations showed that the weirs are oversized but a small amount of recharge took place from these weirs.

13.3.7.2 Muwaqqar Area, Wadi Mugheir

Along wadi Mugheir, 40 km SE of Amman, with an upstream catchment of 70 km^2, three successive weirs were constructed to collect flood flow water for use in supplementary irrigation. The lakes of the weirs served as infiltration basins for groundwater recharge (Salameh et al. 1991).

The catchment area receives an average long-term amount of rainfall of 150–200 mm/yr (MWI. Open files 2018). Short, intensive floods characterize the area. Clastic sediments mainly composed of thin and compacted clastic sediments of gravel and breccia cover the bottom of wadis. The reservoir capacities of the weirs up to the spillway level and the reservoir surface areas are as follows:

- Weir 1 (upstream): 15,580 m^3 and 7200 m^2.
- Weir 2 (middle): 14,860 m^3 and 6000 m^2.
- Weir 3 (downstream): 16,560 m^3 and 8650 m^2.

Water balances for the three weirs were carried out based on onsite-measured evaporation, incoming flood flows, extraction from the weirs, and drop in water levels. That allowed calculating the infiltration to the groundwater of 50–60 mm/day. The infiltration rates decreased after three years of operation by 10–20% due to siltation (i.e., clogging), a matter that must be considered when planning such weirs.

Another issue to be seriously considered is that the floodwater in the upstream areas may naturally infiltrate if left to flow along the natural watercourse toward the base level, which makes such recharge schemes unnecessary unless otherwise proven or other purposes are targeted at.

13.3.7.3 Rajil Dam

This dam was constructed on the wadi Rajil that drains a total catchment of about 324.3 km^2. The dam is located at a few kilometers upstream of its base level; the Azraq Oasis.

The dam's storage capacity is 3.4 MCM. The calculated infiltration rate from the dam lake bottom is 140 L/s when the dam is filled. The dam partly served as a water-harvesting project, but not as a recharge dam because the dam lakebed proved to be of low permeability allowing only small rates of infiltration and hence MAR potential. Therefore, the dam failed to fulfill its purpose as a recharged dam. Nonetheless, its use as a water-harvesting project is satisfactory.

13.3.7.4 Other Sites

Many other studies were conducted on the highland for MAR and water harvesting purposes such as Wadi Jardaneh in the Jafr basin, Wadi Swaqa, and Wadi Abu Safat within the Dead Sea basin, where it was proven that the high evaporation rates in such areas will not allow successful recharge schemes.

Other sites were also studied intensively for MAR and water harvesting but projects have not yet been implemented. Such sites are Wadi Karak in the Dead Sea basin, Wadi Hadi, a tributary of Wadi Shueib in the Jordan Valley basin, and many others. In such areas, MAR projects have another purpose, and that is minimizing erosion in the high-relief areas of the Jordan Rift Valley Mountains.

13.4 Conclusion

The increasing demands for water in Jordan required sustainable water-saving programs, such as MAR. Therefore, it was of utmost importance to define the high-potential sites for MAR as presented through this study, based on the availability of water for recharge, the need for the recharged water, the suitable aquifers, and the availability of recharge areas and extraction potentials. Here, MAR can add useable amounts of water for the different use sectors. The screening of sites for the case of Jordan shows the strength and importance of applying rigorous feasibility studies in order to guarantee safely recovering the recharged water in quantity and quality. Based on this study, the potential MAR sites in Jordan can be delineated as follows:

- Along the main wadis discharging into the Jordan Valley and Wadi Araba, groundwater recharge schemes are of high potential at the foothills, where the main side wadis enter the Jordan Rift Valley. In the upstream areas, surface water storage has high potential, with some bank storage around the surface water reservoir. Owing to the effluent nature of the groundwater, along the upper reaches of these wadis groundwater recharge is not feasible.
- Wadis discharging into the Gulf of Aqaba possess, along their lower reaches, high potentials for MAR because of the high porosity and infiltration rate of alluvial fans covering these areas.

In general, thick layers of coarse sediment (for instance alluvial fans, beds of ephemeral rivers, wide wadis) are beneficial for MAR. MAR can alleviate the impact of climatic changes on the availability of water in arid and semi-arid areas as the case in Jordan by increasing the stored amounts of groundwater

References

Abdallat G, Harahshah E, Salameh E (2020) Management of treated wastewater and flood water using GIS for environmental protection in Jordan. J. Geosci. Environ. Prot. 8(01):86. https://www.scirp.org/journal/Paperabs.aspx?PaperID=97892

AbdelKhaleq RA, Alhaj Ahmed I (2007) Rainwater harvesting in ancient civilizations in Jordan. Water Sci. Technol.: Water Supply 7(1):85–93

Abu-Taleb MF (2003) Recharge of groundwater through multi-stage reservoirs in a desert basin. Environ Geol 44:379–390

Al-Alami H (1996) The potential of weir use as dam substitute for surface water storage and artificial recharge in Karak Area (Doctoral dissertation, University of Jordan)

Al-Alami H (1999) Watershed Management in Wadi Hadi (Jordan) (Doctoral dissertation)

Al-Amoush H, Salameh E, Al-Raggad M (2012) Hydro-geoelectrical investigation for the potential of underground water storage along the lower reaches of King Abdullah Canal—Deir Alla Area/Jordan. J Water Resour Prot 4(7):545

Al-Kharabsheh A (1995) Possibilities of artificial groundwater recharge in the Azraq Basin: potential surface water utilization of five representative catchment areas (Jordan) (Doctoral dissertation, Lehr-und Forschungsbereich Hydrogeologie und Umwelt)

Asano T (ed) (2016) Artificial recharge of groundwater. Elsevier, Butterworth Publishers, Boston, London
Bandel K, Salameh E (2013). Geologic development of Jordan: evolution of its rocks and life. University of Jordan Press, p 277. Accessed 1 June 2020 at https://www.researchgate.net/publication/275946617_Geologic_Development_of_Jordan_-Evolution_of_its_Rocks_and_Life.
Bender F (1968) Geologie von Jordanien. Gebrüder Burnträger, Berlin
Bender F (1975) Geology of the Arabian Pennisula, Jordan. U.S. Geological Survey Professional Paper 560-I, p 136, 3 plates, https://doi.org/10.3133/pp560I
Bouwer H (2002) Artificial recharge of groundwater: hydrogeology and engineering. Hydrogeol J 10(1):121–142
Bundesanstalt für Geowissenschaften (BGR) (1965) Bundesanstalt für Bodenforshung. In: Bender F, Flate H, Futian A (eds) Results of the hydrogeological and geoelectrical survey inthe Southern Jordan Valley, Hannover and Amman
Bundesanstalt für Geowissenschaften (BGR) und Rohstoffe 1977 National Water Master Plan of Jordan, Hannover
Chowdhury A, Jha MK, Chowdary VM (2010) Delineation of groundwater recharge zones and identification of artificial recharge sites in West Medinipur district, West Bengal, using RS, GIS and MCDM techniques. Environ Earth Sci 59(6):1209
De Laat JM, Nonner JC (2012) Artificial recharge with surface water a, a pilot project in Wadi Madoneh. Jordan. Environ Earth Sci 65:1251–1263
Department of Meteorology (JDoM), Jordan, open files (2016). https://doi.org/10.3133/ofr20211089
EXACT (1998) Overview of Middle East Water Resources. Compiled by the U.S. Geological Survey for the Executive Action Team (EXACT) of the Middle East Water Resources Working Group, 44 p. ISBN 0-607-91785-7,. Accessed 1 June 2021: https://www.researchgate.net/publication/317833058_Overview_of_Middle_East_Water_Resources.
El-Hassan O (1999) Geology and hydrogeology of the area between Deir Alla and Suleikhat. A geologic-geophysical study for the potential of groundwater artificial recharge. Unpublished M.Sc. Thesis, University of Jordan, Amman
El-Naqa AR (1989) Hydrogeochemistry of groundwater resources in the area between wadi El-Yabis and Yarmouk River/Jordan Valley area. Unpublished M.Sc. Thesis, University of Jordan, Amman
Fehdi C (2000) Geology and hydrogeology of the area between Suleikhat and Waqqas. A geologic-geophysical study for the potential of groundwater artificial recharge. Unpublished M.Sc. Thesis. University of Jordan, Amman
Gibb A, Partners (1993) Storage Facility in Wadi Mallaha, Karameh Dam Project, prepared for the Ministry of Water and irrigation, Jordan
GIBB report (1994) Inception Report, Karamah Dam, Amman
Eggleston J, Valk M, Kagone S, Velpuri N, Senay G, ElHariry N, ..., Goode D (2021) Managed aquifer recharge suitability–a regional screening and case studies in Jordan and Lebanon. United States Agency for International Development
Hawi MS (1990) Hydrogeology and groundwater flow system in the area between wadi Yabis and Yarmouk River/Jordan Valley area, Unpublished M.Sc. Thesis, University of Jordan
Heviánková S, Marschalko M, Chromíková J, Kyncl M, Korabík M (2016) Artificial ground water recharge with surface water. In: IOP conference series: earth and environmental science, vol 44, No 2. IOP Publishing, p 022036
Mansour Z, Chik Z, Taha MR (2010) Predicting the geotechnical performance of Lisan Marl using finite element analysis, Dead Sea area. Nat Hazards 54(3):931–947
Ministry of Water and Irrigation (MWI) (2016a) National Water Strategy 2016–2025: Water Sector Policies. Ministry of Water and Irrigation, Amman, Jordan
Ministry of Water and Irrigation (MWI) (2016b) Amman, Jordan, Open Files. Available online: http://www.mwi.gov.jo/sites/en-us/default.aspx
NRA: Natural Resources Authority Amman, Jordan, Open files

Pierson WA (2021) Spatial assessment of urban growth in cities of the decapolis; and the implications for modern cities. Graduate Theses and Dissertations. https://scholarworks.uark.edu/etd/3972
Salameh E (2001) Sources of water salinities in the Jordan valley area/Jordan. Acta Hydrochim Hydrobiol 29(6–7):329–362
Salameh E (2001b) The potential of groundwater artificial recharge in the Jordan valley area/ Jordan: selected contributions to applied geology in the Jordan rift valley. Freib Forsch Hefte C 494:63–81
Salameh E, Abdallat G, van der Valk M (2019a) Planning considerations of managed aquifer recharge (MAR) projects in Jordan. Water (Basel) 11(2):182. https://doi.org/10.3390/w11020182
Salameh E, Abdallat G, van der Valk M (2019b) Planning considerations of managed aquifer recharge (MAR) projects in Jordan. Water 11(2):182. https://www.mdpi.com/2073-4441/11/2/182
Salameh E, Abdallat G (2020) Potential areas for managed aquifer recharge in the eastern lower Jordan valley area. J Water Resour Protect 12:330–357. Open access. https://doi.org/10.4236/jwarp.2020.124020
Salameh E, Rimawi O, Bannayan H (1991) Groundwater artificial recharge in Jordan. Water research and Study Center, University of Jordan National Library Dep. No. (174/4/1991), p 197
Salameh E, Shteiwi M, Al Raggad M, Salameh E, Shteiwi M, Al Raggad M (2018a) Water politics. Water resources of Jordan—political, social and economic implications of scarce water resources (World Water Resources Book 1). Springer Publishing Co. https://doi.org/10.1007/978-3-319-77748-1
Salameh E, Shteiwi M, Al Raggad M, Salameh E, Shteiwi M, Al Raggad M (2018b) Water politics. Water resources of Jordan: political, social and economic implications of scarce water Resources. pp 121–141
Steinel E (2012) Stormwater reuse via aquifer storage and recovery: risk assessment for sandy aquifers. In: Climate change and the sustainable use of water resources. pp 17–42
Water Authority of Jordan (WAJ), Open files
Wolf L, Werz H, Hoetzl H, Ghanem M (2007) Exploring the potential of managed aquifer recharge to mitigate water scarcity in the Lower Jordan River Basin within an IWRM approach. In: Proceedings of the 6th international symposium on managed artificial recharge of groundwater, vol 28. ISMAR6, Phoenix, AZ

Chapter 14
Groundwater Artificial Recharge in the Marj Sanour Watershed-Palestine

Sayel Wishahi

Abstract Groundwater forms the main water source in the Marj Sanour Watershed. While domestic water is mainly abstracted from the deeper aquifers, shallow aquifer is mainly used for agricultural purposes. Due to drought and the over-pumping of the agricultural boreholes drilled in the area, the groundwater level of the shallow aquifer has dropped by more than 80 m during the last ten years causing a serious reduction of discharge of the agricultural wells. This has compelled the farmers to look for alternatives to increase the availability of water to sustain their agriculture. Artificial recharge is one of the most promising alternatives that can contribute to enhancing groundwater supplies. Groundwater recharge wells are suggested to be one of the most appropriate techniques that best suites Marj Sanour area due to the following reasons: 1. Availability of floodwater on an annual basis in the Marj area, which submerges large parts of the Marj for a few months. 2. Availability of good storage potential within the shallow aquifer. Accordingly, the Palestinian Hydrology Group (PHG) has proposed a pilot project to drill ten artificial recharge wells under the Land and Water Resources Management Project funded by the Netherland Representation Office (NRO). The aim is to assist in arresting the accelerated decline of the groundwater level and minimize the submerged agricultural area. Among the ten proposed wells, six recharge wells have already been drilled in the part of the watershed which remains flooded for a few months. The source water is filtered through a filtration system before being put into aquifers. The estimated recharge through these six wells is about 20,249 m^3/day.

S. Wishahi (✉)
Senior Hydrogeologist, Palestinian Hydrology Group, Ramallah, Palestine
e-mail: hydrogsayel@gmail.com

D. Saha et al. (eds.), *Managed Groundwater Recharge and Rainwater Harvesting*, Water Resources Development and Management,
https://doi.org/10.1007/978-981-99-8757-3_13

14.1 Introduction

Artificial Recharge is considered one of the main means of enhancing groundwater resources, regulating surface flow, providing underground storage of water, and avoiding its loss to seas or in salt depressions (FAO 2000). With the accelerated overexploitation from the shallow unconfined aquifers, artificial recharge of groundwater can be an effective solution to tame overexploitation. Depending on the local hydrogeological conditions, there are several methods that can be adopted for artificial recharge like infiltration basins and canals, water traps, cutwaters, surface runoff drainage wells, septic tank effluent disposal wells, diversion of excess flows from irrigation canals into sinkholes, etc.

In Palestine, groundwater forms the main water source, where many wells have been drilled in the area targeting the different aquifers. Overexploitation of groundwater has resulted in a considerable decline in groundwater levels with time in some aquifers in the area (Arab Organization For Agricultural Development 2008). Surface water in the area is not used mainly due to the political situation and rainwater mostly drains during rainy months toward the Mediterranean to the west or to the Jordan River and Dead Sea to the east. Thus, rainwater is a potential source of water, which can be harnessed to recharge the depleting aquifers in the area.

14.2 Marj Sanour Watershed

Marj Sanour is located in the northern part of the West Bank within the Jenin governorate area (Fig. 14.1). It forms a unique closed watershed in the West Bank. The catchment area of Marj Sanour covers about 59 km^2 including 16,000 donums forming the Marj Lake area that is considered one of the most fertile agricultural areas in the Palestinian territories, while the other areas are mostly hilly (Palestinian Hydrology Group 2010; Palestinian Hydrology Group, IUCN, UAWC 2013). The watershed has nearly a rectangular shape with northwest-southeast elongation; the length of Marj Lake from northwest to southeast reaches 5–8 km and is 2–4 km in width.

The lake region occupies the area between the 350 and 375 m above msl contour lines. Rainfall precipitates in the area from October to May. The annual rainfall ranges from 400 to 1200 mm with an average of 634 mm (Wishahi 2007).

Carbonate rocks are mostly outcropped in the surrounding mountainous area, while in the lake area, these rocks are overlain by a thick soil and weathered zone reaching more than 30 m at places (Arab Organization For Agricultural Development 2008; Wishahi et al. 2004a). The soil is dark brown in color and is composed of clay, silt, sand, and gravel. This profile is characterized by a high ability for storing water and low permeability. Rain-fed crops are mostly cultivated in the plain area of the lake, while irrigated crops occupy limited areas. Rain-fed trees such as olives are mostly cultivated in the surrounding hilly area.

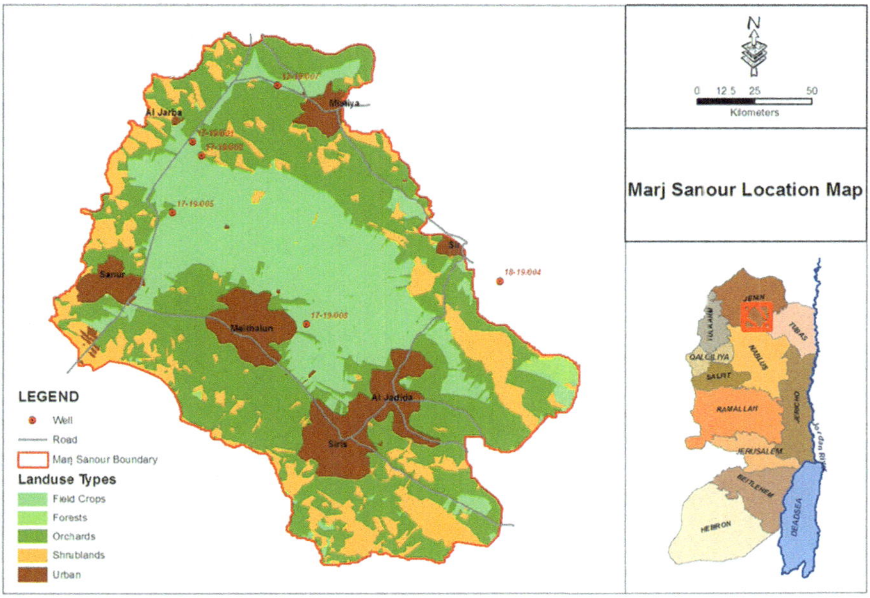

Fig. 14.1 Marj Sanour watershed; location and land use

14.3 Hydro-Geological Setting

Hydro-geologically, the Marj Sanour watershed area is located within the North-eastern groundwater basin (Jenin–Nablus basin). This basin covers about 1050 km^2. Annual recharge of the Northeastern groundwater basin is estimated to be about 145 MCM; 116 MCM is coming from the West Bank area, while 29 MCM is coming from outside areas. Within the watershed area, the groundwater moves from southwest to northeast (Fig. 14.2). The main potential aquifers in the Marj Sanour watershed area are:

- Jenin Formation (Eocene Formation): Due to its lithological composition, this unit forms a local aquifer in the area. Several wells have been drilled in this formation and the hydraulic properties have been found to vary widely due to marl and chalk intercalations. There is a significant fluctuation of water levels of this aquifer. This aquifer is mainly exploited by several wells distributed in the region. At present, the depth of water level in this aquifer ranges between 130 and 150 m below the ground surface.
- Turonian-Upper Cenomanian aquifer system: The three formations, Jerusalem, Bethlehem, and Hebron, form one aquifer unit characterized by its high ground-water potential. This aquifer system is separated from the overlying Jenin aquifer by Senonian aquicludes. The Senonian aquicludes contribute to the aquifer's confined nature and give it natural protection from surface pollutants. The

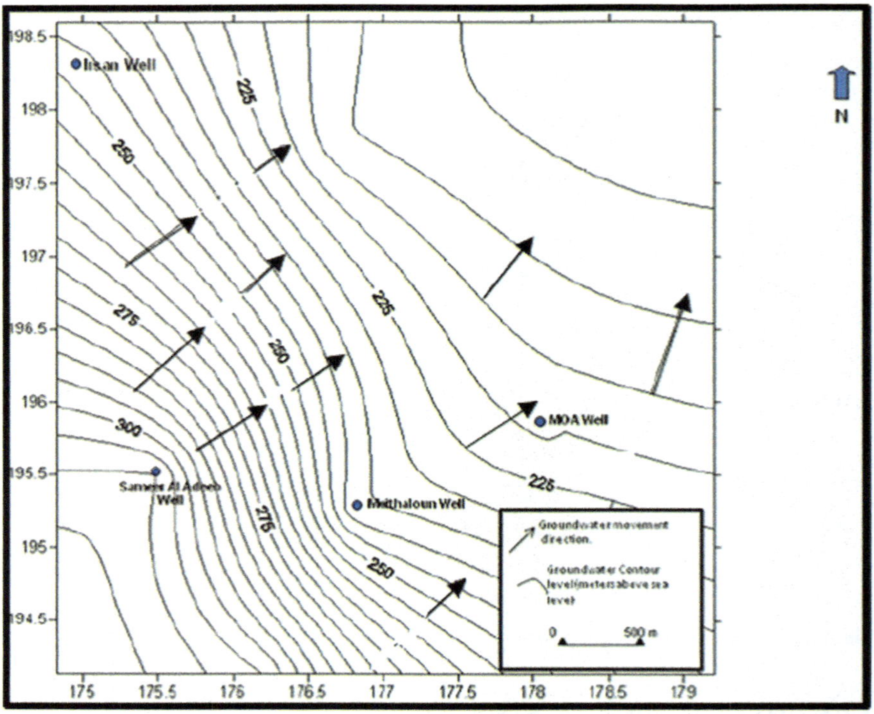

Fig. 14.2 Groundwater movement of Jenin aquifer system

Turonian-Upper Cenomanian aquifer system in the Jenin Governorate area, particularly in the synclinal area, occurs at great depths, even greater than 700 m at certain locations. At Meithaloun, a deep well (1302 m) has been drilled recently targeting this aquifer system. The discharge of the well is found as 200 m^3/hr and is used for domestic purposes for Marj Sanour communities (Wishahi et al. 2004a; Wishahi and others 2004b).

The study by the Palestinian Hydrology Group (2008) reveals that the quality of groundwater in the Marj Sanour watershed is of calcium–bicarbonate type. Results of the analyses show the Total Dissolved Solids (TDS) range between 370 and 1049 mg/l, nitrate ranges from 15 to 80 mg/l, and chloride between 57 and 264 mg/l. The sodium hazard ratio (SAR) of the analyzed samples is within the low range. However, the study showed that all collected samples are biologically contaminated and need to be treated if it is to be used for drinking. Flood water analyses results show TDS range as 230–260 mg/l, nitrate as < 20 mg/l, sodium between 28 and 34 mg/l, and potassium ranging from 3 to 5 mg/l (Palestinian Hydrology Group 2009; Palestinian Hydrology Group 2010). In general, the chemical results show that the flood water is of good quality and is close to the rainwater in composition.

14.4 Problem Statement

Marj Sanour watershed faces the twin problems of water scarcity and flooding. These two problems originate from the same root: mismanagement of the land and water. Due to the drought years witnessed in the region and to the increased groundwater over-pumping from wells tapping the shallow aquifer (mostly used for agricultural purposes) during the past few years, groundwater levels show a drop of about 70 m during the period between 1969 and 2013 (Fig. 14.3) except a rise in 1991 due to above normal rainfall. As a result of the severe water level decline, pumping from the wells has reduced along with deterioration in the water quality, negatively impacting the agricultural sector.

Furthermore, as a result of its nature (without any natural outlet to drain water, and thick clayey layers at the top), Marj Sanour forms a closed watershed. In heavy rainfall years, Marj Sanour Lake is flooded by rainwater that drains from surrounding wadis, and also from rain directly precipitating on the lake area itself. Flooding in the area leads to reduction in the cultivatable land, abetting downward leaching of soil minerals and fertilizers, growing of harmful and undesired grass, increasing anaerobic reactions and limiting aerobic bacteria activity (Arab Organization For Agricultural Development 2008).

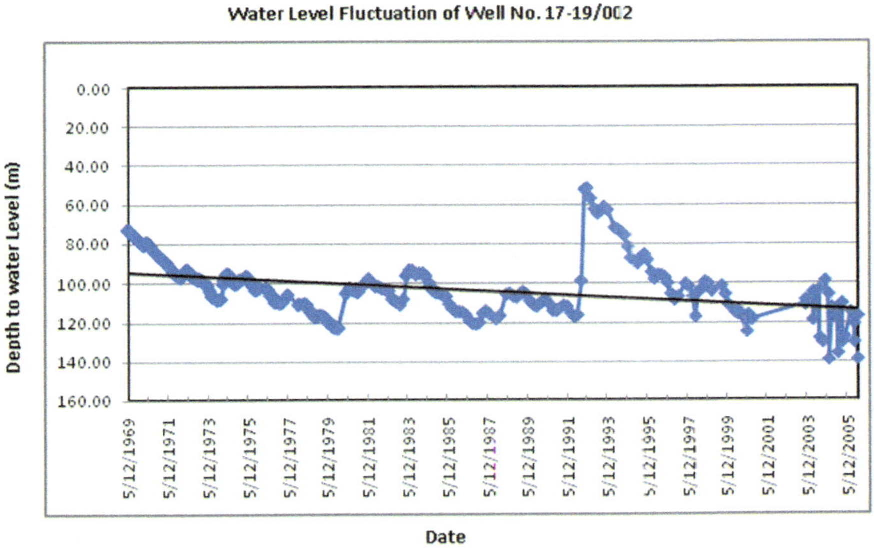

Fig. 14.3 Fluctuation of groundwater level for one of the agricultural wells in the watershed

14.5 Groundwater Artificial Recharge

The idea of groundwater artificial recharge has been studied and discussed further with stakeholders through a regional European Union-funded project "The Social, Ecological, Agricultural Resilience in the face of climate change (SEARCH) project (2011–2014)" that was carried out in the Marj Sanour area. The project developed a resilience framework for adaptation to climate change that includes among other measures; the implementation of artificial recharge interventions.

Palestinian Hydrology Group (PHG) in cooperation with watershed stakeholders proposed a plan of drilling ten recharge wells in the Marj area to recover the declining water levels of the shallow Eocene aquifer. It was agreed to implement a pilot project to enable joint learning with stakeholders, through local action planning and testing of interventions designed to increase climate change resilience.

The pilot project was presented for funding within the framework of the NRO-funded Land and Water Resources Management Project in 2013.

A workshop was organized at the united municipality of Marj Sanour involving the Water Authority, Ministry of Agriculture, Palestinian Hydrology Group (PHG), Union of Agricultural Work Committee (UAWC), farmers, Marj Sanour Watershed Association, and well owners to discuss the proposed location of the pilot artificial recharge wells. As stakeholders agreed, the Watershed Association will supply the required land, while an association with the support of all stakeholders will follow and manage the project. The location of the boreholes was chosen in areas frequently exposed to flooding and also away from the streams so as to avoid strong runoff loaded with huge quantities of suspended materials.

PHG was then asked to design the wells and develop the technical specifications in a way that meets the aim of the project and ensures that the aquifer is rejuvenated while its water quality does not deteriorate. PHG determined the depth of the recharge wells and designed an outer filtration system surrounding the wells in order to reduce the turbidity and the amount of suspended solids in the source water. The filtration system comprises of the following:

- Two layers of square-shaped stone gabion were built around the borehole. The outside gabion consists of stones of 15–20 cm dia, while, the inner one was built by using 7–15 cm dia. stones diameter (Fig. 14.4).
- The 16" dia casing pipe that was installed into the well to support well walls was perforated against the porous rocks and one meter above the ground surface to act as another filter system of the water before entering the well.
- The depth of each well (113–127 m) is kept as less than the depth of the groundwater level, which is at around 140 m below ground. This also enables the percolating to filter as the water passes through unsaturated rocks of at least 20 m, which will act as an additional filter before it mixes with the groundwater. The design of a well (Northern Well is given in Fig. 14.5).

Through the land and water resources program donated by the Kingdom of the Netherlands through the representative office in Palestine, six wells were drilled in

Table 14.1 Depth and recharge capacity of the drilled wells

Well no.	Total depth (m)	Recharge capacity (m^3/day)
1	127	1104
2	124	8033
3	120	1416
4	120	1392
5	113	5952
6	113	2352
Total		20,249

Marj Sanour area. The recharge capacities of the wells range between 1104 and 8033 m^3/day with a total recharge of 20,249 m^3/day (Table 14.1). The test was done by injecting clean water at a capacity of 37 m^3/hr, while the water level was measured by M. scope. The process continued until the water level completely stabilized in each well. Thus, the recharge capacity of each well was estimated based on the equation (Fig. 14.4).

$$\text{Recharge Capacity of Well}(Qr) = 2.75 * \text{h} * \text{d} * \text{k}$$

where, d = diameter, h = Aquifer thickness, and k = Permeability

Fig. 14.4 Filtration system—Gabions

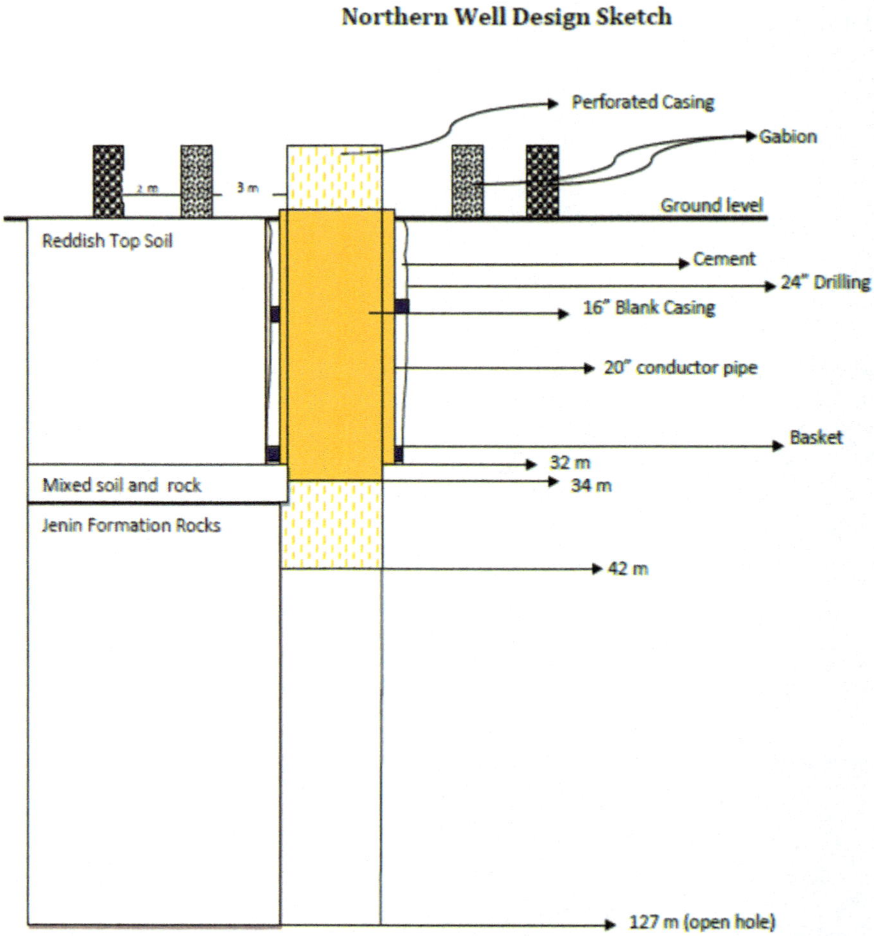

Fig. 14.5 Schematic design of the Northern groundwater Recharge well

14.6 Conclusion

Artificial groundwater recharge can be adopted in areas witnessing severe decline in water levels in Palestine. This technique can be used to recover the declining water levels of the overexploited water-bearing formations, dilute the salinity of groundwater, and can also help in reducing the negative impacts of flooding, as observed in the Marj Sanour area. The data and information that will be generated through this project may help a lot in the scaling up and planning of other artificial groundwater recharge projects in the area that in turn will enhance the groundwater resources (Fig. 14.5).

References

Arab Organization For Agricultural Development (2008) The Technical study for Marj Sanour Development Project

FAO (2000) Irrigation and Drainage Paper No. 56, Crop Evapotranspiration (guidelines for computing crop water requirements)

Palestinian Hydrology Group (2009) Watershed Management Model at Marj Sanour, Water Resources Assessment, Palestinian Hydrology Group (PHG), Palestine

Palestinian Hydrology Group (2010) Environmental Impact Assessment Report for Marj Sanour Watershed, Palestine

Palestinian Hydrology Group, IUCN, UAWC (2013) Vulnerability Assessment Report: Marj Sanour Watershed

Wishahi S (2007) Watershed of Marj Sanour, PHG, Palestine

Wishahi S et al. (2004) Water Resources Assessment in Meithaloun town, Empowers Project, Palestine

Wishahi S and others (2004) Water Resources Assessment in Meithaloun town, Empowers Project, Palestine

Chapter 15
Rainwater Harvesting for Groundwater Recharge: Experience From Sri Lanka

Tanuja Ariyananda and C. Shanthi de Silva

Abstract Almost 80% of the rural populations in Sri Lanka rely on groundwater for their domestic needs. Shallow wells are prone to be affected by climate change. During the monsoon season, the water levels of the shallow wells generally rise near the ground surface as recharge to the aquifer takes place. During dry season, groundwater level goes down due to abstraction, evapotranspiration, and other losses such as seepage and percolation. The quality of groundwater has deteriorated over the years due to various anthropologic activities. Necessary steps need to be taken so that the water levels in the well rise and also to improve the quality. Rainwater harvesting is an age-old practice, which is revived to ease domestic water supply in water-scares areas. Lanka Rain Water Harvesting Forum conducted three studies in different agro-ecological zones of Sri Lanka to monitor the effectiveness of different types of groundwater recharging structures. The traditional groundwater recharge structures such as the garden pond (*Pathaha*) as well as runoff tanks play effective roles in maintaining the water levels in the nearby wells. Rainwater harvested from the roof and directly put into the well is found to have a higher recharging effect than the rainwater harvesting tank. Improvement of groundwater quality is also recorded in wells near the recharge structures.

Keywords Rainwater harvesting · Groundwater · Recharge · Water quality · Sri Lanka

T. Ariyananda (✉)
Lanka Rain Water Harvesting Forum, Battaramulla, Sri Lanka
e-mail: tanuja@sltnet.lk

C. S. de Silva
The Open University of Sri Lanka, Colombo, Sri Lanka
e-mail: csdes@writeme.com

D. Saha et al. (eds.), *Managed Groundwater Recharge and Rainwater Harvesting*, Water Resources Development and Management,
https://doi.org/10.1007/978-981-99-8757-3_14

15.1 Introduction

Groundwater is one of Sri Lanka's most precious natural resources. Many people depend on it for their sustenance. Almost 80% of the rural populations in Sri Lanka rely on groundwater for their domestic needs today because of its excellent natural quality and sustained availability throughout the year (Panabokke and Perera 2005).

Communities in the rural areas of Sri Lanka use either shallow wells of 6–8 m deep or deep wells of 40–50 m deep. In shallow wells, during monsoon/post-monsoon (*Maha* season), groundwater levels generally remain near the ground surface as recharge to the aquifer takes place from monsoon. During dry season (*Yala* season), groundwater level goes down due to abstraction, evapotranspiration, and other losses such as seepage and percolation. The stress on aquifers is aggravated when discharge rates are greater than their recharge rates. Seasonal fluctuation of groundwater is significantly correlated with precipitation, as it is found that recharge into the groundwater system is entirely from rainfall. Due to climate variation, this process has been disturbed. High-intensity rains, limited to a few months of the year, cause both unexpected floods and droughts. As a result of the high-intensity rainfall, the groundwater recharge is less and runoff is high. The runoff often exceeds the capacities of the water bodies creating floods.

The quality of groundwater varies widely depending upon the climate, season, and geology of the area and also anthropogenic activities. In addition to the natural rock–water interaction, groundwater gets enriched with harmful chemical constituents from human activities such as industrial, sewage, and domestic waste disposal, fertilizers, pesticides, etc. The environmental impact of human activity on the groundwater is considered one of the major hazards in the modern days. Rapid urbanization and increased agricultural activities have resulted in the degradation of the water quality. Overdose of fertilizers, pesticides, and effluents discharged from industries and sewage water are the main sources of contamination of groundwater. The chemical budget of major ions and heavy metals is important in determining the quality of groundwater. Total Dissolved Solids (TDS) values are considered important in overall determining the usage of water and groundwater (Rajasooriya et al. 2002).

Rain is the primary source of water in Sri Lanka, with an average annual rainfall of 1800 mm. However, it varies from 900 mm in the dry zone and 5000 mm in the wet zone. The rainfall is bi-monsoonal and varies both seasonally and spatially. Out of the total water received by the island, around 60% escapes to the sea as runoff (Ratnayaka 2001). Therefore, the simple solution to combat water scarcity is to increase the retention of water that is received during the rainy periods to be used during the non-rainy periods. This can be done by rainwater harvesting. Rainwater harvesting, which is traditionally practiced in Sri Lanka is being revived since the last 25 years to address the rising water shortage (Ariyananda 2020). Lanka Rain Water Harvesting Forum (LRWHF), which is a non-government organization (NGO) formed by a small group of people in 1996 has been actively engaged in promoting rainwater harvesting in the country. As a result of influencing and lobbying carried out, LRWHF was able to influence the Government of Sri Lanka to pass a National Policy on rainwater harvesting in 2005.

As a pioneer organization promoting rainwater harvesting in Sri Lanka, LRWHF has been involved in conducting research in collaboration with various research institutes to showcase rainwater harvesting technologies and good practices. Pilot projects have been constructed in three districts, Kurunegala, Badulla, and Killinochchi in collaboration with the Open University of Sri Lanka. The objective was to monitor and assess the effectiveness of various groundwater recharging structures.

15.2 Case Study in Kottewehara-Kurunegala

Kottawehara, the Divisional Secretariat is situated in the intermediate zone in the Kurunegala district in the North Western of Sri Lanka (Fig. 15.1). The annual rainfall in the study area is about 1200 mm, which is distributed in the months of April, and May, and then, from October to December. Generally, about 4 months of the year do not receive rainfall, as a result, people face drought situations. Water shortage is a major problem and the people are mainly dependent on the groundwater for both domestic and irrigation uses. There are few irrigation tanks (*wewa*), which even do not hold enough water to cater for the dry season. Groundwater resource availability is limited in the study area because of the low potential hard rock aquifers. The wells get dried up during the dry months and the rural population is unable to maintain even a small vegetable garden for their own consumption. Therefore, this study was focused on the possibility of raising the groundwater levels during non-monsoon months, through runoff collection tanks, ponds, and artificial recharge structures. In the study area, eight dug wells were selected to study the impact to the existing village tanks (*Wewa*) and constructed runoff collection structures such as traditional gardens pond (*Pathahas)* (Fig. 15.2), runoff collection tanks (Fig. 15.3), and artificial recharging structure (Fig. 15.4). Water levels were monitored daily in all the runoff collection structures and also from the wells. In addition to the water levels, rainfall and evaporation were also measured daily.

The impact of recharge was studied based on the water levels monitored from the wells. The data shows a significant recharge in the vicinity of the structures (Fig. 15.5). Recharge contribution is reflected more in shallow wells than the deep wells (Fig. 15.6). But it is also a factor that seeping water may take longer time to contribute to the deeper aquifers. It is also evident that runoff collection structures such as *pathaha* and runoff collection tanks play a role in maintaining the water levels in the nearby wells. All the shallow wells in the study area maintained an average water level depth of 4 m below ground, which is desirable for cultivation.

The estimated recharge from the tank (*wewa*) and runoff collection tanks was found to be 130 mm. The modified soil balance method estimated recharge and runoff as 190 mm and 157 mm, respectively. The estimated recharge as found by the water table fluctuation method ranged from 130 to 280 mm, which was higher than

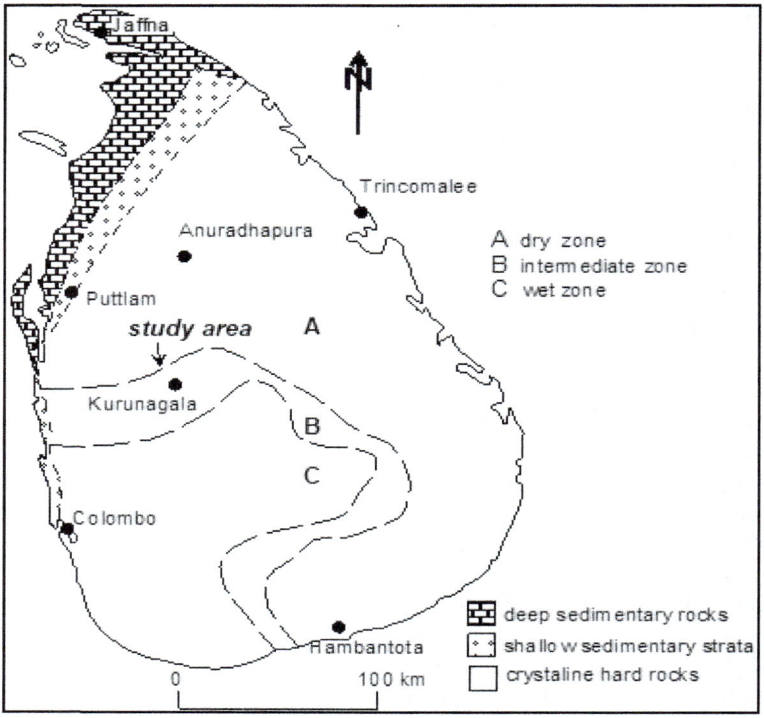

Fig. 15.1 Project location, climatological zones, and aquifer types in Sri Lanka

Fig. 15.2 A typical traditional garden pond (pathaha)

the average recharge in the study area (De Silva and Rushton 2007; De Silva and Ariyananda 2006).

Fig. 15.3 Runoff collection tank

Fig. 15.4 Artificial recharging structures

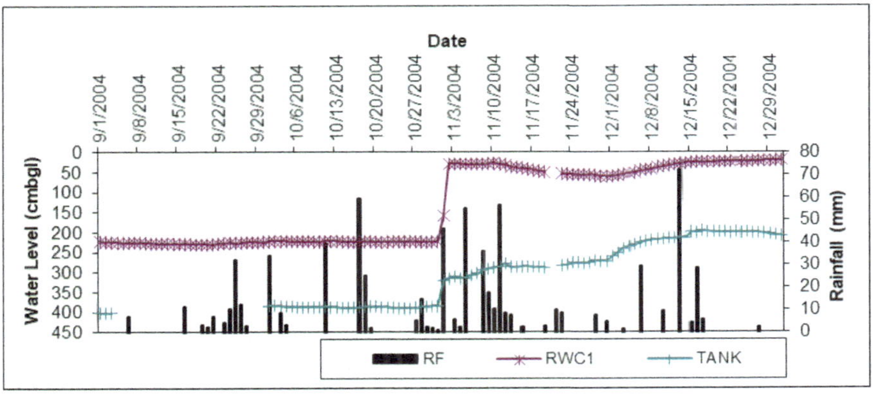

Fig. 15.5 Water level fluctuation in recharging structures and tank in cmbgl *Note* RF: Rainfall, RWC1-Rainwater collection (Pathaha 1), TANK-Surface water storage tank (wewa)

15.3 Case Study in Kilinochchi

LRWHF under the project titled "Safe Disaster-Resilient Drinking Water to Floods and Drought Prone Areas in Sri Lanka" (supported by USAID) installed well recharge structures in schools, hospitals, and primary health care units in Kilinochchi district in

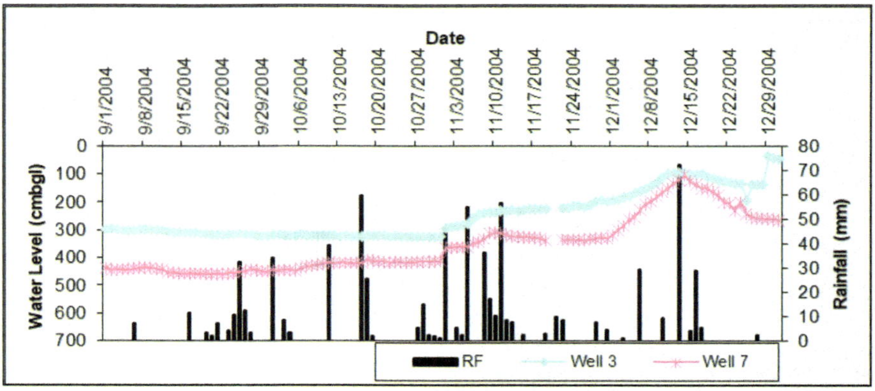

Fig. 15.6 Water level fluctuation in shallow wells (less than 7 m) no 3 and 7 cmbgl

the Northern Province of Sri Lanka (Fig. 15.7). Located in the dry zone, Kilinochchi receives an average annual rainfall of 1325 mm and 75% of the rainfall is received during the period of September to December by North–East monsoon.

Artificial recharge structure using rainwater as a source was installed in a total of 11 places, near/within schools and primary medical care units (PMCU). These wells were holding less amount or little water during the dry season and some wells get completely dried up during the dry season. Further, the quality of the groundwater was not good because of the high TDS during the dry season. The wells were monitored to study the impact on the groundwater resource and quality over 2-year period.

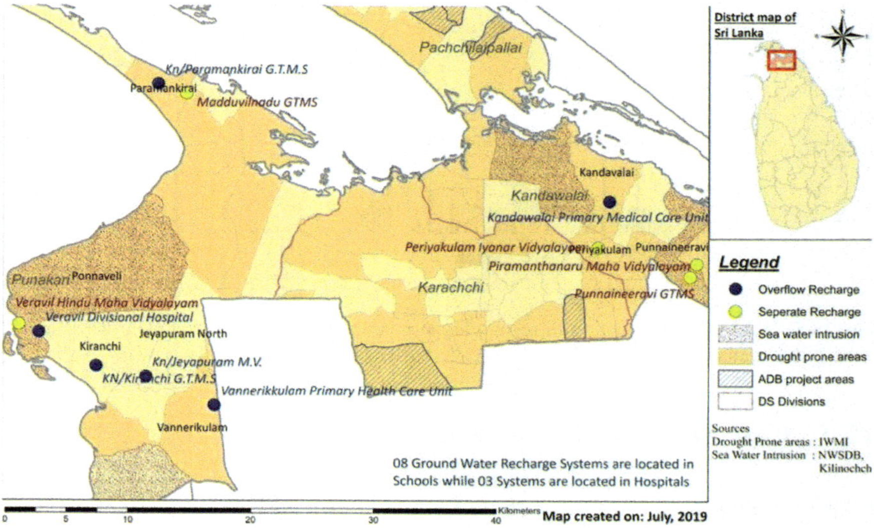

Fig. 15.7 Locations of the recharge wells in Kilinochchi district

According to the study by Pankaj Kumar et al. (2016), the area is predominantly covered by four main soil types, namely, immature brown loams (dry zone); reddish brown earths; regosolic alluvial soil; and solodized solonetz and solonchaks. The depth of these wells varies from 4 to 7 m with the highest of 10 m and a diameter range between 2 and 3 m, except one with less than 2 m diameter (Table 15.1).

Table 15.1 Aquifer types and artificial recharging method in the study area

Well name	DSD/GND	Well no	Well depth (m)	Well diameter (m)	Aquifer type	Recharge method
Shallow wells >4 m *depth*						
Madduvill Nadu GTMS	Poonakari	1	4.1	2.36	Red earth red and brown sand	Direct
Paramankirai GTMS	Poonakari	2	4	2.03	Red earth red and brown sand	Overflow
Kallaru Tamil Vidyalayam	Kandawallai	4	4.06		Alluvial lagoonal clay, Sand	Overflow
Medium depth wells 4–5.5 m *depth*						
PeriyakulamIyanar Vidiyalayam	Karachchi	3	5.2	1.3	Alluvial Lagoonal clay, Sand	Direct
Kandawallai PMCU	Kandawallai	6	4.52	3.07	Alluvial lagoonal clay, Sand	Overflow
Veravil Divisional Hospital	Poonakari	10	5.1	3.05	Red earth red and brown sand	Overflow but no water
Veravil Hindu MahaVidyalayam	Poonakari	11	5.1	2.6	Red earth red and brown sand	Overflow but no recharge
Deep wells 5.5–7 m						
Vannerikkulam PMCU	Karachchi	7	5.7	2.14	Red earth red and brown sand	
Piramanthanaru Maha Vidayala	Kandawallai	8	5.95	2	Alluvial lagoonal clay, Sand	Direct
Punnaineeravi GTMS	Kandawallai	5	6.2	2.14	Alluvial lagoonal clay, Sand	Direct

(continued)

Table 15.1 (continued)

Well name	DSD/GND	Well no	Well depth (m)	Well diameter (m)	Aquifer type	Recharge method
Deep well >7 m						
Jeyapuram Maha Vidayala	Poonakari	9	10.4	2.64	Red earth red and brown sand	Overflow

Note GTMS: Government Tamil Medium School, PMCU: Primary Medical Care Units

Two recharge methods adopted were;

(i) Direct Recharge: (Fig. 15.8a) when the wells are located near the building and the rainwater run-off from the roof is directly diverted to the well.
(ii) Overflow Recharge: (Fig. 15.8b) when the wells are located far away from the roof, the rainwater from the roof is first collected into a tank and then taken to the recharge well.

Both Paramankirai GTMS and Madduvill Nadu GTMS wells are located in red earth and red-brown sand. The well has shallow depth of 4 m. The Madduvil Nadu GTMS well was recharged directly from the roof from 15/11/2017 to 31/5/2019 (one and half year) (Fig. 15.9). Paramankirai GTMS well was recharged by the overflow of the rainwater harvesting tank, from 15/11/2017 to 31/05/2019 (Fig. 15.10). The groundwater level rise observed in Madduvil Nadu GTMS was 3.27 m in the year 2018 and it was 1.21 m higher than 2017, before the construction of the recharge structure. Whereas, at Paramankirai GTMS, the water level rise was 2.76 m during 2018 and it was 0.56 m higher than the non-recharging year of 2017. Therefore, it

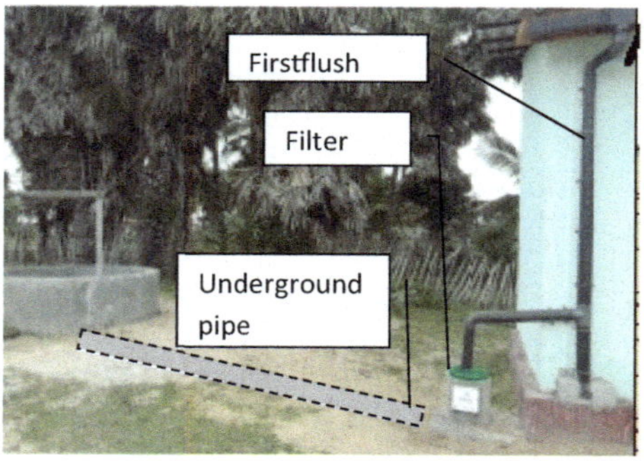

(a)

(b)

Fig. 15.8 **a** Direct recharge method. **b** Overflow recharging method of closed dome rainwater tank

appears that the well in Madduvill Nadu GTMS has recharged more water than that of the Paramankirai GTMS with overflow method. The water level in the well at Madduvil Nadu GTMS was almost near to the ground surface (0.3 mbgl) facilitating the availability of groundwater during the dry season.

Based on the water table rise, the specific yield of the aquifer was calculated (De Silva and Rushton 2007; Mikunthan and De Silva 2012, Mikunthan and De Silva 2009; Pankaj Kumar et al. 2016). Table 15.2 shows the calculated recharge in each

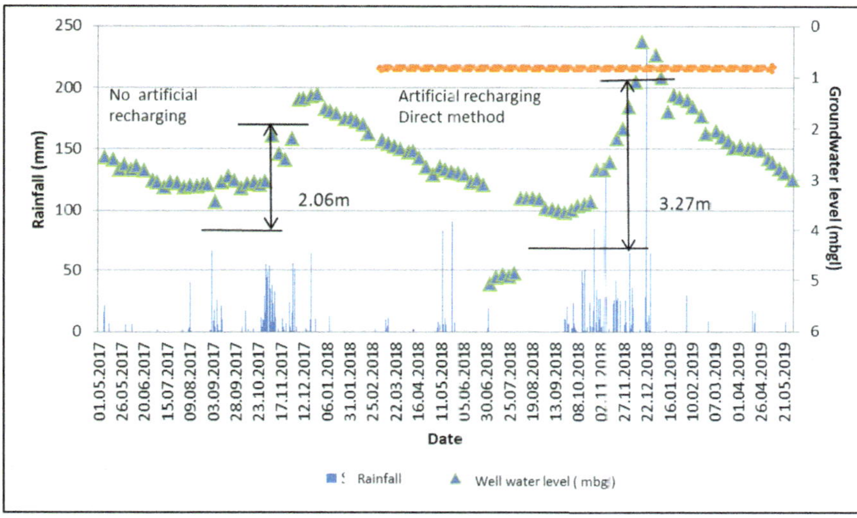

Fig. 15.9 Artificial recharging by rainwater harvested using direct method monitored at Maddivill Nadu GTMS well

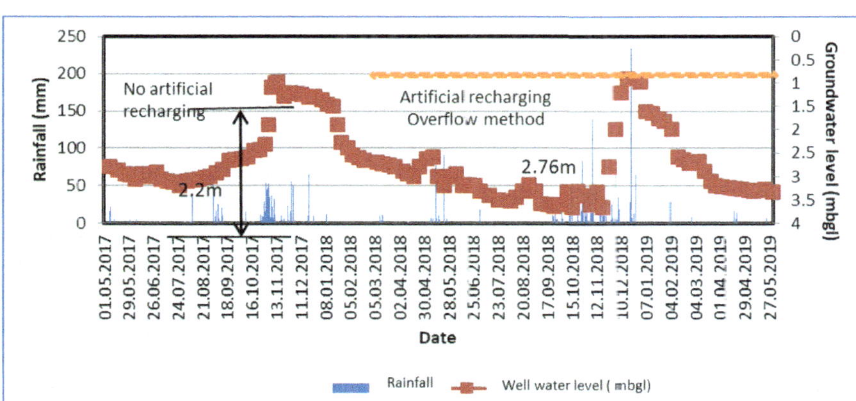

Fig. 15.10 Artificial recharging by rainwater harvested using overflow method monitored Paramankirai GTMS well

Table 15.2 Recharge in groundwater wells from artificial recharging by direct method or overflow method

Well name	Recharge method	Well no	Recharge 2017 (mm)	Recharge 2018 (mm)
Madduvill Nadu GTMS	Direct	1	124	196
Paramankirai GTMS	Overflow	2	132	166
PeriyakulamIyanarVidiyalayam	Direct	3	237	260
Kandawallai PMCU	Overflow	6	–	205
Veravil Divisional Hospital	Overflow	10	–	180
Veravil Hindu MahaVidyalayam	Overflow	11	–	192
Vannerikkulam PMCU	Overflow	7	–	295
Punnaineeravi GTMS	Direct	5	–	301
Jeyapuram M.V.	Overflow	9	–	170

location during the years 2017 and 2018. Recharge during 2018 was higher than that in 2017, mainly due to the impact of artificial recharge. Even within 2018, the direct method of artificial recharge contributed better than the overflow method of artificial recharging. Therefore, it can be concluded that direct recharge from the roof is effective in drought-prone areas of Sri Lanka for groundwater recharge. Average recharge varies from 196 to 301 mm per year.

Water quality from the rainwater harvesting tanks remained within the safe limits for potable consumption (6–8.5 of pH, 1500 μS/cm of EC, and 500 mg/L TDS). The source water quality was found to be superior than the groundwater in the study area. Thus, the groundwater quality was improved due to the dilution process by artificial recharging using rainwater (De Silva and Ariyanada 2020a, 2021).

15.4 Case Study in Badulla

Badulla district is located to the east of the central highlands and is a district of the Uva Province of Sri Lanka. It is a region with mountain ranges, dissected plateaus, escarpments, and narrow valleys. The average annual rainfall in the district is around 1885 mm but varies from 900 mm in the northern and southern-most extremities to over 2500 mm in the eastern flanks of the central highlands. The northeast monsoon provides a comparatively higher component of rainfall, between December and February. During the first inter-monsoon season (March to mid-May), the whole district receives about 300–500 mm of rain. Drought is an important climatic characteristic in the Badulla District. Partial drought occurs very often during February to July in the Intermediate Zone. Badulla soil is named as red yellow Podsolic soil according to local classification (Dharmasena 2014). Surface soil is dark yellowish brown in color with a sandy clay loam texture and medium size granular structure.

Fig. 15.11 Rainwater harvesting system at Orubendiwewa Maha Vidyalaya, Badulla

LRWHF installed rainwater harvesting and recharge well systems in schools to combat water scarcity in the district (Fig. 15.11). In total, seven wells located in schools were selected for recharge (Fig. 15.12). The depth of the well varies from 5.2 m to 9.75 m and the diameter from 1.0 m to 5.10 m. The source water for recharge was overflow water of the rainwater harvesting tank of 30,000-L capacity. The rainwater harvesting tanks get rain run-off from the roof.

A rise in water level was recorded at Mahagama Maha Vidyalaya (Fig. 15.13) and Orubendiwewa Maha Vidyalaya (Fig. 15.14) in the year 2018. Since the volume of overflow rainwater is limited, there is not much difference in the recharge. Accordingly, recharge during the years 2017 and 2018 were 132 mm and 140 mm, respectively, even though the annual rainfall in 2018 (1827.9 mm) was less than the annual rainfall in 2017 (1924.5 mm).

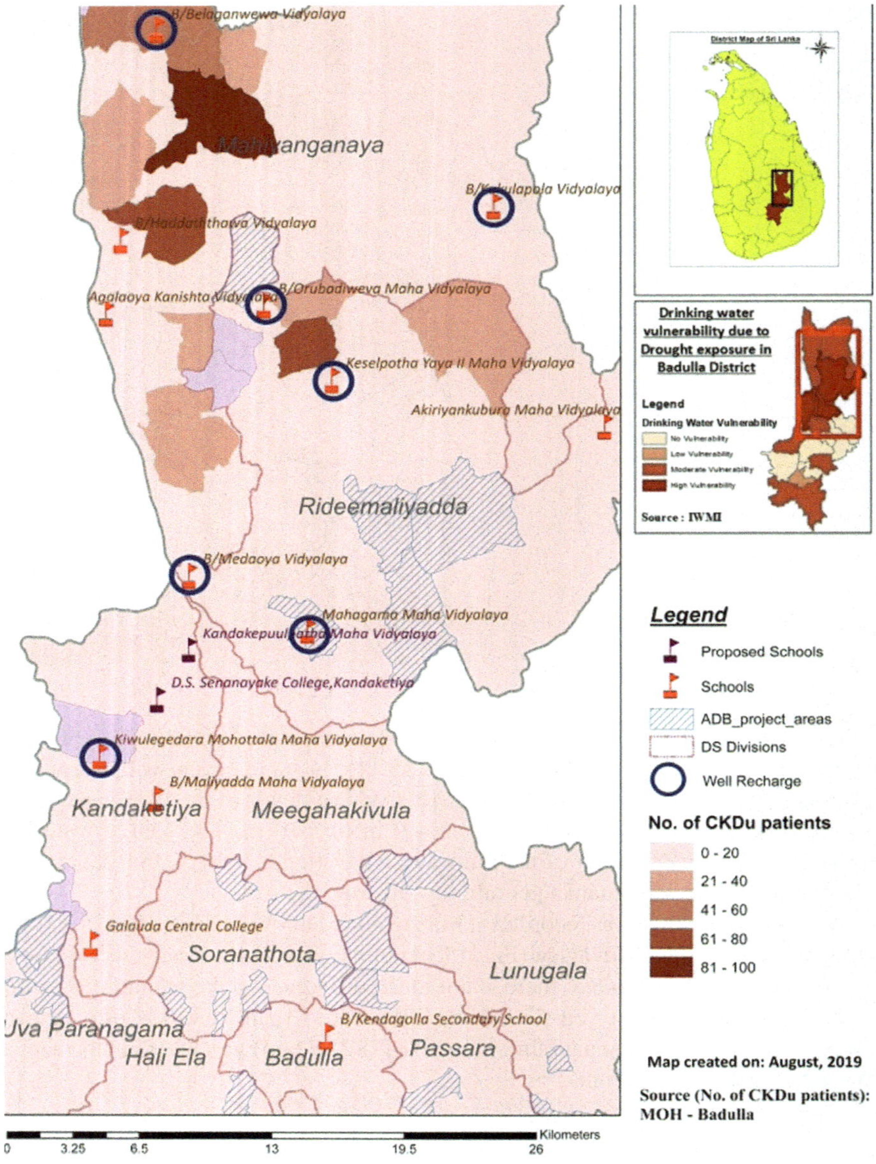

Fig. 15.12 Location of well water recharging at Badulla

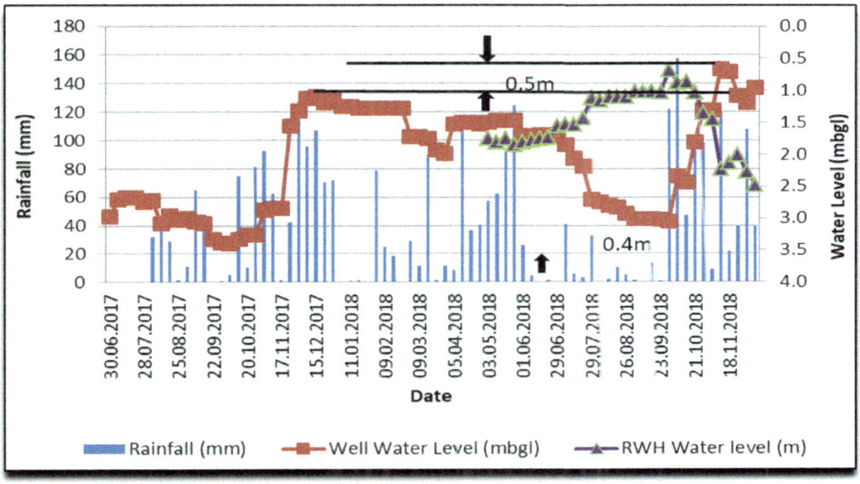

Fig. 15.13 Well water level (mbgl) and rainfall (mm) in Mahagama Maha Vidyala well in Ridimaliyadda, Mahagama

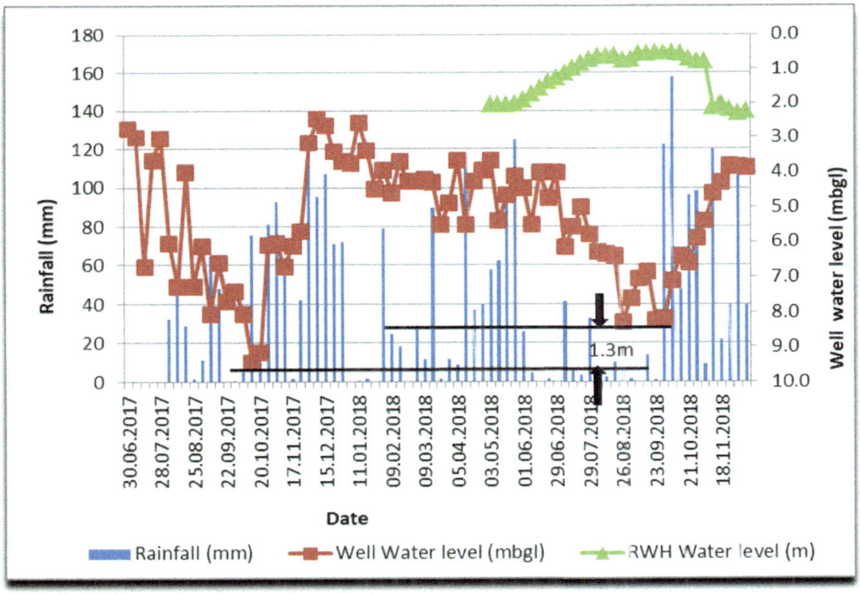

Fig. 15.14 Well water level (mbgl) and rainfall (mm) in Orubendiwewa M. V. Ridimaliyadda, Sagabopura

Groundwater quality analysis showed that pH, EC, and TDS remained within the safe limits of 6–8.5, 1500 μS/cm, and 500 mg/L, respectively. Rainwater harvesting tank water quality remained within the safe limits. EC and TDS values were less in the tank than in the well water. Low TDS values (less than 100 mg/L) in the rainwater harvesting tank are preferred by Chronic Kidney Disease (CKD) patients in the study area (De Silva and Ariyananda 2021).

15.5 Conclusion

Groundwater recharge through the structures such as *pathahas* and runoff collection tanks has shown a significant recharge to aquifers and built up of water level in the well in the near vicinity. Recharge contribution to the wells of shallow depth was better than the deep wells. The recharge to the shallow wells by the rainfall runoff collected in the structures is effective to keep the water level high during the dry season.

Roof rainwater harvested during the wet season to recharge the groundwater wells is found to be effective and helps to keep the well water levels shallow during the dry periods. This simple low-cost method effectively increases the well water levels as well as improves the groundwater quality. Direct method of artificial recharge using rainwater is effective in drought-prone areas.

Rainwater harvesting is a solution to meet the challenge of providing water for all by 2030 (Sustainable Development Goal 6) amidst rising demand, increasing cost, pollution of water resources, frequent drought, and other climate change effects.

15.6 Recommendations

Based on the study and analyses of the results, the recommendations are as follows;

- Introduce rainwater harvesting tanks such as *pathaha*, runoff collection tanks, or any suitable similar structures to hold runoff so that it will contribute to the groundwater recharge.
- Groundwater recharge would improve the groundwater levels and they could be kept near the ground surface even during the dry season.
- Introduce roof rainwater harvesting system as a safe drinking water source in all households in the dry zone, especially in areas where groundwater quality is poor.
- Introduce direct recharge method for the wells using rainwater harvested from roofs.
- Artificial recharge using rainwater is strongly recommended for areas affected by Chronic Kidney Disease of unknown etiology (CKDu), from Northern, North Central, Southern, and Uva provinces as the quality of groundwater would improve significantly.

Acknowledgements Authors acknowledge the USAID Grant No: AID—383-A-16-00001 of the project, Safe Disaster-Resilient Drinking Water to Floods and Drought Prone Areas of Sri Lanka given to Lanka Rainwater Harvesting Forum to undertake these research studies in Kilinochchi and Badulla.

References

Ariyananda T, Aheeyer M (2011) Effectiveness of Rain Water Harvesting (RWH) systems as a domestic water supply option

Ariyananda T (2010) Domestic rainwater harvesting as a water supply option in Sri Lanka. Hydro Nepal J Wa 27–30

Ariyananda T (2020) Transforming tradition of rainwater harvesting in Sri Lanka. In: International rainwater catchment systems experiences. IWA Publication, pp 41–53

De Silva CS, Ariyananda T (2006) Impact of rainfall runoff harvesting in drought prone areas-a case study in Kotawehara of Sri Lanka. In: Proceedings of the international workshop on rainwater harvesting. Kandy, Sri Lanka, pp 96–102

De Silva CS, Rushton KR (2007) Groundwater recharge estimating using improved soil moisture balance methodology for a tropical climate with distinct dry season. Hydrol Sci J 52(5):1051–1067

De Silva CS, Ariyananda T (2020) Groundwater responses to artificial recharge of rainwater in Kilinochchi District in Sri Lanka. J Eng Technol 8(1):77–101. The Open University of Sri Lanka

De Silva CS, Ariyananda T (2021) Artificial groundwater recharge using rainwater in Kilinochchi District in Sri Lanka. In: International Conference on Looming water crisis and climate change in South Asia: Rainwater Harvesting as an adaptation

De Silva CS, Ariyananda T (2020) Groundwater responses to artificial recharge of rainwater in Badulla district in Sri Lanka. J Dry Zone Agric 6(2):15–29. University of Jaffna

Dharmasena PB (2014) Current status of land degradation in Badulla district. https://www.researchgate.net/publication/355186317_Current_Status_of_Land_Degradation_in_Badulla_District

https://doi.org/10.1007/s40899-016-0072-5

Mikunthan T, Shanthi De Silva C (2009) Estimation of groundwater recharge in limestone aquifer using an improved soil moisture balance method: a case study in Jaffna district. J Trop Agric Res 21(1):51–61

Mikunthan T, Shanthi De Silva C (2012) Assessment of groundwater resources in Jaffna limestone aquifer, tropical agricultural research, vol 23

Panabokke CR, Perera AP (2005) Groundwater resources of Sri Lanka. Water Resources Board, Colombo, Sri Lanka

Pankaj Kumar SH, Avtar R, Takeuch K (2016) Mapping of groundwater potential zones in Killinochi area, Sri Lanka, using GIS and remote sensing techniques. Sustain. Water Resour Manag 2:419–430

Rajasooriyar L (2002) Groundwater quality in the valigamam region of the Jaffna Peninsula. SriLanka Geol Soc Spec Publ 193(1):181

Ratnayaka R (2001) Sri Lanka water vision 2025 framework fro action water fro food and agriculture-Sri Lanka, 109

Chapter 16
Groundwater Recharge in the Kabul Plain (Afghanistan) Through Rainwater Harvesting

Abdulhalim Zaryab, Mohammad Zia Jamal, Hamid Zaki, Zamen Jafari, Asadullah Farahmand, and Mohammad Salem Hussaini

Abstract The groundwater of Kabul Plain aquifers is the main source of water for the Kabul city for drinking, and also extensively used for agricultural and industrial sectors. The exploitation of groundwater from Kabul Plain aquifers is unsustainable and the groundwater storage has significantly decreased over the last decades. Therefore, rainwater harvesting and implementing managed aquifer recharge (MAR) projects are very important to reduce stress on the Kabul Plain aquifers. In this study, we evaluate the groundwater conditions in the Kabul Plain and present the current state of knowledge associated with the rainwater harvesting (RWH) in the Kabul Plain. A systematic RWH has not been adopted in any cities in Afghanistan. However, in Central Afghanistan, the farmers normally collect the rain and spring waters in traditional ponds and use the collected water for irrigation and livestock consumption. A MAR pilot project has been carried out in the Kabul Plain between 2017 and 2020. Lately, the Ministry of Energy and Water (MEW) of Afghanistan performed a

A. Zaryab (✉)
Department of Engineering Geology and Hydrogeology, Faculty of Geology and Mines, Kabul Polytechnic University and Highland Groundwater Research Group, Kabul, Afghanistan
e-mail: a.zaryab@kpu.edu.af

M. Z. Jamal · H. Zaki · Z. Jafari
Department of Engineering Geology and Hydrogeology, Faculty of Geology and Mines, Kabul Polytechnic University, Kabul, Afghanistan
e-mail: ziajamal@kpu.edu.af

Z. Jafari
e-mail: z.jafari@kpu.edu.af

A. Farahmand
Department of Hydrogeology, Ministry of Energy and Water, Kabul, Afghanistan and Portuguese Chapter of the International Association of Hydrogeologists, Porto, Portugal

M. S. Hussaini
Department of Engineering Geology and Hydrogeology, Faculty of Geology and Mines, Kabul Polytechnic University, Kabul, Afghanistan and Portuguese Chapter of the International Association of Hydrogeologists, Porto, Portugal

D. Saha et al. (eds.), *Managed Groundwater Recharge and Rainwater Harvesting*, Water Resources Development and Management,
https://doi.org/10.1007/978-981-99-8757-3_15

pilot RWH project in the campus of the Ministry of Agriculture and Livestock. The MEW and Municipality of Kabul encourage the Kabul residents to store rainwater in the subsurface during the rainy season. This study provides a synoptic view regarding rainwater harvesting that could be helpful for sustainable groundwater management in the Kabul Plain where the people suffer from a shortage of drinking water.

Keywords Groundwater · Over-abstraction · Rainwater harvesting · Managed aquifer recharge · Kabul · Afghanistan

16.1 Introduction

Groundwater is one of the most important natural resources in the globe. It provides the primary source of human water needs for domestic consumption, agriculture, industrial applications, and a wide range of other services (Jac 2022). Many large cities around the globe primarily depend on groundwater resources (Zhou et al. 2012; Dash et al. 2013; Ramos-Leal et al. 2010, Mukherjee et al. 2015). Groundwater levels are declining and its storage in many aquifers, particularly urban aquifers are decreasing due to overexploitation (Maliva 2020; Saha et al. 2013). Therefore, it is really important to reduce stress on urban aquifers through implementing RWH and MAR projects (Julius et al. 2013). However, the availability of adequate sources of water with suitable chemical and physical quality is an important issue (Kresic 2023).

The Kabul Plain is located in the Kabul River Basin (KRB) in the eastern part of the country. The Kabul city is classified as the fifth fastest-growing city in the world. The KRB is so sensitive to population growth and climate variations, which a slight increase in average temperature can change the hydrologic conditions and water resources (Bromand 2015). Groundwater is the main source of water for domestic, agricultural, and industrial use in the Kabul Plain (Zaryab et al. 2021a). But, the population growth and rapid urbanization have created a high pressure on the groundwater resource of the Kabul Plain (Hussaini et al. 2022; Zaryab et al. 2021b). The groundwater levels have severely declined in most parts of the Kabul Plain due to overexploitation (Noori and Singh 2021). Presently, Kabul residents suffer from a shortage of drinking water for almost all seasons of the year. From a land-use perspective, urban areas in the Kabul Plain increased significantly, while agricultural areas considerably decreased over the past two decades.

Although the MAR techniques are utilized in many countries in the world, but it is a new topic in Afghanistan. A pilot MAR project entitled Kabul Managed Aquifer Recharge (KMARP) funded by USAID through the Asian Development Bank, implemented by Landell Mills and Danish Committee for Aid to Afghan Refugees (DACAAR) under the supervision of the Ministry of Energy and Water of Afghanistan was successfully conducted in the Kabul Plain from 2017 to 2021. This pilot project investigated the efficacy of MAR techniques to raise the water table and increase the groundwater storage. During this pilot project, various MAR technologies were examined. As a major result of this project, 40 monitoring wells were drilled and equipped to monitor the water table using data loggers (Landell Mills 2020). Implementing the next phase of the project is important and may reduce stress on the Kabul Plain aquifers. Since the recharge water is not accessible in all four seasons (in a year) in the region, the MAR project could be efficient in the rainy seasons at least (Zaryab et al. 2021b). Masoom (2018) has conducted a study to increase groundwater storage in the Kabul Plain to obtain a sustainable yield. In addition, Hussaini et al. (2022) carried out a study on site selection of aquifer recharge the Kabul Plain using geographic information system (GIS) and multi-criteria decision analysis (MCDA) techniques to identify the most suitable places for a MAR project in the plain.

Rainwater harvesting systems are classified as either earthwork (passive) or storage (active) systems. Earth systems are employed to harvest and hold water in aquifers and storage systems are utilized to harvest and store rainwater in tanks or reservoirs (Maliva 2020). The RWH is commonly classified as a small-scale recharge of groundwater (Alley et al. 2022). Rainwater harvesting is increasingly being implemented by individuals and organizations in developed countries. In addition, the RWH is suitable for developing countries as it has the advantages of being small-scale and simple operations with low costs and high adaptability (Maliva 2020). Although the RWH is often viewed as a new technology, but it actually is an ancient technique (Younos et al. 2016). The annual estimated RWH potential per household in many villages and cities such as Gayama village of Nigeria and Sylhet city of Bangladesh can meet the water demand (Ishaku et al. 2011; Alam et al. 2011). Since the groundwater storage in the Kabul Plain has declined significantly over the past decade, therefore, the RWH could be an effective and suitable approach for groundwater recharge in the Kabul Plain. Lately, the Ministry of Energy and Water of Afghanistan has encouraged the residents of Kabul city to collect rainwater from their rooftops and recharge it to aquifers. The authors believe this manuscript will render an useful outline for researchers who want to work on RWH and MAR in Afghanistan, and particularly, in the Kabul Plain.

The main objective of this study is (1) to review the relevant studies, which have been conducted in the Kabul Plain, (2) to assess recharge systems and techniques through rainwater harvesting that could be applicable for Kabul Plain, and (3) to highlight and make visible the importance of rainwater harvesting in Afghanistan.

16.2 Historical Background of RWH in Afghanistan

As known, any systematic study on RWH has not been carried out in any cities of Afghanistan in the past. However, in some parts of Afghanistan, particularly in Central Afghanistan, the farmers collect rain and spring waters in traditional man-made ponds (called *Nawor*) and use them for irrigation and livestock consumption. In 2020, the government of Afghanistan constructed infiltration trenches in some hills and mountain foots around the Kabul Plain to capture rainwater. But, the constructed infiltration tranches might not operate for a long time due lack of maintenance and rehabilitation.

During the rainy season, some individuals in Kabul city collect rainwater on their rooftops and channelize it to the subsurface through dug wells to avoid water logging in their houses. A few dissertations related to RWH have been completed by the Department of Engineering Geology and Hydrogeology at Kabul Polytechnic University (Zaki 2022). Lately, the Ministry of Energy and Water of Afghanistan has launched an RWH pilot project on campus of the Ministry of Agriculture and Livestock of Afghanistan in western Kabul Plain. In addition, the Ministry of Energy and Water of Afghanistan and the Municipality of Kabul City encourage Kabul city's residents to collect rainwater from their rooftops and recharge it to the subsurface via dug or drill wells.

16.3 The Needs for RWH in Afghanistan

Water utilization in Afghanistan is mostly for agriculture and approximately 85% of the country's crops are produced under traditional irrigation systems. Irrigation systems in Afghanistan use mostly traditional methods and distribute water in traditional ways throughout the country with 84.6% of irrigation water being tapped from rivers. Water from springs and *Karezes* make up 7.9% and 7.0% of irrigation water, respectively. A small amount (0.5%) of water is tapped from dug wells (Aini 2007).

According to an estimate, all traditional groundwater irrigation systems have reduced or dried up completely. About 60–70% of the *Karezes* are not in use and 85% of shallow wells are dried out, the main reason being low precipitation and consequently low recharge to groundwater. The population dependent on these systems has suffered badly. In addition, the boring of deep wells in the vicinity of *Karezes* and shallow wells has adversely affected on production of the shallow wells and yield of the *Karezes*. This has threatened the sustainability of these systems in the future (USAID 2021).

In most urban areas of Afghanistan, shallow wells are used for drinking and other household activities. The Kabul Plain experienced a severe groundwater decline(0.8 m/year on average) in the past two decades (Zaryab et al. 2021b). The poorer families in Kabul city are unable to dig their wells deeper and thus are forced to get water from community wells. Many of these wells are already dried up and people (often women and children) are forced to walk miles to meet their daily water demands (Qureshi 2002).

Groundwater availability and accessibility vary significantly across Afghanistan due to geologic and climate conditions. Most groundwater recharge is linked to infiltration along rivers. Overexploitation of groundwater in the Harirud–Murghab and Helmand Basins is reducing the water supply for irrigation. Increasing water demand and climate change will further reduce their long-term viability. The exploitation of groundwater from Kabul Plain aquifers is twofold higher than the rate of natural recharge (Zaryab et al. 2021b), which Saffi (2019) claims to be more than sixfold.

Agricultural withdrawals also added to the problem in the Kabul Basin and the eastern Helmand Basin.

On the other side, Afghanistan has low long-term average precipitation (327 mm/year). The precipitation rates are highest in the mountainous northeast (over 1,000 mm/year) and lowest in the southwestern deserts (less than 110 mm/year) (USAID 2021). Over 80% of annual precipitation is snowfall received between January and March in the central and northeastern mountains. Snow melts in the spring and summer contribute to peak river flows in June/July. Snowpack and glaciers are a natural inter-annual and inter-seasonal water storage mechanism that sustains perennial watercourses and recharges aquifers. Precipitation in the March–May period is mostly rain rather than snow. Unlike snow, which is stored in the mountains, rainwater washes away quickly, with lower capture for agricultural needs and groundwater recharge, and exits the country swiftly. In the absence of reservoirs to retain rainfall, most of the water is lost quickly. According to the streamflow lag time map, surface water leaves Afghanistan in two to 14 days, depending on the length of the river basin (Mayar 2022).

In water-stressed regions, the availability of freshwater for human consumption is significantly affected by inadequate runoff collection infrastructure, particularly in arid and semi-arid regions. The insufficient water supply throughout seasonal dry periods and drought disturbs economic growth with substantial influence on the agricultural sector. Rainwater and runoff harvesting through recharge pits, recharge trenches, and recharge wells can be employed for groundwater recharge of the Kabul aquifer during rainy seasons (Zaryab et al. 2021b).

16.4 MAR Study in Kabul Plain

16.4.1 Description of the Study Area

Kabul Plain is situated in the center of the Kabul province in the eastern Afghanistan (Fig. 16.1) and lies between 69°1′ E and 69°25′ E longitude and 34°25′ N–34°37′ N latitude, covering an area of about 381 km^2. The Kabul Plain is enclosed by low but quite steep mountain ranges made up of different types of metamorphic and sedimentary rocks, which include schist, amphibolite, sedimentary rocks, limestone, and dolomite (Bohannon 2010). The Kabul Plain is divided into western and eastern plain by the Asmai (TV) and Share Darwaza mountains. Paghman, Kabul, and Logar are the major rivers that flow from the west and southwest to the east of the plain. The Kabul Plain experienced a rapid unplanned urbanization in the past two decades and the population living in Kabul city increased from about one million in 2001 to 5.38 million in 2021 (Zaryab et al. 2021b).

The rapid population increase continues in Kabul city, and the population is estimated to increase to 6.5 million in 2025 (Rahimi and Murakami 2018). Groundwater is the main source of water for domestic, agricultural, and industrial usage in Kabul

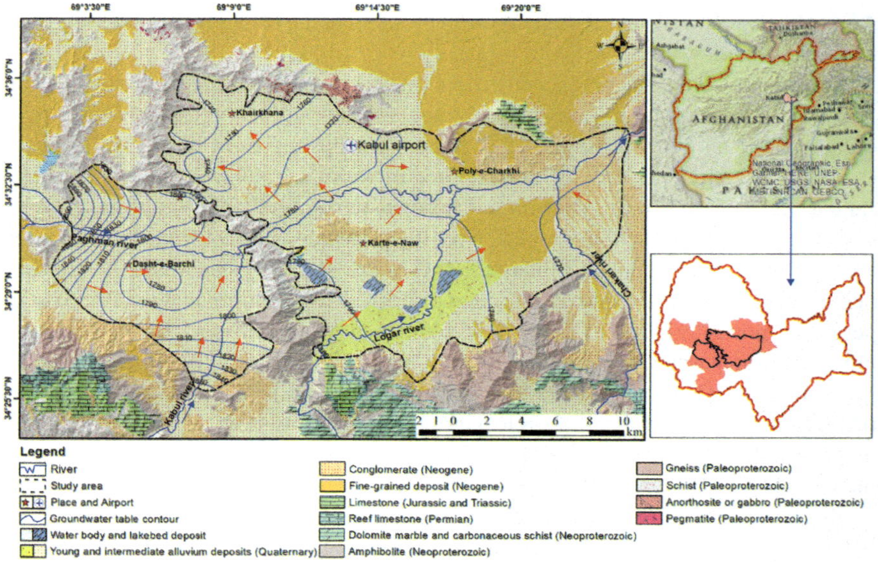

Fig. 16.1 Location of the Kabul Plain aquifer with simplified geological map (Houben et al. 2009; Bohannon 2010). Groundwater table contour lines (July 2021) and flow direction (red arrows) are depicted. The labeled contour lines exhibited the elevation of the groundwater level

(Zaryab et al. 2022). Many households in Kabul depend on shallow wells as municipal water is only distributed intermittently to 20% of the population. It is estimated that there are over 120,000 shallow wells in Kabul and over 40% are projected to be seasonally dry or completely dry by 2057 due to overexploitation (USAID 2021). The Kabul Plain aquifers are the main source of water supply for Kabul residents and socioeconomic development in the city. The groundwater tables have severely declined in some places of the Kabul Plain due to overexploitation. There is a significant imbalance between groundwater recharge and discharge in the Paghman and Central Kabul aquifers thus; they are in crisis because of over-exploitation.

Overall, natural groundwater recharge within the Kabul Plain has been altered as a consequence of rapid urbanization, and natural recharge within the urban area has decreased due to the construction of buildings and transportation infrastructures. Zaryab et al. (2021b) estimated groundwater recharge rates from precipitation in Kabul city is only 25.7% (28.1 MCM/year) of total natural recharge. Immediate actions are required to reverse the further depletion of groundwater resources and deterioration of groundwater quality in the plain. Therefore, effective measures and integrated approaches for safeguarding the water supply should be implemented to decrease groundwater deterioration and increase groundwater recharge. Furthermore, increasing groundwater recharge by using artificial recharge should be seriously taken into consideration (Zaryab et al. 2021b).

16.4.2 Climate

The Kabul Plain is characterized by an arid to semi-arid climate and receives approximately 315 mm of precipitation yearly with the highest precipitation occurring during winter months as snow and rain (Zaryab 2022). The snowy and rainy season, accounts for 68% of the annual precipitation from January to April. The temperature in the area ranges from an average monthly low of −7 °C in January to a mean monthly high of 32 °C in July. The rate of evaporation in the plain is approximately 1,600 mm/yr. The winter snowpack in the mountains adjacent to the Kabul Plain is a crucial source of recharge in the Kabul Plain (Mack et al. 2010).

16.4.3 Hydrogeological Setting

The Kabul Plain is underlain by Quaternary and Neogene sediments (Houben et al. 2009). The Quaternary aquifer which forms shallow aquifer, consists of gravel and sandy beds partially covered by loess silt and clay with a thickness of up to a few meters. The average thickness of shallow aquifer estimated around 45 m and it is maximum can be up to 80 m, with hydraulic conductivity (K) varies from 1 to 34 m day^{-1} (Zaryab 2022). The Quaternary alluvial aquifers along the major river channels are the most productive aquifers (Broshears et al. 2005). Thousands of shallow wells have been installed tapping the shallow aquifers over the past decades. However, a considerable number of them in the western plain and northwest of the eastern plain dried up because of heavy exploitation over the past decade (Saffi 2019). Therefore, currently, shallow aquifer plays as an unsaturated zone in most parts of the Kabul Plan.

The general groundwater flow direction in the plain is from the west and southwest to the east (Fig. 16.1). However, in recent years, the groundwater flow direction locally changed due to over-abstraction at places (Zaryab et al. 2021b).

The Neogene sediments in the Kabul Plain is divided into lower Neogene (N1) and upper Neogene (N2). The upper Neogene dominantly consists of consolidated clay with partially intercalated layers of sandstone and some gravel (JICA 2011). JICA (2011) considered upper Neogene an aquiclude, however, lately, Landell Mills (2020) considered the N2 layer as an aquitard. The thickness of the upper Neogene is estimated from 260 to 360 m (JICA 2011). Lower Neogene consists of hard clay with gravel, sand, and conglomerates. The thickness of the lower Neogene is estimated from 160 to 255 m. The Lower Neogene formation is known as deep aquifer (JICA 2011; Landel Mills 2020). The hydraulic conductivity of the deep aquifer is significantly lower than shallow aquifer (JICA 2011; Houben et al. 2009). Thus, the deep aquifer is not considered a major productive aquifer in the Plain.

16.4.4 Aquifer Storage Depletion

As shown in Fig. 16.2, during the past 13 years (from 2007 to 2020), groundwater levels have declined all across the Kabul Plain, except in a small area in Bagrami district, where the water level rose about 2 m during the period. Rising groundwater level in Bagrami (Fig. 16.5) district is most likely to be a result of recharge from the river and irrigation return flows. The maximum groundwater table decline is observed in the southwest of the western plain and in the northwest of the eastern plain. These two places are the most densely populated areas of the Kabul city. Thus, population growth and rapid unplanned urbanization have played a major role in the water table decline in the Kabul Plain (Farahmand et al. 2021). However, the riverbeds covered by a thick mud/clay layer that prevents significant infiltration to the subsurface. Mahdwai et al. (2022) has reported that the areas between 50 and 300 m from the rivers in the Kabul Plain are suitable for recharging aquifers.

Overall, 12 land-use classes are identified in the Kabul Plain (Fig. 16.3). The main land-use in Kabul is residential (34% of the total area), followed by agriculture (17%), institutional (13.8%) vacant plots (12.3%), barren land (7.3%), transport network/transport infrastructures (6.3%), and industrial (2.4%). The area covered by residential area is about 56 km^2. The area under cultivation in the Kabul Plain covers approximately 45 km^2. In recent years, rainfall patterns have changed and

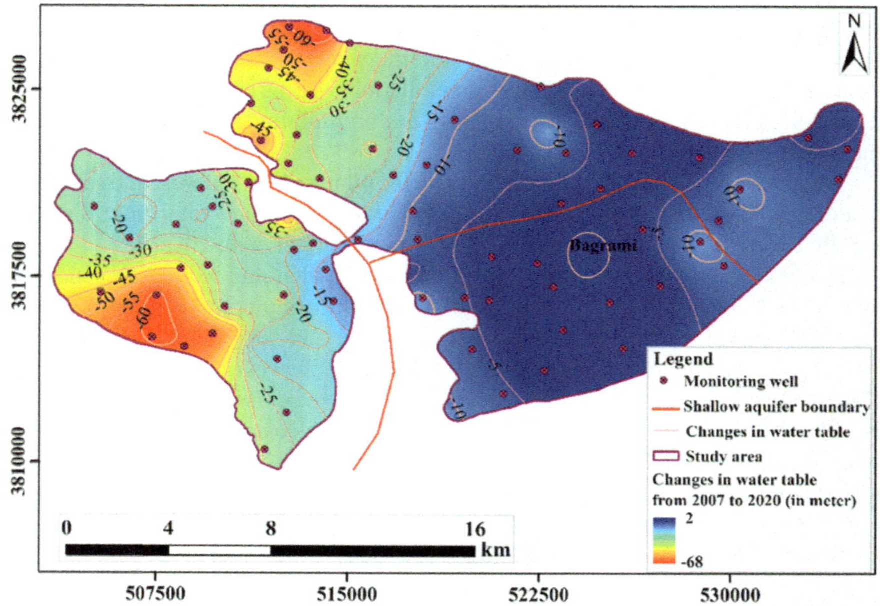

Fig. 16.2 Changes in groundwater level from November 2007 to November 2020 (Zaryab et al. 2021b)

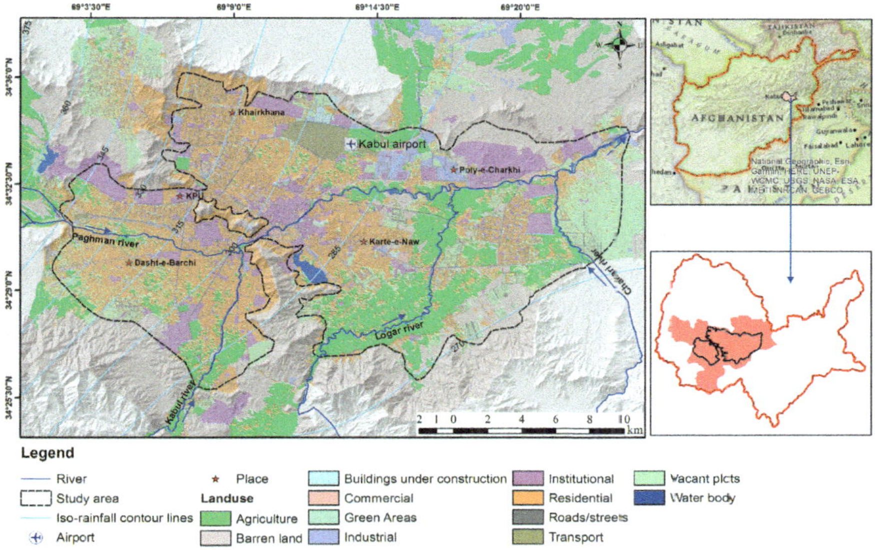

Fig. 16.3 Land-use map of the study area (Kabul city, Afghanistan) and iso-rainfall contour lines

most part of the precipitation occurs as rain. The average annual iso-rainfall contour lines in Kabul Plain are exhibited in Fig. 16.3. There is a positive relationship between precipitation and altitude.

16.4.5 Water Availability in the Kabul Plain

The primary water resources within the Kabul Basin are the Panjshir River, Loger River, Maidan River, Salang River, and Panjshir fan aquifer (Zaryab et al. 2017; Noori and Singh 2021). Over the past decade, a number of studies on surface water and groundwater resources have been conducted by the government of Afghanistan with the technical and financial cooperation of international organizations such as JICA, USIAD, BGR, Landel Mills, DACAAR, etc., in the Kabul Basin. However, nothing has been carried out to supply drinking water to Kabul City's residents in a planned manner from other available water resources within the basin. Over the past two decades, groundwater storage was remarkably depleted in the Kabul Plain due to over-exploitation. Zaryab et al. (2021b) stated that the groundwater exploitation is more than twofold than the rate of natural recharge. Thus, an urgent action is needed to supply drinking water to Kabul city's residents. Artificial recharge of the Kabul aquifers by surplus river flow and snow melt water also could be a good option for raising groundwater table in the Kabul Plain (Zaryab et al. 2017).

16.5 Managed Aquifer Recharge History in Afghanistan

Managed aquifer recharge (MAR) is defined as the purposeful recharge of water to aquifers for subsequent recovery or environmental benefit. It is a water resources management tool that encompasses a wide variety of water sources, recharge methods, and storage management practices. MAR technologies have a long history. For instance, harvesting and storage of monsoon rains, including techniques that recharged local aquifers, have been continuously practiced in many places like in India since at least the third millennium BC (Agarwal and Narain 1997). However, the adoption of MAR as a decentralized water management solution has been indicated and fallen out of favor in some areas in preference to large-scale solutions, such as major dams and desalination systems (Maliva 2020). MAR is likely to see increased use as growing populations create greater demand for water and as a strategy to adapt to increased variability of water supplies due to climate change (Alley et al. 2022). MAR is most commonly employed to either increase the volume of freshwater stored in an aquifer or reduce the rate of decline in water storage caused by excessive groundwater pumping (Maliva 2020; Braune and Israel 2021; Saha et al. 2021).

Although the MAR technology is in use in numerous countries, it is new to Afghanistan and there are not many designs and practices available. A few scientific studies regarding MAR aspects have been carried out in the Kabul basin recently, mostly in urban and populated areas, of which one was conducted by Masoom (2018). The objective of his study was to increase groundwater storage in the basin to reach a sustainable yield. In his study, a direct surface recharge method employing a recharge basin was modeled. Results of this study indicate that a daily diversion of 75,000 cubic meters per day (m^3/day) from the Kabul River to the recharge basin during the rainy season, increases the groundwater storage by an annual amount of 9 million cubic meters. Hussaini et al. (2022) integrated geographic information system (GIS) and multi-criteria decision analysis (MCDA) techniques to identify the most suitable areas for a MAR project in the Kabul city area. A suitability map was generated from analytical network process (AHP) and analytical hierarchy process (ANP) techniques separately based on a combination of all thematic layers. The result of the ANP was found to be more precise and reliable compared with that of the AHP. As a result, seven sites have been prioritized for the design and implementation of MAR projects the location of three of these seven sites is matched with the MAR sites of the pilot project (KMARP). Therefore, MCDA techniques integrated with GIS found an effective method for the selection of the MAR site. Generally, a very suitable area for MAR is located in the western part of the study area in the upper Kabul sub-basin close to the upstream of the Paghman River. Suitable areas for the application of MAR are located in the southwest, western, northern, and central parts of the study area but the larger portion of suitable areas is in the western and central parts.

In addition to the mentioned studies, a pilot MAR project entitled Kabul Managed Aquifer Recharge (KMARP) funded by USAID through the Asian Development Bank, and implemented by Landell Mills and DACAAR were successfully carried out in Kabul city under the supervision of Ministry of Energy and Water of Afghanistan from 2017 to 2020. This project investigated the use of MAR techniques to improve groundwater levels and quality, as well as to access the drinking water in Kabul City. It includes testing pilot infrastructure as well as the design of a follow-on loan investment project.

The milestones of the pilot project were; Project preparatory, Kabul MAR feasibility study, capacity building and demonstration, detailed design, and procurement documentation for the next phase of the project (ADB 2015). During this project, different MAR technologies were examined such as injection wells (pumping), injection wells (gravity), above-surface receiving well, and spreading basins at the four selected sites.

As a main project activity, 40 piezometers were constructed to monitor the water table and temperature. Water level sensors and data loggers have been installed in all piezometers. The monitoring system could record water levels and groundwater temperature hourly, and when GPRS is available, the recorded data are transmitted (pushed) to the external server in an internet portal for visualization. Due to signal strength being highly variable in Kabul City when the signal is low, data transmission (i.e., communication) is not possible. At the end of the project, the recording, transmission, and communication were working properly, but at the moment of writing this paper, all dataloggers installed could not transmit data and all were deactivated. This problem negatively affects the water level data accessibility and monitoring of the groundwater level changes could not be possible. The suggestion for the next phase of the project was to implement 12 spreading basins, 60 above-ground intake boreholes, 100 dug-wells with horizontal galleries, 50 recharge boreholes, 120 km rehabilitation of unlined canals, and land acquisition, which may take time between 5 and 7 years (Landell Mills 2020). The outcomes of the pilot MAR project (KMARP) indicate that groundwater tables have been raised locally in the Kabul Plain. Therefore, implementing such projects may reduce stress on the Kabul aquifers. Since the recharge water is not accessible in all four seasons in the region, the MAR project could be efficient in the rainy seasons at least (Zaryab et al. 2022).

16.6 The RWH Pretreatment System Design

The designing of the recharge well in the RWH structure was carried out based on three key points, i.e., (i) considering the location and available space, (ii) the rate of recharge should be as rapid as possible, and (iii) the impurities in the recharging of groundwater should be minimum. The surface water runoff flows into an underground tank (6 ft deep and 9 ft wide). This tank is equipped with a 2 ft deep coarse filter mechanism to avoid impurities like leaves and plastics. The horizontal filtration mechanism provides clay and slit-free water to recharge the groundwater. The water will be recharged under gravity. The typical design with all details is shown in Fig. 16.4 (Hussain et al. 2019).

The advantages of this design are that the groundwater recharge rate through recharge well is high. The inflow rate will be approximately equal to the recharge

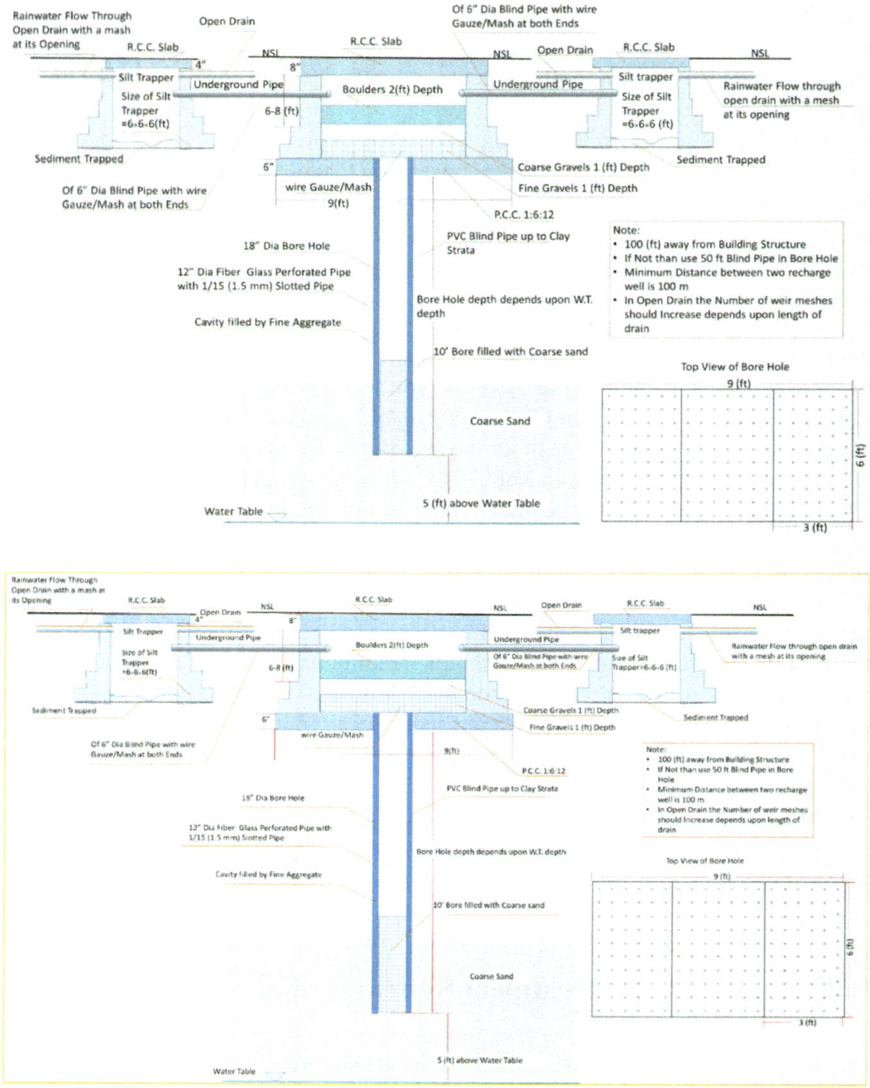

Fig. 16.4 Typical design of a recharge well employed for RWH is adopted from Hussain et al. (2019)

rate and the structure is cost effective and easy to maintain. While, the disadvantages are that the heavy metals, if present in the source water, will face little hindrance in making their way to the aquifer.The filtration material requires periodic cleaning to keep the filtration process efficient (Hussain et al. 2019).

The design further takes care of the quality issues through some measures like, (i) the recharge well casing ends 5 ft above the water table, where the percolating water would further be filtered as it flows through a 5-feet unsaturated zone; (ii) The silt trapper structures known as settlement chambers (provided where necessary) will remove silt and other floating impurities from rainwater. Moreover, in the case of excess rainfall, the recharging rate of bore wells might not match with the rate of rainfall. In such situations, these chambers hold the excess amount of water till it is soaked up by the recharge structure (Hussain et al. 2019).

The schema of groundwater recharge and rising of the water table are displayed in Fig. 16.5a. The components of the filter are shown in Fig. 16.5b. The design of recharge well is according to the volume of water accumulating in the recharge well. The water will infiltrate and percolate to the subsurface in a downward direction which tends to raise the existing groundwater level. The percolation rate depends upon the hydraulic conductivity of subsurface materials (Table 16.1) (Ahmed et al. 2022).

16.7 Conclusion

The groundwater in Kabul Plain from both the shallow and deep aquifers are widely utilized for drinking, agricultural, and industrial purposes. Groundwater tables considerably declined throughout the Kabul Plain due to over-abstraction. The groundwater depletion will likely continue in the future. In recent years, the Kabul urban areas have suffered from shortage of drinking water round the year. In this study, we presented the groundwater conditions in the Kabul Plain and reviewed the studies on RWH and managed aquifer recharge.Rivers and rainwater are the only sources of recharge to the aquifers but most of the rainfall is drained out, which need to be captured and recharged. In recent years, the Ministry of Energy and Water of Afghanistan and the Municipality of Kabul City encouraged the Kabul residents to collect rainwater and recharge it and also implemented an RWH pilot project. The RWH has been successfully practiced in many parts of the globe and it can be successfully implemented in Afghanistan as well. The MAR and RWH offer a promising approach in Kabul Plain as well as in other parts of Afghanistan affected by water level decline to arrest the groundwater resource depletion. Govt. may come out with a policy and more pilot projects are needed to showcase and assess the efficacy of such structures.

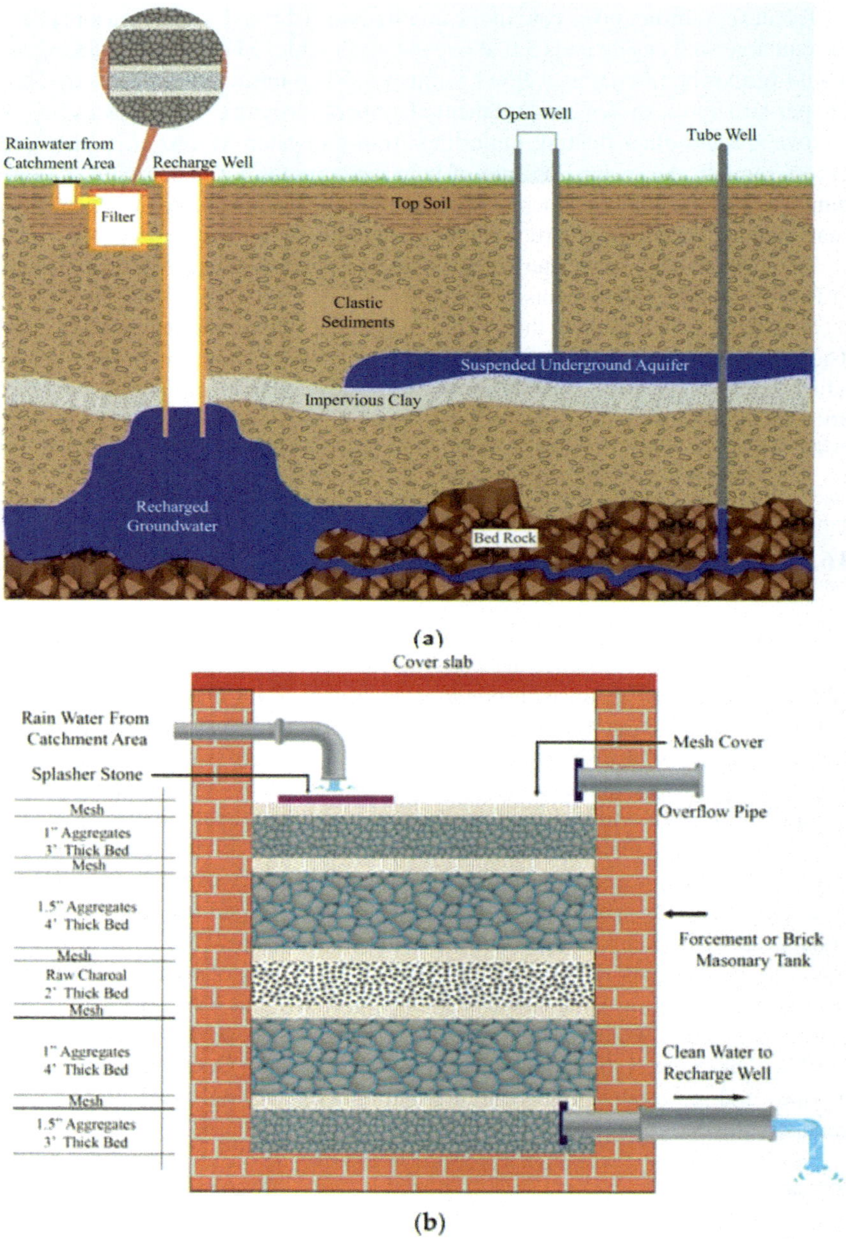

Fig. 16.5 **a** Recharge takes place by recharge well, and **b** Schematic diagram of filter is adopted from Ahmed et al. (2022)

Table 16.1 Typical ranges of hydraulic conductivity for different soil types (Humphreys et al. 2010)

No	Soil type	Coefficient of permeability (mm/s)
1	Clean gravel	$10^{+1}-10^{+2}$
2	Coarse and medium sands	$10^{-2}-10^{+1}$
3	Fine sand, loose silt	$10^{-4}-10^{-2}$
4	Dense silt, clayey silts	$10^{-5}-10^{-4}$
5	Silty clay	$10^{-8}-10^{-5}$

Acknowledgements The authors would like to thank the Ministry of Energy of Water of Afghanistan for access to groundwater table and managed aquifer recharge data of the Kabul Plain.

Author Contributions All authors contributed to the study's conception and design. Material preparation, data collection, and analysis were carried out by Abdulhalim Zaryab, Mohammad Zia Jamal, Zamen Jafari, Asadullah Farahmand, Hamid Zaki, and Mohammad Salem Hussaini. All authors reviewed and commented on previous versions of the manuscript. All authors read and approved the final manuscript.

Funding This research did not receive any specific grant from funding agencies.

Conflict of Interest The authors declare no conflict of interest.

References

Agarwal A, Narain S (1997) Dying wisdom. Rise, fall, and potential of India's traditional water harvesting systems. India Centre for Science and Environment, New Delhi

Ahmed A, Valyrakis M, Ghumman AR, Muhammad Arshad M, Pasha GA, Rashid Farooq R, Janjua S (2022) Assessing the rainfall water harvesting potential using Geographical Information Systems (GIS). Civil Eng 3(4):895–908

Aini A (2007) Water conservation in Afghanistan. J Dev Sustain Agric 2:51–58

Alam R, Munna G, Chowdhury M, Sarkar M, Ahmed M, Rahman M, Jesmin F (2011) Feasibility study of rainwater harvesting system in Sylhet city. Environ Monit Asses

Alley W, Dillon P, Zheng Y (2022) Basic concepts of managed aquifer recharge, in managed aquifer recharge: overview and governance. IAH Special Publication. https://recharge.iah.org/. ISBN 978-1-3999-2814-4

Asian Development Bank (ADB) (2015) Preparing the Kabul managed aquifer recharge project. Kabul-managed aquifer report, ADB, Mandaluyong, Philippines

Bohannon RG (2010) Geologic and topographic maps of the Kabul North 30 × 60 quadrangle, Afghanistan: U.S. Geological SurveyScientific Investigation Map 3120, 34 p. pamphlet, 2 map sheets, scale 1: 100,000. http://pubs.usgs.gov/sim/3120

Braune E, Israel S (2021) Managed aquifer recharge: Southern Arica. The Groundwater Project, Gulph, Ontario, Canada

Bromand MT (2015) Impact assessment of climate change on water resources in the Kabul River Basin, Afghanistan. Ritsumeikan University, Japan

Broshears RE, Akbari MA, Chornack MP, Mueller DK, Ruddy BC (2005) Inventory of groundwater resources in the Kabul Basin, Afghanistan: U.S. Geological Survey, Survey, Scientific Investigation Report 2005–5090, U.S. Geological Survey, p 34

Dash JP, Sarangi A, Ruggieri G (2013) Spatial variability of groundwater depth and quality parameters in the National Capital territory of Delhi. Environ Manag 45:640–650

Farahmand A, Hussaini MS, Zaryab A, Aqili SW (2021) Evaluation of hydrogeoethics approach or sustainable management of groundwater resources in the upper Kabul sub-basin, Afghanistan. Sustain Water Resour Manag 7:48. https://doi.org/10.1007/s40899-021-00525-9

Houben G, Niard N, Tünnermeier T, Himmelbach T (2009) Hydrogeology of the Kabul Basin (Afghanistan), part 1: aquifer and hydrology. Hydrogeol J 17(3):665–677. https://doi.org/10.1007/s10040-008-0377-z

Humphreys E, Kukal SS, Christen EW, Hira GS, Sharma RK (2010) Halting the groundwater decline in north-west India—which crop technologies will be winners? Adv Agron 109:155–217

Hussain F, Hussain R, Wu R-Y, Abbas T (2019) Rainwater harvesting potential and utilization for artificial recharge of groundwater using recharge wells. Processes 7(9):623

Hussaini MS, Farahmand A, Shrestha S et al (2022) Site selection for managed aquifer recharge in the city of Kabul, Afghanistan, using a multi-criteria decision analysis and geographic information system. Hydrogeol J 30:5978. https://doi.org/10.1007/s10040-021-02408-x

Ishaku H, Majid MR, Johar F (2011) Rainwater harvesting: an alternative to safe water supply in nigerian rural communities. Water Resour Manag 295–305

Jac V (2022) Large aquifer systems around the world. Groundwater Project, Guelph, Ontario, Canada

Japan International Cooperation Agency (JICA) (2011) The study on groundwater resources potential in Kabul Basin, JICA

Julius JR, Prabhavathy RA, Ravikumar G (2013) Rainwater Harvesting (RWH)—a review. Int J Sci Eng Res 4(8). ISSN 2229-5518

Kresic (2023) Hydrogeology 101: introduction to groundwater science and engineering. Blue Ridge Press LLC. VA, USA, p 564

Mack TJ, Akbari MA, Ashoor MH, Chornack MP, Coplen TB, Emerson DG, Hubbard BE, Litke DW, Michel RL, Plummer LN, Rezai MT, Senay GB, Verdin JP, Verstraeten IM (2010) Conceptual model of water resources in the Kabul Basin, Afghanistan. U.S. Geological Survey Scientific Investigation Report 2009–5262, p 240. https://pubs.usgs.gov/sir/2009/5262/

Mahdawi Q, Sagin J, Absametov M, Zaryab A (2022) Water recharge suitability in Kabul aquifer system within the upper indus basin. Water 14:2390. https://doi.org/10.3390/w14152390

Maliva RG (2020) Anthropogenic aquifer recharge, WSP methods in water resources evaluation series No. 5. Springer Hydrogeology. https://doi.org/10.1007/978-3-030-11084-0

Masoom MF (2018) Artificial recharge of groundwater as a management tool for the Kabul Basin, Afghanistan. LSU Master's Theses. 4653. https://digitalcommons.lsu.edu/gradschool_theses/4653

Mayar MA (2022) Global warming and Afghanistan: drought, hunger and thirst expected to worsen, Afghanistan analysts network. https://www.afghanistan-analysts.org/en/reports/economy-development-environment/global-warming-and-afghanistan-drought-hunger-and-thirst-expected-to-worsen/

Mills L (2020) The feasibility of scaling-up Managed Aquifer Recharge (MAR) Technology in Kabul City. AFG: Kabul Managed Aquifer Recharge Project Preparation, Landell Mills, Kabul

Mukherjee A, Saha D, Harvey CF, Taylor RG, Ahmed KM, Bhanja SN (2015) Groundwater systems of the IndianSub-Continent. J Hydrol: Regional Stud. https://doi.org/10.1016/j.ejrh.2015.03.005

Noori AR, Singh SK (2021) Spatial and temporal trend analysis of groundwater levels and regional groundwater drought assessment of Kabul, Afghanistan. Environ Earth Sci 80:698

Premchand M, Bhavani S (2018) Rooftop rainwater harvesting technique for Anits campus. J Emerg Technol Innov Res (JETIR) 5(5). ISSN-2349-5162

Qureshi AS (2002) Water resources management in Afghanistan: The Issues and Options, Working paper 49, Pakistan Country Series No. 14, International Water Management Institute

Rahimi O, Murakami K (2018) Analysis of rooftop rainwater harvesting in Kabul New City: A case study for family houses and educational facilities. Open J Civil Eng 8:155–165. https://doi.org/10.4236/ojce.2018.82013

Ramos-Leal JA, Noyola-Medrano C, Tapia-Silva FO (2010) Aquifer vulnerability and groundwater quality in mega cities: case of the Mexico Basin. Environ Earth Sci 61:1309–1320. https://doi.org/10.1007/s12665-009-0434-5

Saffi H (2019) National alarming on groundwater natural storage depletion and water quality determination of Kabul city and immediate response to the drinking water crises. DACAAR, Kabul. DACAAR, Kabul, pp 1–73

Saha D, Dwivedi SN, Singh RK (2013) Aquifer system response to intensive pumping in urban areas of the Gangetic plains, India: the case study of Patna. Environ Earth Sci. https://doi.org/10.1007/s12665-013-2577-7

Saha D, Sikka AK, Goklani R (2021) Artificial recharge endeavors in India: a review. Water Secur. https://doi.org/10.1016/j.wasec.2022.100121

USAID (2021) Afghanistan Water Resources Profile by USAID's Sustainable Water Partnership. https://winrock.org/wp-content/uploads/2021/08/Afghanistan_Country_Profile-Final.pdf

Younos T, Parece TE, Barcelo D, Kostianoy AG (2016) Sustainable water management in urban environments. Springer Nature, AG Switzerland

Zaki H (2022) Study the geological structure and artificial groundwater recharge using rainwater harvesting from the roof-tops of the Kabul districts. Kabul Polytechnic University

Zaryab A, Nassery HR, Alijani F (2021a) Identifying sources of groundwater salinity and major hydrogeochemical processes in the lower Kabul Basin aquifer, Afghanistan. Environ Sci Process Impacts 23:1589–1599. https://doi.org/10.1039/D1EM00262G

Zaryab A, Nassery HR, Knoeller K, Alijani F, Minet E (2022) Determining nitrate pollution sources in the Kabul Plain aquifer (Afghanistan) using stable isotopes and Bayesian stable isotope mixing model. Sci Total Environ 823:153749. https://doi.org/10.1016/j.scitotenv.2022.153749

Zaryab A, Noori AR, Wegerich K, Kløve B (2017) Assessment of water quality and quantity trends in Kabul aquifers with an outline for future drinking water supplies. CAJWR 3(2):3–11

Zaryab A, Nassery HR, Alijani F (2021b) The effects of urbanization on the groundwater system of the Kabul shallow aquifers, Afghanistan. Hydrogeol J. https://doi.org/10.1007/s10040-021-02445-6

Zaryab A (2022) Determination of the source of nitrate in Kabul aquifer using isotopic indicators and simulation of contaminant transport, PhD thesis, Shahid Beheshti University, Tehran

Zhou Y, Wang L, Liu J, Li W, Zheng Y (2012) Options of sustainable groundwater development in Beijing Plain, China. Phys Chem Earth 47–48:99–113. https://doi.org/10.1016/j.pce.2011.09.001

Chapter 17
Combating Urban Waterlogging with Support from Underlaying Over Exploited Aquifer: A Case Study from India

Arunangshu Mukherjee, Nidhi Didwania, Sneha Rai, Sandeep Kumar, Priya Pahil, N. C. Wadhwa, and Dipankar Saha

Abstract Urban waterlogging and groundwater depletion are two diverse but major challenges of Indian cities under changing climatic conditions. The enhanced extreme events of rainfall in recent years along with rapidly altered hydrological conditions in the urban environment pose a conducive situation for urban waterlogging. On the other hand, intense and large withdrawal of groundwater, higher than the natural annual recharge has depleted the groundwater level severely in many Indian cities. A pilot project is executed in Faridabad Smart City of National Capital Region India, to combat waterlogging and also to rejuvenate groundwater resource. In the study, all steps were undertaken meticulously, beginning with hydrogeological study, site selection, rainfall analysis, calculations of runoff generation, framing well design after identification of suitable recharge zones within depleted aquifer, and determination of its intake capacity. It further elucidates the estimation of the suitable dimension of the desilting chamber, fixing a suitable dose of ferric chloride for coagulation, and assessment of recharge volume. The constructed recharge system is tested whether it is working effectively as per the feedback obtained from independent sources. It has high scalability in similar hydrogeological situations in other parts of India.

Keywords Faridabad · Groundwater · Over-exploitation · Recharge · Flash food · Urban hydrogeology

A. Mukherjee (✉) · N. Didwania · S. Rai · S. Kumar · P. Pahil · N. C. Wadhwa · D. Saha
Manav Rachna Center for Advance Water Technology and Management, Manav Rachna International Institute of Research and Studies, Faridabad 121001, India
e-mail: arunangshu.fet@mriu.edu.in

D. Saha et al. (eds.), *Managed Groundwater Recharge and Rainwater Harvesting*, Water Resources Development and Management,
https://doi.org/10.1007/978-981-99-8757-3_16

17.1 Introduction

Incidence of prolonged waterlogging in areas of cities and towns are increasingly reported worldwide and particularly from India in the light of the changing global climate. There are several origins of such urban waterlogging creating frequent threats for infrastructure damage, economic loss, and social unrest (Chen et al. 2021; Subham et al. 2021; Wei et al. 2021). Environmentalists and town planners are looking for effective interventions. On the other hand, several Indian/world cities are hugely dependent on groundwater sources for meeting up their daily requirement of water (Howard and Israfilov 2002; Schirmer et al. 2013; Casanova et al. 2016; Saha et al. 2018). It exerts unprecedented stress on the underlying aquifers.

India is critically dependent on its aquifers, with an annual groundwater extraction, as per the latest estimation of 244.92 billion m^3 (CGWB 2021), which is currently the world's highest annual draft from a single country. As a result, in India, nearly 16% groundwater assessment units are categorized as *over-exploited* where the annual draft exceeds the annual dynamic groundwater recharge. A large part of these over-exploited units is concentrated in the north & north-western parts of the country and also in the south & south-eastern states of the country (Fig. 17.1 CGWB 2021). Further, nearly 20% of the other assessment units are grouped under *semi-critical, critical*, and *saline* categories. Therefore, only 64% of assessment units still remained under the *safe* category having less than 70% extraction of annual dynamic recharge. Many of them are even potentially low in groundwater resources. The distribution of *over-exploited* assessment units shows that the aquifer underlying them are largely alluvial in nature in the northern and north-western parts of India and occupied lithologically by deposits of Quaternary age. Due to excessive irrigational and domestic extraction, groundwater levels in these *overexploited* units have depleted severely. To overcome such an environmental challenge, rainwater harvesting and artificial groundwater recharge has been practices as supply-side intervention along with demand-side interventions like, crop rotation, crop diversification, behavioral change, and adopting improved irrigation technology.

The paper highlights the pilot project on artificial recharge of the depleted aquifers in the Faridabad urban area. The city suffers from waterlogging during monsoon, the frequency of which has increased due to the climate change. The Pilot Project offers a scalable ecofriendly solution to combat waterlogging by diverting the water into the underlaying over-exploited aquifers.

17.1.1 Study Area

The site is located in one over-exploited assessment unit, i.e., Administrative Block Faridabad of Haryana state, India. The intervention site is located at Officers' Colony of Sector-15A, Faridabad (28°23′45.50″ N, 77°19′6.96″ E, 53H/7). The Colony is

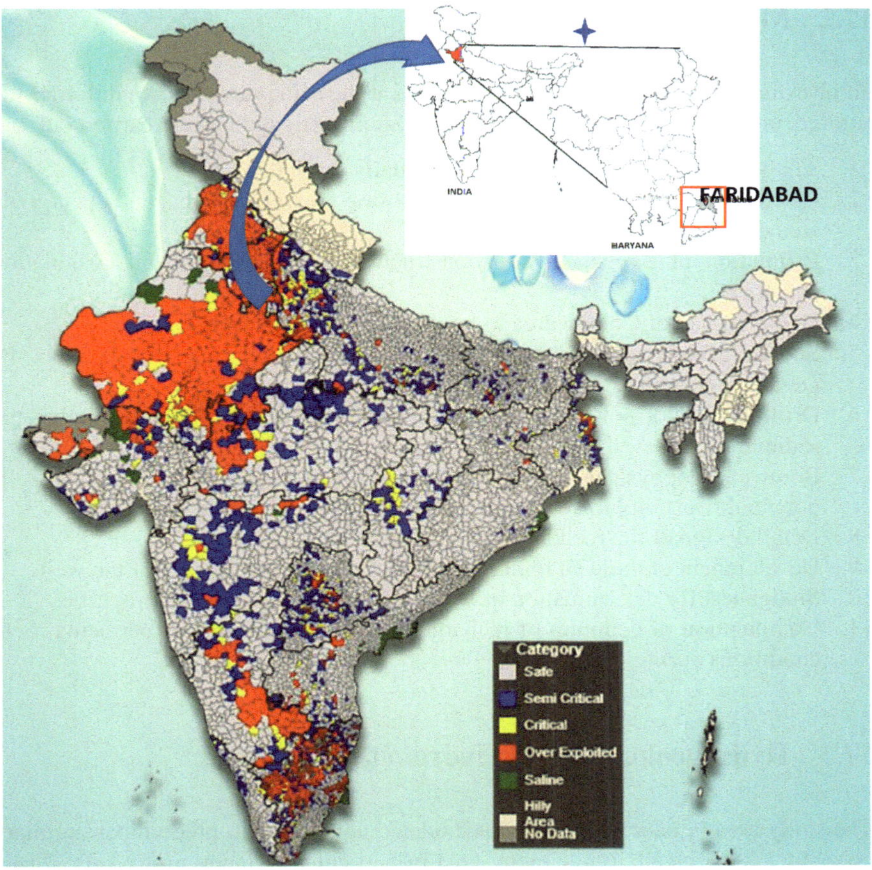

Fig. 17.1 Location map of study area with groundwater over exploited units in India in red

having 3.25-hectare open area. Geomorphologically, the Faridabad area is represented by an alluvial plain, bounded by residual hills of Aravalli in the West and the Yamuna River (flows from North to South) in the East. The regional slope of the area is toward the south-east. The elevation of the area is 201 m above mean sea level (msl). Geologically, the area is occupied by Quaternary Older Alluvium, which has a thickness of over 200 m underlain by quartzite of Alwar Group, Delhi Super Group (Palaeo-Proterozoic age) (Saini et al. 2017).

In Faridabad urban area, the settlements are aligned along the north–south running railway line, Agra-Gurugram canal, and National Highway (NH-19). The spread of the township is also constrained by the Aravalli ranges on the western side and the Yamuna River on the eastern side. The population density along the NH-19 is higher and gradually reduces toward the eastern and western sides.

17.2 Methodology

To investigate the waterlogging and its cause at the same time, the condition of the aquifers underlying in the study area, following investigations were carried out.

1. Analysis of long-term, short-term, and daily rainfall
2. Calculation of runoff generated from rooftop area, paved area, and open/green area of the Colony.
3. Estimation of dimension and construction of de-siltation and coagulation chambers.
4. DGPS survey of Colony area to locate pinpoint site for the structure.
5. Surface geophysical investigation to generate first-hand knowledge about the geometry of aquifer.
6. Drilling of pilot hole by rotary method, preparing litho-logs, examining the sediments
7. Borewell geophysical logging applying short normal, long normal, and self-potential resistivity to confirm the lithological log prepared.
8. Detail design of the recharge well and the filter material
9. Development of the well to make it mud-free and construction of the well.
10. Sludge test for determination of well performance and aquifer properties.
11. Computation of quantum of recharge and rate of recharge as per actual field conditions during Monsoon.

17.3 Hydrogeological and Hydrological Setup

The study area is underlain by unconsolidated aquifers with predominant primary porosity. Lithologically, the Colony area in particular is underlain by silty clayey sand, silty sand, clayey silt, and medium sand with silt up to 60 m depth. The silty clay sand prevails down to 21 m and a clay horizon occurs between 33 and 36 mbgl. The lithology made from the drill cut samples and the sub-surface geophysical log is given in Fig. 17.2. The annual water level in the Colony area is below 50 m. The transmissivity of the aquifer ranges between 145 and 875 m^2/day and the storativity ranges from 1.03×10^{-5} to 3.85×10^{-5}. The discharge varies from 1060 to 2300 lpm with a drawdown of 4 to 40 m (CGWB 2015).

The upper part of the lithological section in the study area is clay and thus acts as an aquitard. This aquitard prevents infiltration of runoff and promotes water ponding in the area. The Colony area is located in the densely populated part of the city and has higher paved area/rooftop that further reduces infiltration. Thus, geogenic and anthropogenic factors together create a conducive condition for waterlogging. Further, due to changing climatic conditions extreme events of rainfall have increased in recent years. Hydrogeologically, the aquifers in the area remains semi-unconfined to a confined condition where the stage of groundwater development remains >100%.

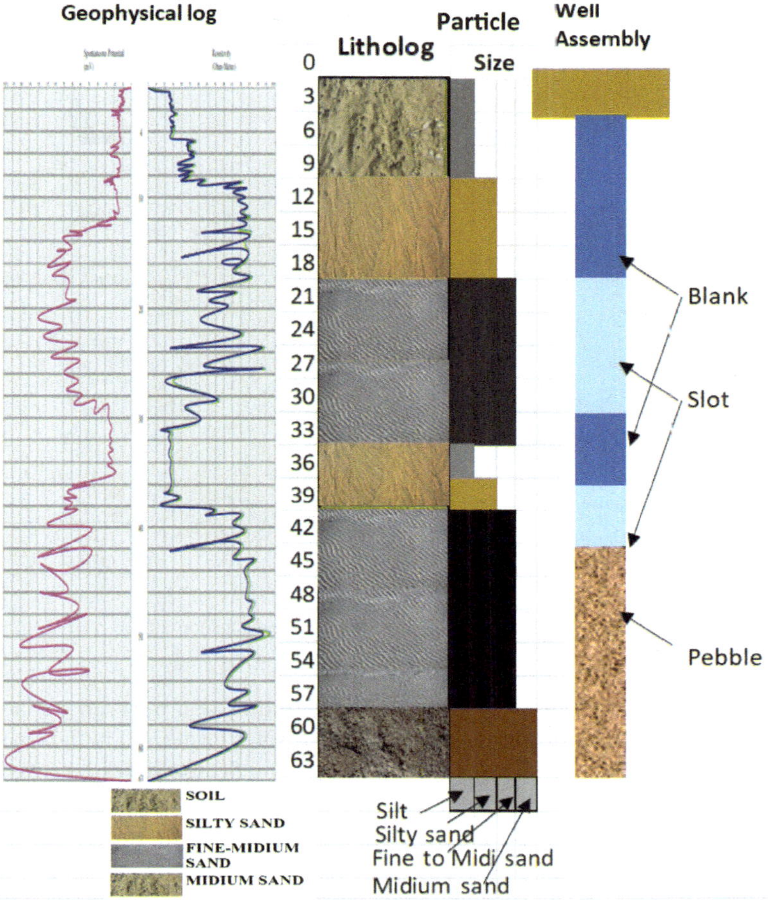

Fig. 17.2 Composite graph of borehole geophysical log, lithology, grain size, and well assembly

17.3.1 Depth of Water Level

Water level within the City area and in the adjacent parts of the Faridabad block varies between 20 and 77 m bgl. The DTW map (Fig. 17.3) clearly depicts that the deepest water levels are observed within the township area of Faridabad, along the National Highway. The levels gradually become shallower in the east and west sides. In the <3 m zone, the water level, however, remains between <1 and 2.75 m bgl. Further, the area having a water level between 3 and 10 m bgl is observed surrounding <3 m zone. Largely in the Aravalli hills area, the water level is found between 20 and 40 m bgl, whereas along the River Yamuna, it varies from 10 to 40 m bgl during the pre-monsoon period. Deeper groundwater level, >40 m bgl occurs in the central part of the township in an elongated stretch in north–south direction (up to Gurgaon canal)

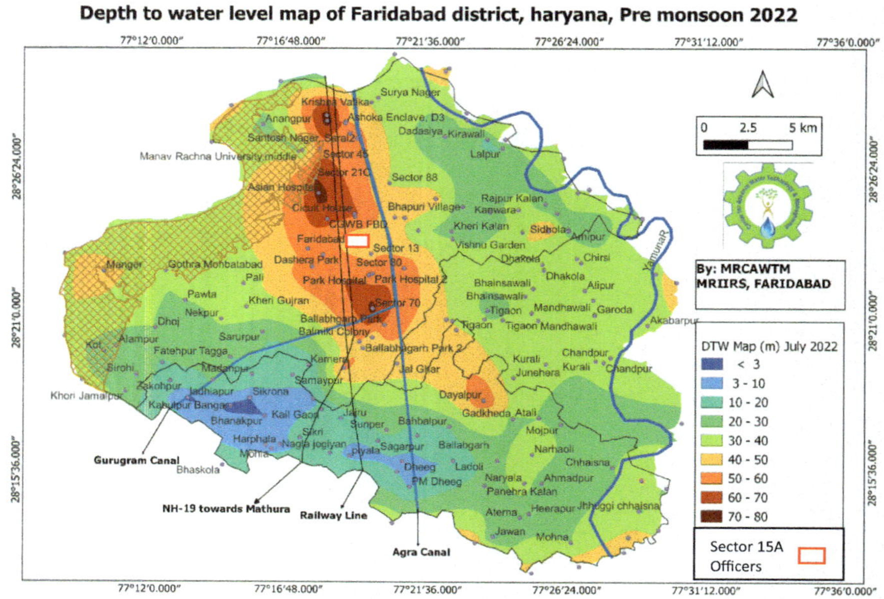

Fig. 17.3 Depth to water level map of Faridabad district, Pre-Monsoon 2022

and then the alignment turns toward south-eastern direction between Ballabhgarh and Dayalpur. In three detached patches, deepest-water level (>60 m bgl) has been identified, around Sharma Colony, Asian hospital, and Sector-70 (Fig. 17.3).

17.3.2 Groundwater Flow Map

The water level contour map reveals the formation of groundwater trough along the north–south aligned National Highway, extending toward the Dayalpur area. It also reveals recharge from Aravalli hills and the flow from waterlogged area (along Yamuna?) toward the central trough. On a regional scale, groundwater flow direction is towards the Yamuna River but locally within the city area, it has been totally altered. Recharge of groundwater from river Yamuna towards the central trough zone can be seen (Fig. 17.4).

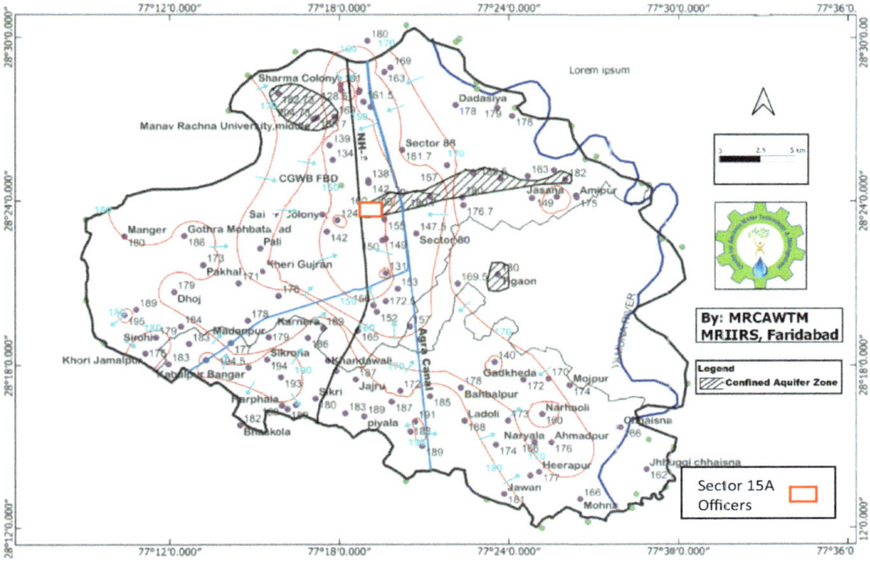

Fig. 17.4 Water level contour map of Faridabad district, Pre-Monsoon 2022

17.3.3 Distribution of Electrical Conductivity

It has been observed, largely the electrical conductivity (EC) of groundwater remains within 1500 µs/cm (Fig. 17.5), however, a concentrated high EC zone has been found around shallow groundwater level area EC 2250 to >9000 µs/cm).

Figures 17.3, 17.4, and 17.5 reveal that due to the formation of a central trough along the National Highway the groundwater movement is directed toward the central trough. Since the south-western part predominantly has higher EC, therefore, along with the groundwater flow the high EC plume started migrating toward the north-east direction. The salinity has been aggravated due to over-exploitation and a zone of 1500–2500 µs/cm of EC is created covering a large area of the southern side of the central trough. However, after the Gurgaon canal turns toward Dayalpur, the EC is <1500 µs/cm, indicating recharge from Yamuna. The situation is alarming as not only there is a possibility of drying up of aquifers along the National Highway but also the water quality will deteriorate rapidly unless corrective measures are taken.

17.3.4 Rainfall Analysis

Faridabad area falls under semi-arid region, and monsoon generally strikes around 1st week of July and retrieves around mid-October. Long-term trend shows a marginal decline in annual rainfall in Faridabad-NCR area (Fig. 17.6). To understand the

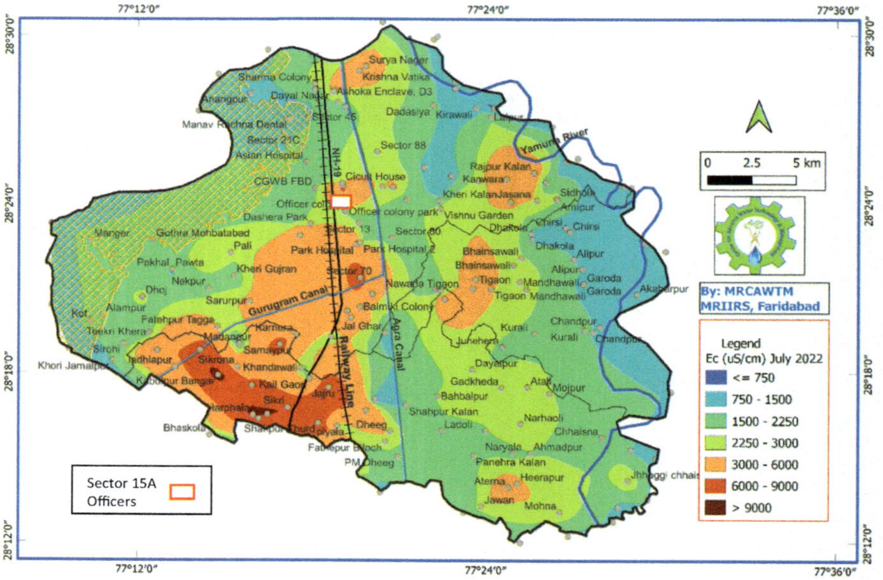

Fig. 17.5 Electric conductivity map of Faridabad district, Pre-Monsoon 2022

impact of changing climate, long and short-term meteorological data were analyzed to find out the changing patterns of rainfall and temperature indices (Chauhan et al. 2022). Further, a recent analysis by IMD of the past 30 years of data (1989–2018) shows Faridabad district having 29–34 rainy days with an annual mean of 618.8 mm (Pulak et al. 2020).

Rainfall data for 2020 and 2021 were collected from a station located at 2.25 km from the site (Fig. 17.7). It has been observed that the daily rainfall remains by and large around 25 mm/day, and occasionally up to 60 mm barring occasional extreme

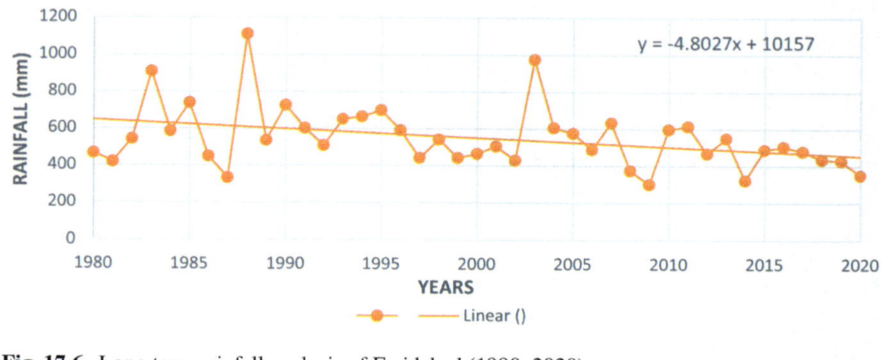

Fig. 17.6 Long term rainfall analysis of Faridabad (1980–2020)

events.. Extreme events were observed thrice in a year when rainfall crossed 80 mm/ day mark and have even exceeded 140 mm/day in 2021.

This rainfall analysis data was used for calculating the potential runoff generation as given in the table. Based on the rainfall data, it has been calculated that from 3.25-hectare area of Officers' Colony, a maximum of 2000 m^3 (Table 17.1) water per day can be generated (Fig. 17.8). For computing the dimensions of desiltation chamber (considering the mean grain size of the sediments) and considering the LULC pattern, the various categories of runoff generation have been tabulated (Table 17.1).

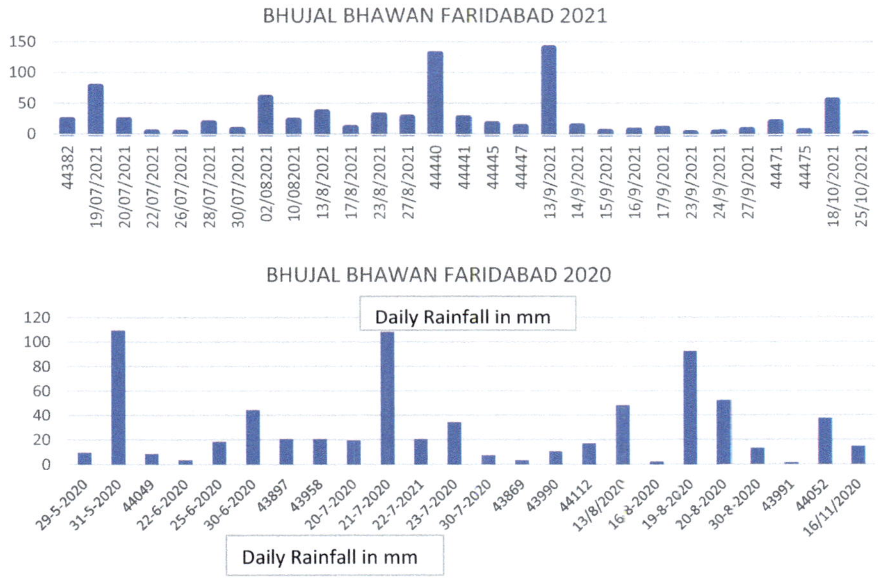

Fig. 17.7 Analysis of daily rainfall data of Faridabad city for two years (2020 and 2021)

Table 17.1 Computation of runoff from officers colony, Sector 15A, Faridabad

SN	Rainfall in m	Roof Area runoff @ 80%	Paved Area/ runoff @ 65%	Green Area/ runoff @ 15%	Total ham	Runoff m^3
Area in Ha		0.75	1.65	0.85	3.25 ha	–
Annual Rainfall	0.6	0.36	0.6435	0.0765	1.08	10,800
Normal per day Rainfall	0.025	0.015	0.0268125	0.0031875	0.045	450
General Rainfall/Day	0.061	0.0366	0.0654225	0.0077775	0.1098	1098
Maximum Rainfall/Day	0.11	0.066	0.117975	0.014025	0.198	1980

Fig. 17.8 Google earth image showing location of construction in officers colony, Faridabad

Fig. 17.9 Photographs of actual waterlogging in some locations of Faridabad and of the colony area

17.3.5 Magnitude of Waterlogging

The entire area of Sector 15 and 16, and large parts of Sector 19, 21, 22, 23, and 28 of Faridabad experience waterlogging during a high rain event (>40 mm/day). The water remains stagnated for several days and creates immense difficulties for the citizens, traffic jams, etc. It is estimated nearly 17 km^2 area of Faridabad City gets frequent waterlogging (Fig. 17.9).

17.3.6 Calculation of Dimension for Desilting Chamber

As the area is having silty clay/clay at the top, therefore, it has been considered that substantial runoff will be generated with considerable suspended and colloidal material. To prevent this suspended and colloidal load to move toward the recharge well, treatment chambers are constructed. Finally, the treated water passes through the sand filter around the recharge well. The dimension of desilting chamber has been calculated as (Table 17.2).

17.4 Designing the Recharge System

The pinpointing of site was done based on the results of surface geophysical investigations—resistivity profiling and sounding and DGPS survey of the campus.

17.4.1 Construction of Civil Structures

A dug cum recharge well structure is adopted (Fig. 17.10). A 60 m deep and 200 mm dia pilot hole was drilled first by using a rotary rig. Borehole logging was carried out when long normal, short normal, and self-potential logs were generated for a 60 m deep pilot hole (Fig. 17.2). The groundwater level in the area is observed around 50 m bgl thus deliberately the screen is kept down to 45 m bgl only (Fig. 17.2) and the lower 15 m is backfilled with 6–10 mm size pea gravel. Then, the annular space between 150 mm screen and 200 mm pilot hole is gravel-packed. After completion of lowering of assembly, a well was developed with compressed air and over-pumping. A 3.0 m dia and 3.0 deep dug well is constructed around the recharge well and filled with 2.0 m thick filter bed. The filter bed is made of 1.0 m thick bottom pebble, 0.5 m thick gravel, and 0.5 m thick top coarse sand so as to keep 1 m blank space for water. About lower 1.0 m screen of the recharge well within the dug well section is kept slotted for the entrance of filtered water. A small gutter is created in front of desilting chamber for bringing water to the system. The desilting chamber (6.5 * 2.5 * 2 m) is having a baffle wall of 1.4 m height to restrict course material and slow down the velocity of water entering the system. The desiltation chamber is connected to the coagulation chamber (3 * 2.5 * 2.5) through a 150 mm dia inlet pipe from 1.0 m below the top of the chamber and outlet at 1.5 m below the top. Both the chambers were provided with air vents, steps, and cleaning openings. A slab was casted over the system to cover the chembers. The coagulation chamber is dosed with ferric chloride as coagulation agent.

17.4.2 Establishing Well and Aquifer Efficiency

To know the intake capacity of the recharge well, slug tests were performed twice, first with 320 m^3 water and then with 450 m^3 water (Fig. 17 11). It is calculated that the intake capacity of the well is 10 m^3/min. Therefore, even with 110 mm/day rainfall, about 2000 m^3 of runoff can easily be accommodated within the system.

Table 17.2 Desilting chamber dimensioning and hydraulic calculations

Discharge Q	Normal/day	450 m^3
	Max/day	2000 m^3
	Annual	11000 m^3
Size of sediment particles say		<0.0020 mm
Settling velocity through Desilting Chamber (DC) Vs = C*d2		0.0358 m/s
Consider particle size to be settled 'd'		0.0010 mm
Settling velocity thru DC, Vs =		0.00895 m/s
Flow-through velocity < Critical Velocity		
Critical velocity Vc = a*sqrt(d); a = 44 constant		1.3914 cm/s
Let us consider Flow through velocity V		1.30000 cm/s
=		0.01300 m/s
(Considering max pracipitation of 2000 m3/day, we assume av. rain fall duration of 4 hrs)		
Precipitation/hr		500.000 m^3
Peak/Design Discharge		0.1389 m^3/s
Velocity thru DC for Qmax will be governed by the size of outlet to Coagulation Chamber		
Required Cross sectional area of outlet		0.107 m^2
Providing 2 nos. circular pipes with dia d =		0.300 m
Area of cross section A=2*Pi*d2/4		0.141 m^2 > 0.107
Provide Free board		0.300 m
Total height of chamber		2.000 m
Center of pipe from bottom of chamber		1.550 m
Depth of chamber for silt deposition dead storage		1.400 m
Length of DC is taken as 6.5 m and height as 2.0 m		
Width of DC, W = V*D/Vs		

(continued)

Table 17.2 (continued)

Discharge Q	Normal/day	450 m^3
	Max/day	2000 m^3
	Annual	11000 m^3
Where; W = Width of DC		
V = 0.0130 m/s (flow through velocity)		
D = 1.700 m (depth for settlement)		
Vs = 0.00895 m/s (settling velocity)		
Width = 2.46927 m		
Finally, the dimensions of desilting chamber:		
	Length	6.50 m
	Width	2.50 m
	Height	2.00 m

Fig. 17.10 Photographs of constructed desilting and coagulation chambers and dug cum recharge well at officers colony

17.5 Discussions

The urban flooding is caused by a combination of factors which is getting, aggravated due to climate variability, overburdened drainage, unregulated construction, no regard to the natural topography, drainage, and hydrogeology all make urban floods a man-made disaster.

The rapid urbanization and related paving of open lands by concretization and road construction has aggravated waterlogging problems in many cities in India and

Fig. 17.11 Photograph showing slug testing arrangements at officers colony, Faridabad

the World. Roads and streets of cities during heavy downpour profoundly overflow and mix with sewage and stormwater, contaminate the soil and water also. In the older parts of the cities in India, where drains and sewerages constructed decades back; many-a-time unable to carry the enhanced load, faces such a situation commonly with heavy rainfall. Such waterlogging problems are more acute in low-lying areas of the cities (McCartney and Smakhtin 2010).

Faridabad City (Haryana) in the backyard of the capital city of India is no exception to such a problem. The Municipal Corporation of Faridabad (MCF) responsible for water and sewage amenities of the city has to face such waterlogging-related issues every year, particularly in the Older Faridabad Constituency area. Now, part of the Older Faridabad Constituency area is covered under the Faridabad Smart City Project. Water supply of the entire Faridabad urban area is totally by groundwater (Faridabad Development Plan 2031 2016). This has resulted in continuous decline (Figs. 17.3 and 17.4) in groundwater level (Mukherjee and Shekhar 2016; Ibkar et al. 2022). The city area is categorized under *Over-exploited* (Fig. 17.1).

The abovementioned problems of Faridabad city are deteriorating the environment and creating potential health risks for the citizens. Technological interventions addressing these water issues can provide a solution to eradicate the waterlogging problem as well as address the need of Artificial Recharge of groundwater (Steinel 2012; Chadha 2014; Dillon 2005; Paul et al. 2015; Tuinhof et al. 2005) to minimize the potential health risk to the citizens of Faridabad and improve the amenities of Faridabad Smart City.

The present pilot project is an attempt toward a sustainable and ecofriendly solution to these problems. Fast-depleting groundwater levels in urban area is imposing water security challenges. Intervention through action research has been taken up to demonstrate to the local authorities a solution to combat waterlogging and groundwater depletion (Fig. 17.12) through a single intervention.

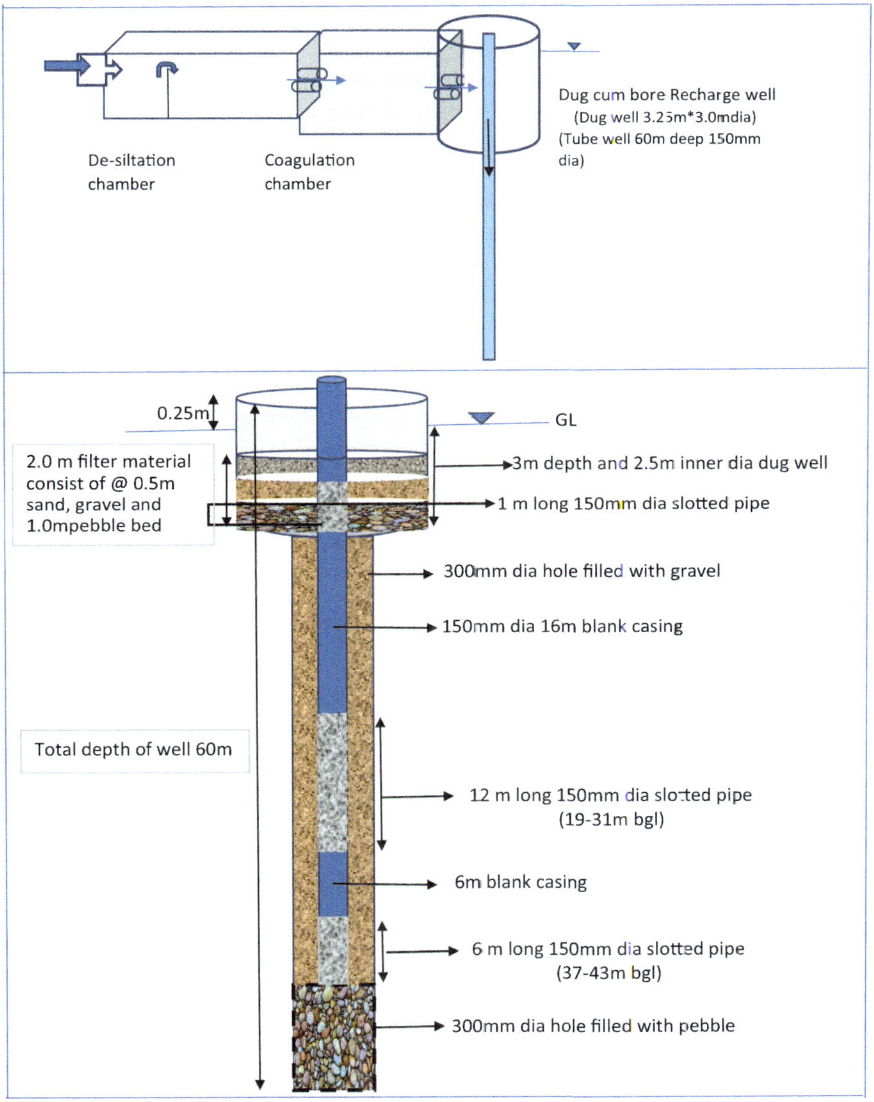

Fig. 17.12 Design of groundwater recharge system at officers colony, Sector 15A Faridabad

17.5.1 Source Water Availability and Suitability

The stormwater is used as source water and needs pre-treatment to meet the requisite quality criterion for recharge. Removal of silt, colloidal particles, and dissolved pollutants using a de-siltation method, coagulation process in conjunction for removing settleable particles, is being practiced in several countries (Steinel 2012; Wei et al. 2021).

An investigation has demonstrated that coagulation by adding ferric chloride is an effective method to reduce the turbidity, though it depends on pH, temperature, and flow rate. 85–98% reduction of turbidity from raw water can be achieved by using the optimum coagulant dosage (8 ppm) of ferric chloride in the optimum pH 9.2 at a temperature of 20 °C, (Koohestanian et al. 2008; Salim 2015). The primary disadvantage of clogging is that it leads to reduced infiltration over time.

17.5.2 *Identification of Suitable Recharge Zone*

The analysis of lithology, results of borehole logging and slug test show that the zone between 21 and 33 m depth is suitable for recharge and can be designated as zone 1. Similarly, zone between 40 and 60 m (Fig. 17.2) can be considered as zone 2. As the groundwater level is below 50 m in this area, the slotted casing is lowered down to 45 m depth so that the recharge water can be released into the unsaturated zone. The advantages are: (1) the rate of vertical infiltration is higher than horizontal so more recharge would take place; (2) any suspended particle if still left in the source water will pass through the unsaturated zone and be further filtered. The zones 1 and 2 together can absorb about 10 m^3/min.

17.5.3 *Effectiveness of Constructed Recharge System*

After completion of all constructions of the recharge system, two heavy rainfall episodes occurred in the area on 30th Aug and 22–23rd Sept 2022 (74 mm and 150 mm rainfall). On both the occasions, the system worked very well and the waterlogging of the Colony area was cleaned within 3–5 h, and no water was noticed over the filter bed around the recharge well then after. Several residents of the officers' Colony have given positive feedback about the working of the system.

17.6 Conclusions

The waterlogging in Faridabad City area is a combined effect of near-surface lithological condition and urbanization related concretization and pevmentization. Waterlogging is aggravated by climate change. The area is categorized as over-exploited and the current rate of groundwater depletion is estimated nearly 1 m/yr. These are underlain by a two-tier aquifer system. The top aquifer (21–33 mbgl) has gone dry as the present groundwater level is >50 mbgl. It has been estimated maximum 2000 m^3 runoff can be generated with about max. 110 mm/day rainfall from the 3.25 ha study area. The cumulative intake capacity of the recharge well (150 mm dia 60 m deep) tapping 12 m and 20 m thick two recharge zones, is found as 10 m^3/min. Calculations indicate for suspended load generated from silty-clay surface with max. 2000 m^3/day runoff, a 6.5 * 2.5 * 2 m size desilting chamber is required and for further coagulation, 8 ppm dosing of $FeCl_3$ is enough in a 3 * 2.5 * 2.5 m size coagulation chamber. A sand filter of 0.5 m thickness around the recharge well is further able to attenuate remaining particles before the water finally enters the recharge well. To minimize coagulation of the casing slot of well, the bottom of casing is kept open to a 15 m thick back-filling gravel zone below the well assembly within the drilled pilot well. The system is tested and found working efficiently during two consecutive rainfall episodes of 74 and 150 mm/day. The designed system has high scalability within the Faridabad Smart City area and in the other townships in similar hydrogeologic framework.

Acknowledgements The authors thankfully acknowledge DST, Government of India for the grant of Project "Co-solving waterlogging and groundwater depletion issue in parts of Faridabad Smart City using Underground Taming of Flood water for Aquifer Storage and Recovery" (DST/TMD/ EWO/ WTI/ 2K19/EWFH/2019/237 (G)) to corresponding authors. Special thanks to Sh H K Sahu, Former Member CWC for helping in calculating the dimension of the desilting chamber. The authors also thankfully acknowledge the Management of MRIIRS for providing logistic support and allowing authors to publish the paper. Authors are grateful to the editors of this book for their generous invitation to contribute a chapter to this publication. The view expressed in this paper is the authors' own and may not be of its Institution and funding agency.

References

CGWB (2015) Aquifer Mapping and Formulation of Aquifer Management Plan for the National Capital Region (NCR), Haryana (Volume II), Central Ground Water Board Ministry of Water Resources, River Development & Ganga Rejuvenation Government of India North Western Region, Chandigarh

CGWB (2021) National compilation on dynamic groundwater resources of India. CGWB Department of Water Resources, River Development and Ganga Rejuvenation, Ministry of Jal Shakti, Government of India

Casanova J, Devau N, Pettenati M (2016) Managed aquifer recharge: an overview of issues and options. In: Book integrated groundwater management. https://doi.org/10.1007/978-3-319-23576-9_16

Chadha DK (2014) Water conservation and artificial recharge of aquifers in India. Boletin Geology Minero 125(2):203–218

Chauhan AS, Singh S, Maurya RKS, Rani A, Danodia A (2022) Spatio-temporal and trend analysis of rain days having different intensity from 1901 to 2020 at regional scale in Haryana, India. https://doi.org/10.1016/j.ringps.2022.100041

Chen W, Dong J, Yan C, Dong H, Liu P (2021) What causes waterlogging? explore the urban waterlogging control scheme through system dynamics simulation. Sustainability 13(15):546. https://doi.org/10.3390/su13158546

Dillon PJ (2005) Future management of aquifer recharge. Hydrogeol J 13(1):313–316

Faridabad Development Plan 2031 A.D. (2016) Town and Country Planning Department bearing notification No. CCP(NCR)/FBC(DDP-2031)/2016/3303, dated the 7th December, 2016, the Governor of Haryana hereby publishes the Final Development Plan 2031 A.D. of Faridabad
Howard KWF, Israfilov RG (2002) Current problems of Hydrogeology in urban areas, urban agglomerates and industrial clusters. In: NATO science series, IV earth and environmental science, vol 8
Ibkar A, Mukherjee A, Didwania N, Rai S (2023) Impact of urbanization on groundwater in changing climatic scenario: a case study. In: Thambidurai P, Dikshit AK (eds) Impacts of urbanization on hydrological systems in India. Springer, Cham, pp 323–345
Koohestanian A, Hosseini M, Abbasian Z (2008) The separation method for removing of colloidal particles from raw water. American-Eurasian J Agric Environ Sci 4(2):266–273
McCartney M, Smakhtin V (2010) Water storage in an era of climate change: addressing the challenge of increasing rainfall variability. In: Blue Paper. Colombo, Sri Lanka: International Water Management Institute (IWMI), p 14. Available at http://www.iwmi.cgiar.org/Publications/Blue_Papers/PDF/Blue_Paper_2010-final.pdf
Mukherjee A, Shekhar S (2016) Hydrogeology of Smart cities of India (Ed.) INC-IAH Publication, p 160
Paul P, Karthikeyan B, Amarnath G, Eriyagama N, Muthuwatta L, Smakhtin V, Gangopadhyay PK, Malik RPS, Mishra A, Sharma BR, Hanjra MA, Reddy RV, Mishra VK, Verma CL, Kant L (2015) Controlling floods and droughts through underground storage: from concept to pilot implementation in the ganges river basin IWMI Research Report 165: International Water Management Institute (IWMI). Colombo, Sri Lanka. https://doi.org/10.5337/2016.200
Pulak G, Nitin N, Preetha M, Ashwini Kumar P, Neha S (2020) Observed rainfall variability and changes over Haryana State, Met Monograph No.: ESSO/IMD/HS/Rainfall Variability/09(2020)/33, IMD
Saha D, Marwaha S, Mukherjee A (2018) Groundwater Resources and Sustainable Management Issues in India. Clean Sustain Groundw India. https://doi.org/10.1007/978-981-10-4552-3_1
Saini HS, Kaura LPS, Mujtaba SAI, Verma K and Pant NC (2017) Assessment of the soils around Faridabad, NCR in the context of changing land use. Indian Journal of Geosciences 70(3&4) and 71(1): 219–226
Salim MS (2015) Effect of pH on the turbidity removal of wastewater. Open Access Libr J 02(12), Article ID:69015,9. https://doi.org/10.4236/oalib.1102283
Schirmer M, Laschink S, Musolff A (2013) Current researches in urban hydrogeology-a review. Ad Water Res 51:280–291
Steinel A (2012) Storm water reuse via aquifer storage and recovery: risk assessment for sandy aquifers. In: Book: Climate Change and the Sustainable Use of Water Resources. https://doi.org/10.1007/978-3-642-22266-5_2
Subham R, Singha BA, DebanjanBasak RC, Iandrajit, (2021) Urban waterlogging risk as an undervalued environmental challenge: An Integrated MCDA-GIS based modeling approach. Environ Chall. https://doi.org/10.1016/j.envc.2021.100194
Tuinhof A, Olsthoorn T, Heederik JP, de Vries J (2005) Groundwater storage and water security: making better use of our largest reservoir. Water Sci Technol 51(5):141–148
Wei M, Lin S, Xue-Yi Y (2021) Establishment of urban waterlogging pre-warning system based on coupling RBF-NARX neural networks. Water Sci Technol 82(9):1921–1931. https://doi.org/10.2166/wst.2020.477

Chapter 18
Water Harvesting and Managed Aquifer Recharge to Combat Water Scarcity in a Country of Water Abundance

Maria Jose Iturbide-Chang, Héctor Francisco Espinoza García, and Angela María Méndez Mora

Abstract In Guatemala, rainfall ranges between 700 and 5,000 mm per year depending on the region, with two seasons marked as dry and rainy. It is a country whose meteorological data show that there is a good quantity of water to satisfy the demands of its different uses and sectors, but a weak regulatory framework and inefficient management cause many regions to face water scarcity amid this abundance. The situation is more visible in the metropolitan region where the availability of water is 191 m^3/person/year, well below the 1.700 m^3/person/year that has been established as a reference for avoiding water stress. Currently, there is no specific law that regulates the use and exploitation of water in the country. Despite the multiple law proposals during the last 30 years by different sectors, none have been approved by the country's Congress. This lack of legal framework coupled with the high contamination of surface water sources due to deficient sanitation, makes the supply from private wells the primary alternative option at the municipal level and for companies and residential areas that can afford it, since they are allowed to drill their own well without considering the proximity to each other or the overexploitation of the underground water resources. Since 2018, FUNCAGUA has been working on monitoring piezometric levels in the metropolitan region of Guatemala, generating data that has made it possible to observe the decrease in groundwater levels over time. In this context, different techniques for the supply of water and/or mitigation of the extraction of underground water resources have emerged, such as rainwater harvesting applied in schools and in municipal markets, a new architectural vision for the use of water from rain in private homes, a pilot project for the managed aquifer recharge through dry wells that seek to generate technical guidelines for this type of intervention at the local level; incorporating the learning of empirical experiences that were not documented in the country and the impulse that begins to acquire more technical and engineering solutions, provided by private companies that venture into providing

M. J. Iturbide-Chang (✉) · H. F. Espinoza García · A. M. Méndez Mora
FUNCAGUA, Guatemala City, Guatemala
e-mail: mjiturbidef@gmail.com

D. Saha et al. (eds.), *Managed Groundwater Recharge and Rainwater Harvesting*, Water Resources Development and Management,
https://doi.org/10.1007/978-981-99-8757-3_17

advice and services for the management of rainwater and urban runoff in different projects in the country. These experiences have shown that: (a) rainwater harvesting systems are a functional alternative in both rural and urban areas, whose implementation can be accelerated thanks to public–private partnerships, given that the cost of materials can be high; (b) the use of shallow wells to capture rainwater following certain constructive recommendations should be socialized and taught at the municipal level, adding the understanding of how rainwater runoff can reach the aquifer for recharge purposes and allowing to reduce contamination risks. In addition, better technical guidelines for recharge monitoring and estimation will transform current empirical practice into a truly managed aquifer recharge approach; and finally, (c) technical options for stormwater management combined with groundwater recharge are available in Guatemala and begin to attract attention for different developments. Soon, these kinds of projects will gain visibility and be more valued as they address critical water scarcity and climate change issues by enhancing water supply from the harvesting of rainwater as well as incorporating adapted water recharge approaches.

18.1 Introduction—Guatemala: A Country of Abundant Water Availability

18.1.1 Bio-Physical Context

Located in Central America, The Republic of Guatemala is bounded to the north and west by Mexico, northeast by Belize, and (along a short coastline) the Gulf of Honduras (leading to the Caribbean Sea and the Atlantic Ocean) east by Honduras, and by El Salvador and the Pacific Ocean to the south and southeast. Guatemala has a total extension of 108,900 km^2. It has four major topographical features characterizing its surface: Southern Guatemala is composed of a chain of 27 volcanoes extending for about 300 km between Mexico and El Salvador in a northwestern to southeastern tract. Between the volcanoes and the Pacific Ocean lies a fertile plain of approximately 50 km. The Northern region, a large low-lying area, is part of the Yucatán Peninsula, a limestone platform shared with Mexico and Belize. The high mountain ranges and valleys in the center of the country, part of the American Cordillera, are inserted between the volcanic landscape and the Northern region.

Guatemala's hydrological division integrates 38 watersheds distributed like this: the Pacific (18 watersheds), the Caribbean Sea (10 watersheds), and the Gulf of Mexico (10 watersheds) (Fig. 18.1) (MAGA 2009a, b). Of these watersheds, 22 are transboundary: 20 are shared binationally and 2 tri-nationally (MARN 2013). Consequently, Guatemala is one of the Central American countries with the highest percentage of the national territory (64.6%) lying within international watersheds (Hernández-Ulate et al. 2000). In most of these territories, the highland is in Guatemala, so the country can be considered a net exporter of water (GWP 2015) and an important water tower.

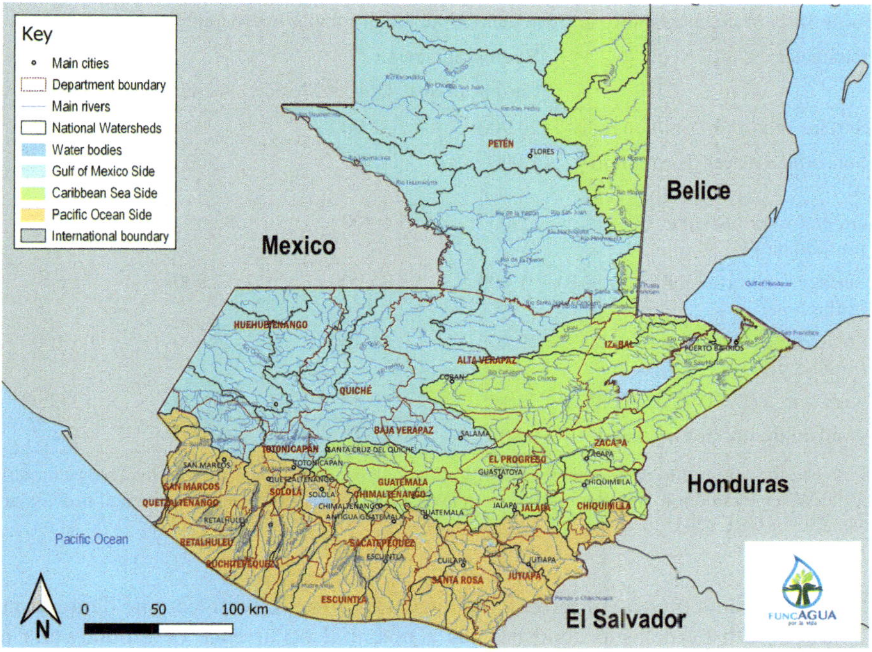

Fig. 18.1 Major hydrographic units of Guatemala

Guatemala has abundant water resources as it can be seen from Table 18.1, which estimates water availability in 2015.

The average annual rainfall in Guatemala is approximately 2,000 mm, with a geographic variation ranging from 700 mm to 5,000 mm. The country experiences two seasons: rainy and dry. The rainy season occurs between June and September, with less precipitation between July and August, known as the "canícula." In dry regions, which are mainly in the central highlands and the northern parts of the lowlands in the north of the country, a lack of rain occurs for six months, from November to April, while in the wettest regions, rain is reduced for two or three months, without being clearly defined (URL 2005).

Based on the groundwater recharge, the total estimated annual groundwater recharge potential is 20,500 million m^3. This availability does not mean that all this volume can be safely used. The data is only a reference to estimate a reasonable number of m^3 that could be sustainably used.

The alluvial aquifers of the Pacific Plain are estimated to have the highest yield, while aquifers of the Altiplano region represent the most used source of supply (Castañeda et al. 2000; URL 2005).

Hydrogeologically, the country is divides into four hydrogeological regions:

1. The quaternary alluvial plains of the Pacific coast are the formations with the most significant groundwater potential.

Table 18.1 Water availability in Guatemala at the level of major hydrographic units for 2015

Parameter	Hydrographic unit			
	Caribbean Sea	Gulf of Mexico	Pacific Ocean	Combined
Surface runoff (A) (million m^3)	20,900	33,900	16,300	71,000
Subsurface runoff (B) (million m^3)	1,000	1,100	1,500	3,600
Groundwater recharge (C) (million m^3)	5,500	6,400	8,600	20,500
Surface water availability (A + B) (million m^3)	21,900	35,000	17,800	74,600
Total availability (A + B + C) (million m^3)	27,400	41,400	26,400	95,100
Area (km^2) (INSIVUMEH 2022)	34,105	50,803	23,990	108,898
Availability (million m^3/km^2)	0.80	0.81	1.10	0.88

Note The model used to determine the hydrological balance corresponds to the hydrology module of the Water Evaluation and Planning System (WEAP). Details of the methodological process are available in URL (2015)

2. The volcanic plateau of tertiary and quaternary rocks, with tectonic depressions filled with pyroclastic deposits, has a potential occurrence of groundwater at relatively great depths.
3. The mountainous chain of crystalline highlands of granitic and metamorphic igneous rocks is the formation with the lowest occurrence of groundwater in the country.
4. The northern sedimentary karstified region, where groundwater occurs in karstic conduits is of great importance, while its dynamics have been little studied.

In general, it has been determined that only 26% of the territory has a high groundwater potential, while 47% is low or very low (MAGA 2009a, b).

The evaluation of Guatemala's water resources conducted by the United States Corps of Engineers (US Corps of Engineers 2000) agrees that Guatemala has abundant water resources and of good quantity to meet everybody's demands. However, it points out that weak management of resources is the main factor in complying with national strategies to satisfy the population's requirements. The Foundation for the Conservation of Water in the Metropolitan Region of Guatemala (Fundación para la Conservación del Agua de la Región Metropolitana de Guatemala—FUNCAGUA) concluded that Guatemala has adequate water availability, but because of poor integrated management, many regions face water scarcity in the midst of abundance (FUNGUA 2022).

18.1.2 Socioeconomic Context

According to a census of 2018, Guatemala has a population of 14.9 million inhabitants, 6.5 million (43.75%) of whom self-identify as indigenous, primarily from the Maya, Garífuna, and Xinca Indigenous Peoples, or the Creole (Afro-descendants) groups. At a growth rate of 2.9% per year, most Guatemalans live in the valleys of the mountainous regions in the center of the country, characterized by numerous lakes and volcanoes and a temperate climate.

The country's water stress is particularly pronounced where there is a mismatch between population density and water endowment. The population is dense in regions where water availability is low due to altitude and/or rainfall deficit, and the opposite is true in areas where water resources are abundant. Guatemala City is a prime example. The city is home to more than 20% of the population. However, the valley in which Guatemala City is located crosses the Continental Divide, where river discharge is minimal. The lack of adequate infrastructure for water delivery and disposal of waste leads to constant shortages of safe, potable water. They also increase river and possibly aquifer pollution, especially in poorer areas of the cities, leading to popular discontent. Furthermore, since scarce surface water resources of Guatemala City cannot support the requirements of its population, groundwater resources are increasingly required to supplement the needs of the city (US Corps of Engineers 2000).

Guatemala's average water availability per capita is around 6,380 m^3/person/year, which is above the internationally acknowledged standard thresholds of water stress (1,700 m^3/person/year) and water scarcity (1,000 m^3/person/year) (Nepomilueva 2017). In comparison, the water availability in Guatemala City is 190,84 m^3/person/year (URL 2022) indicating the great geographic disparity across the country and the localized water stress.

18.1.3 Legal and Institutional Context

In general, Guatemala's legal and institutional framework is weak and reflects the legacy of the civil war in the period of 1960–1996 and continuous political instability. As a result, its institutions face significant challenges, and according to the Worldwide Governance Indicators (WGI), Guatemala scores in the lowest quartile globally in three of six indicators (government effectiveness, the rule of law, and political stability), with the rule of law indicator scoring in the lowest decile. Only in the regulatory quality indicator does the country approach the Latin America and the Caribbean median (scoring in the 48th percentile). Furthermore, Guatemala is in the bottom third of the rankings regarding control of corruption and voice and accountability. Moreover, the country has not significantly improved in any of the six governance dimensions in the past 20 years (World Bank Group 2018).

The public institutions' pillar of the World Economic Forum's (WEF) Global Competitive Index also shows the weakness of Guatemala's institutions. In nine of the twelve components of the institutions' pillar, Guatemala ranks in the lowest 25th percentile in the world, and in five of those pillars (business costs of crime and violence, organized crime, public trust in politicians, the strength of investor protection, and diversion of public funds) ranks in the lowest decile (FUNDESA 2021).

Over the past decade, the percentile ranking of Guatemala's institutions appears to be deteriorating for most of the indicators of institutional quality. In parallel, the index for municipal public services for 2013 showed that 76% of municipalities (223 of 340) were either medium–low or low in terms of the provision of public services, which undoubtedly explains the gap in coverage of water and sanitation supply for the poor (World Bank Group 2018).

Guatemala is a unitary state with three tiers of government:

- The central government
- The departmental government with 22 departments
- The local government with 340 municipalities.

A governor appointed by the president of the republic is the head of the departmental governments, while local governments are led by a municipal council elected by popular vote. The mayor (alcalde in Spanish) presides over council sessions and exercises legal representation of the municipality (World Bank Group 2018).

The Constitution establishes that municipalities are autonomous institutions. The Municipal Code establishes and reaffirms this responsibility concerning local public services stipulating that municipalities must provide and regulate public services in their respective jurisdictions. Each municipality has the power to develop its management model of service provision, following complementary regulations from national ministries, specifically, the Health Ministry and the Ministry of Environment. There are three models of water service provision in urban areas of Guatemala: direct and delegated municipal public management and private management. In rural areas, community-managed systems dominate. Although the municipalities are legally responsible for providing water services in rural areas, communities generally build, operate, and maintain their systems through local drinking-water committees that are elected locally based on the population that is being benefited by the water service or can be voluntary by community leadership (World Bank Group 2018).

Currently, there is no specific law to regulate water use and exploitation. Despite multiple law initiatives proposed during the last 30 years by different sectors, none have been passed by the Congress of the country. However, in 2021, a legal initiative was released, creating the Water Vice-Ministry as a part of the Ministry of Environment and Natural Resources (MARN). Creating a Vice-Minister was a big step toward the necessary water resources management. This vice-ministry includes a water quality laboratory and watershed committees to monitor water resources. In the same year, another initiative was released, promoting general guidelines to protect watersheds in the country and elaborating diagnosis as a baseline to create and implement natural resources management plans. However, even with these efforts, the vice-ministry lacks the legal authority and mandate to coordinate action, assign rights, and mediate conflict. As a result, policies are drafted with little knowledge of water users and needs and limited power to enforce, and ultimately limited is achieved in terms of capturing the value of and investing in better infrastructure.

In addition to, and partly due to the lack of water regulations and confused institutional commitments with gaps and duplications in roles and responsibilities, there is a lack of systematic information to rely on for decision-making. Public institution analyses differ from national reports when considering water balance. In 2003, the Instituto Nacional de Sismología, Vulcanología, Meteorología e Hidrología (INSIVUMEH) reported a value of 93,300 million m^3 of water supply (GWP 2015). Subsequently, studies of Secretaría de Planificación y Programación de la Presidencia (SEGEPLAN) reported a balance of 97,100 million m^3 (SEGEPLAN 2006); also, as established previously, the Rafael Landivar University (URL) estimated 95,100 million of m^3 (URL 2015), and later, in 2018, the Ministry of Environment and Natural Resources (MARN) reported 154,900 million m^3 (CEPAL et al. 2018). The difference between each data set is due to the different methodologies used for the water balance. Generally, hydrologic data are lacking throughout the country, particularly since the mid-1980s when data collection decreased dramatically. Data are needed to solve water resources problems and to address critical need issues.

The census in 2018 validates the idea of scarcity in the middle of abundance, as it confirms a massive gap in drinking water supply coverage, impacting millions of Guatemalans. In addition, four of ten homes do not have a water supply source inside their house, and if they have it, there is no certainty of its functionality or the water quality (INE 2018).

On this background, this book chapter addresses the water situation in Guatemala and the options for enhancing water security through groundwater use that is managed through new rainwater harvesting and recharge solutions.

18.2 Main Sources of Water Supply and Its Use

18.2.1 Surface Water (Quantity and Quality)

Freshwater (<1,000 mg/L Total Dissolved Solids (TDS)) from streams and lakes provides about 70% of the public water supply in urban areas and 90% of the water supply in rural areas. The area of surface of water bodies amounts to about 1,000 km^2 of land area. However, many streams cease to flow during March, April, and May, giving rise to seasonal water scarcity. Unregulated deforestation has further altered the dynamics of the hydrologic cycle, reducing the availability of surface water. Over the past 40 years, clean surface water availability has decreased by 60–70%, primarily due to deforestation and other human activities (US Corps of Engineers 2000).

Unfortunately, the surface water quality is generally poor due to anthropogenic activities, especially the discharge of untreated wastewater, and many rivers are considered severely polluted and not suitable for water supply (US Corps of Engineers 2000). A recent study conducted by Gil-Rodas et al. (2021) evaluated the quality of six rivers in the Pacific Coastal Region of Guatemala and found them to be moderate to severely polluted due to high concentrations of fecal coliforms, phosphates, and organic matter, with seasonal variations (Gil-Rodas et al. 2021).

Regarding lakes, Guatemala has approximately 20 lakes of different sizes and in different environments. Water in each of them is freshwater but not drinkable. Not unexpectedly, Amatitlán Lake is the most polluted, receiving approximately half of Guatemala City's domestic and industrial sewage water effluents transported to the lake by the Villalobos River (US Corps of Engineers 2000). Even though regulation requires the implementation of wastewater treatment, many municipal treatment plants are out of service because of the high operational costs. The result is a high fecal contamination load, especially in densely populated areas (US Corps of Engineers 2000).

18.2.2 Groundwater (Quantity and Quality)

Due to high levels of surface water pollution, groundwater is increasingly perceived as a more reliable and, therefore, a primary water source. However, due to the high deforestation rate in Guatemala, the amount of water that recharges the aquifers is reduced, resulting in declining groundwater levels (FUNCAGUA 2022). Since most existing hand pumps cannot produce water from depths greater than 90 m (US Corps of Engineers 2000), there is a risk that groundwater may fall out of reach in certain areas that are already hard to access because of the deepness of the groundwater table (MAGA 2009a, b).

Although groundwater is generally safer than untreated surface water supplies, the contamination of many shallow aquifers by biological components due to improper human or animal waste treatment and disposal has been detected near populated areas. Chemical contamination is primarily related to fertilizers and pesticides in the sugarcane and banana plantations of the Pacific and Caribbean coastal plains. Leaching from the main landfill in Guatemala City has severely contaminated the local aquifers (World Bank Group 2018). Generally, only deep confined aquifers should be considered safe from biological and chemical contamination.

Shallow wells (less than 10 m deep) in the interior highlands may go dry intermittently during the dry season. While currently not a problem for the coastal areas, saltwater intrusion may happen if coastal wells are over-pumped. It is acknowledged that water should be thoroughly tested from any well before consumption or use (World Bank Group 2018).

To promote adequate water resources management, especially in the Metropolitan Region of Guatemala, FUNCAGUA, as a water fund for this area, has advanced monitoring of at least 200 municipal wells with depths of up to 500 m since 2018 and has been collecting data each year since then. As a result, data have been generated for the first time, providing science-based information for decision-makers.

18.2.3 Water Usage

According to FUNCAGUA (2022), the distribution of water supply sources for the Metropolitan Region of Guatemala is as follows: 82.92% of the population uses groundwater as their primary source, 17.00% surface water, and 0.08% rainwater. The municipalities provide water to 70% of their population from all these sources, and 30% have private wells.

Domestic use

According to the national census (INE 2018), domestic water supply conditions are:

- 58.9% of the households have a piped water supply connection inside the house,
- 14.8% get water from a piped connection outside the house, but on their own premises
- 12.2% get water through a public or private drilled well,
- 10.9% take water from a river, lake, spring, water tanker truck, plastic barrel, rainwater catchment, etc., and
- 3.2% have water from a public water tap.

In most cases, where tanker trucks or plastic barrels are the primary water supply source, companies that provide these services use groundwater.

Industrial/commercial uses

Industrial and commercial operations are primarily located within the metropolitan region's central municipalities, particularly Guatemala City. Industries operate their own water systems for their operations or rely on public supply. The water comes from both surface and groundwater sources. However, commercial and industrial needs are mostly satisfied using groundwater from own wells, due to the high pollution levels from surface water sources. There is relatively limited information regarding water use in industry (US Corps of Engineers 2000). However, the overexploitation of some aquifers already represents a high cost and increasing investments for different sectors (for additional electricity and deeper drilling). In the medium term, the use could become unsustainable due to a lack of control of abstractions (SEGEPLAN 2006).

Agricultural use

Most irrigation water comes from surface water sources, but this option is depleting due to deforestation. Thus, more groundwater is being exploited (US Corps of Engineers 2000).

18.3 Promoting New Water Supply Solutions and Techniques

As a result of the information on declining groundwater levels generated through groundwater monitoring (FUNCAGUA 2022), FUNCAGUA has been promoting different water supply techniques. The following sections give an overview and explain these techniques.

18.3.1 Rainwater Harvesting

FUNCAGUA (public sector)

FUNCAGUA implemented seven rainwater harvesting systems in peri-urban schools located in Amatitlán (two cases), San Miguel Petapa, Santa Catarina Pinula, Villa Nueva, Guatemala, and Villa Canales. For each school, the available roof surface area for rainwater harvesting and the capacity of the implemented systems were determined (Table 18.2).

In each of the educational establishments, a rainwater harvesting system (or SCALL–Sistema de Cosecha de Agua de Lluviain Spanish) was installed, which required in most cases the improvement or change of the rooftops to enhance water catchment and to ensure good quality water harvested. Furthermore, investment in gray infrastructure, mainly the construction of storage tanks, was required. In addition, an analysis of the quality of collected rainwater and the provision of Ecofiltros® for purification after collection were required. Ecofiltros® are small systems that filter off residues and microbiological elements present in the rainwater allowing the water to be safely consumed.

The project included creating, training, and equipping a water committee with students, parents, teachers, and administrative staff to maintain the SCALL. FUNCAGUA provided a personalized manual that details each of the components

Table 18.2 Available roof surface and rainwater harvesting capacity in seven schools of Guatemala

No	Municipality	School	Roof area (m^2)	Rainwater harvesting capacity (m^3)
1	Amatitlán	San José Humitos	400	20
2	Amatitlán	El Durazno	290	22
3	San Miguel Petapa	Playa de Oro	500	30
4	Santa Catarina Pinula	El Carmen	930	31
5	Villa Nueva	El Tablón	500	42
6	Guatemala	Cebadilla, Zona 25	600	72
7	Villa Canales	Lomas de Rustrían	–	17

[a]Capacity estimated based on the volume of water storage facilities installed

and how to maintain the system properly. The project also implemented an environmental education day. Through an itinerant and interactive exhibition of five educational modules (dynamic and experimental), the beneficiary students received information on the importance of caring for water, its cycle in nature, and human impacts on it. This training system served 480 students in one day. The direct beneficiaries of the overall project since 2018 were 2,932 people. Assuming each person is part of a household of five, the total number of beneficiaries was estimated to around 14,700 people.

In a country with high rainfall such as Guatemala, rainwater harvesting is a sound alternative in both rural and urban areas. However, challenges must be resolved such as: the poor quality of the roofs of many schools (in some cases with asbestos) which could make the cost of projects unfeasible as well as the availability of space for storage. Public–private partnerships and sponsorship of materials by leading companies in the market may be an answer to scale up this type of intervention, making it affordable to more schools.

Municipality of Guatemala City (Public Sector)

In 2013, the Municipality of Guatemala City started the implementation of rainwater harvesting systems as an alternative to supplying water from traditional sources. Currently, Guatemala City has a population of 3.4 million people, and a water demand of around 4.8 m^3/s of water, a need that is only 50% covered by the current supply system.

Rainwater harvesting solutions have been implemented for municipal markets and schools. There are 61 operating spots in the city with water harvesting systems built between 2013 and 2018:41 are in schools and 20 in markets. The Municipality worked together with the organization Fundación Solar to encourage low-income communities to use cost-effective rainwater harvesting at the household level and at the local scale. The initial design and implementation were carried out by these two actors. As markets are part of the municipal administration, the Municipality is taking care of maintaining these systems. In the case of the schools, every school is responsible for the maintenance as part of its existing central government budget; however, the Municipality could give support if required.

The first system was for the municipal market El Amate in 2013. The design applied is modular, making it applicable and scalable in many contexts, depending on the flow rate demanded at each site. The whole market structure covers more than 2,000 m^2, and has a tarpaulin roof of 392 m^2 and a sheet roof of 350 m^2, where these latter two roof options were used to catch rainwater. The market is frequented by around 3,000 people daily.

The system consists of the catchment area (sheet roofs and tarpaulin), pipes leading to two parallel batteries of settling tanks, built-in filters in plastic containers, plastic storage tanks, and a single cistern (underground concrete tank). The captured flow is divided into the settling tanks, then the filters, and finally the storage tanks (Fig. 18.3). The retention time in each filter layer should be approximately one minute. The cistern has a capacity of 60,000 L and allows measurement of the amount of water coming from the catchment area.

There is an initial settling tank in each battery with an overflow system, activated in case the system exceeds its capacity. That water is diverted and discharged to the pluvial drainage system. The settling tank precedes the filters and functions as a 'first flush diverter' since it has a drain in its lower part with a stopcock that must remain open during the first days of the rainy season. This is to clean the roof and 'flush' impurities of various sorts, such as sediment, organic material, and various contaminants. Only after this first flush, does rainwater go through the filters.

The settling tank directs the water to filter 1, interconnected with filter 2. Filter 1 is made of gravel of different grain sizes, and filter 2 has silica sand and activated carbon. The filter system is placed at a high level to work by gravity and does not require external energy. By gravity, the water passes from one filter to the next. The water then passes to the storage tanks, each with a capacity of 5,000 L. Figure 18.2 shows the system, including details of the settling tank and the filter layers.

Monitoring of the system happened in 2013 and 2014. After seeing successful results, new systems were implemented in other sites and adapted to the existing infrastructure of each location. The number of settling tanks and filter batteries is determined according to rainfall, demand, available roof area, and functional space to place them. All 61 systems are still operating and being maintained. The filter material is replaced once a year, taking advantage of the dry season. However, in the case of the silica sand filters, they can simply be backwashed and remain in operation for two rainy seasons before cleaning.

The existing height difference is constantly an advantage, and in case of not having sufficient differential, scaffolding-type platforms with a metallic structure are installed to help water flow easily by the gravity effect. The cost of building a system ranges from USD 2,500, varying to more depending on the need to add these structures to create a height differential.

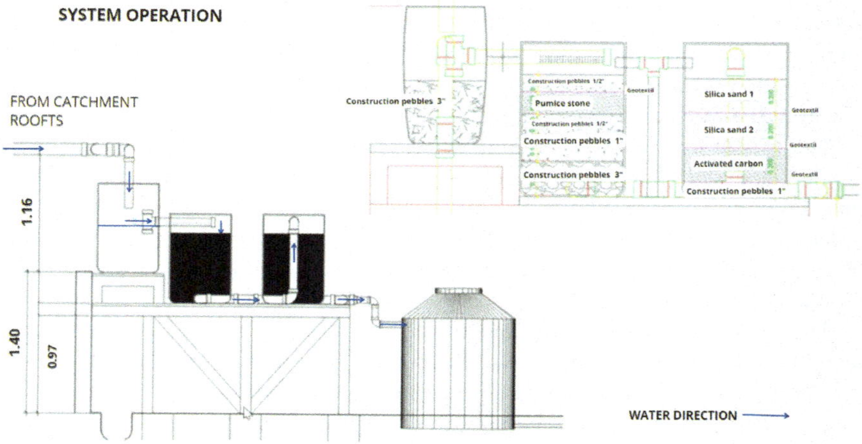

Fig. 18.2 Diagram of the rainwater harvesting system of El Amate market, Zone 1, Guatemala City. *Source* own elaboration with information from Rashjal (2016)

Fig. 18.3 Representation of finished architectural details of a rainwater harvesting system. *Source* Balcárcel Arquitectos (2021)

In total, the rainwater harvesting systems of the Municipality of Guatemala City have benefited 62,234 people (one in 27 people of the whole city) in the municipal markets, 18,534 people in schools (in addition to what is implemented by FUNAGUA described above), and 5,000 people in a demonstration site of La Aurora Zoo, which is built with transparent polycarbonate to visualize the filtering process. This amounts to 85,768 people in total, or 1 out of 40 people in the whole city.

The information gathered from the implemented systems indicates a reduction in water consumption from the traditional sources (municipal service) of between 17 and 68% depending on the rainfall, compared to consumption without any collection system.

The outcome of increased water security for an important number of people in markets and schools of the metropolitan area and the water demand management points to a successful intervention and investment. During this experience, the Municipality of Guatemala City has advised the private sector and followed up on some initiatives. Currently, there is still a budget allocation for the maintenance of the systems already installed, and more resources will be available to continue implementing new systems in this kind of public facility (markets and schools).

Private Investments

Since water is a crucial part of life, it is essential to preserve it. In addition, if conditions allow, architecture can help create attractive, functional, and sustainable means to protect and make better use of water. For the firm Balcárcel Architectos, sustainable architecture is about adapting the housing from a bioclimatic point of view and protecting the environment while protecting ourselves from any risks, like floods and droughts. In this way, carefully designed rainwater harvesting systems can be both functional and aesthetical (Fig. 18.3).

Over the last decade, Balcárcel Architectos has worked on many private high-end residence projects in Guatemala for the hotel and other sectors, conceived by their vision of adapting and incorporating rainwater and runoff harvesting systems.

As an example, large houses have large roof covers, resulting in less rainwater infiltration into the ground. So, these houses require better management of rainwater by using these covers as catchment areas, connecting their respective downspouts directly to the "first rains devices" (filters, see below) and storage tanks. The design of the infrastructure is discreetly and aesthetically incorporated into the architecture.

Systems have a simple sand trap filter in plastic barrels, so-called "first rains devices", which catch the first rain and then allow water to go to the storage tanks. The latter has access possibilities and is typically surrounded by cubicles aesthetically designed to match the environment of the house. If topography and design allow, the filters and cisterns can be placed underground, in basements, or hidden by ground irregularities.

The collected rainwater is used in swimming pools or for irrigating extensive gardens in the dry season. It is also used to elegantly highlight garden or landscape spaces and features with canals and small fountains. One of the projects even directs rainwater into a small infiltration pond, planned for landscape and recreational purposes. The collection and retention of water through these systems can also be used to recharge the aquifer at the macro level and for use at the residential level.

The design of houses with rainwater harvesting systems allows for taking advantage of the qualities and role of water while using it consciously. Another example is green roofs that use rainwater are adapted to be home gardens and incorporate the concept of permaculture. The use of rainwater harvesting systems also applies to vertical projects, green areas, and gardens with automatic pipe systems from the rainwater tank to balconies.

Balcárcel Architectos indicates that rainwater harvesting systems do not have to be ugly or costly. The role of the architect is to promote these alternatives and make them attractive so that clients can incorporate water conservation into their plans. In addition, the rainwater harvest system could be used for economic reasons. The client saves the cost of this water and adds value to the real estate by enabling sustainable design certification and improving its image and publicity.

The current lack of knowledge of this alternative prevents it from being more widely adopted by clients. The challenge is to arouse their interest in taking advantage of a free resource while reducing costs. By making these systems more widely known, e.g., through media, explaining the economic, social, and environmental benefits, more people would opt for the implementation. Doing so would increase the rainwater usage, reducing the use of potable water from existing surface and groundwater sources, while possibly addressing both floods and droughts, if systems are properly designed.

Other Rainwater Harvesting Initiatives

Further illustrating the uptake and out-scaling of rainwater harvesting in Guatemala, Table 18.3 gives some recent initiatives.

18.3.2 Managed Aquifer Recharge

Pilot project (GREMIA-FUNCAGUA)-Technical guide for MAR activities

In 2020, FUNCAGUA was invited by the Gremial de Empresas para el Manejo Integral del Agua (GREMIA) (English: Guild of Companies for Integrated Water Management) to initiate a pilot project on Managed Aquifer Recharge (MAR). The main idea was to equip and record the recharge potential of various sites by equipping the study points with elements such as water meters, digital rain gauges, and data loggers to record variations in groundwater levels. This would help guide the understanding of the potential for MAR using infiltration wells.

The project currently has four recharge wells located in three sites on private property, where there are nearby abstraction wells that could benefit from this type of practice. The sites are in different parts of Guatemala City and Colonia El Búcaro, Villa Nueva (Table 18.4).

One of these infiltration wells captures rainwater collected from nearby roofs, and another receives surface runoff water from the streets of a nearby paved residential area, while one site has both types of collection. The initiative has attracted different actors (private, academic [Universidad de San Carlos de Guatemala (USAC)], and public), allowing entities to work on the issue separately but to share relevant information and criteria that could be useful to enrich the practice and give greater scope to the pilot project. It has also made it possible to learn about MAR experiences not known before, serving as a starting point for knowledge compilation.

This pilot experience is being duly documented and synthesized by FUNCAGUA, to serve as a reference on the subject. Between August and December 2021, it was possible to estimate a volume of 641.6 m^3 of water recharged from the three sites. The one with the best performance is the well that receives recharge only from roofs (Zone 3 of Guatemala City), where, thanks to the digital rain gauge data, an estimated volume of recharged water to the dry well is 459.5 m^3. The better performance in this site may be explained by the fact that all harvested water is directed to the well,

Table 18.3 Ongoing initiatives on implementing water harvesting systems

Project implementer	Sector	Overview	Objective	Roof area (m^2)	Water usage	Rainfall amount (mm/year)	Storage volume (m^3)	Estimated volume captured (m^3/year)	People benefited
Ministry of Environment and Natural Resources (MARN)	Public	The ministry is working on developing a rainwater harvesting system inside its central facilities, located in Zone 13 of Guatemala City	Use the 3,240 m^2 of roof area available within the facilities to supply internal services	3,240	Internal sanitary uses for working staff	907	48	2600	1600
El Manzano School, Santa Catarina Pinula	Public	The necessity of water is essential in schools to keep sanitary services available at any time	Provide the school with better-reliability water access implementing a rainwater harvesting system	288	Internal sanitary uses for students and administrative personnel	1,250	36	324	325
Bottling industry	Private	The bottling industry in Guatemala is known for using large amounts of water for their operations. Knowing this, implementing rainwater harvesting solutions will help economically and socially by securing water access	Implement a water harvesting system to provide working staff with water while taking advantage of the existent infrastructure	940	Internal sanitary uses for working staff	1,250	50	1059	–

Source Own elaboration with information provided by Pedro Ramírez (Envirotech)

Table 18.4 Rainwater harvesting capacity in three pilot sites in Guatemala

Location	Characteristics	Pictures
El Búcaro, Villa Nueva	Water source: surface runoff Number of wells: 2 Depth: 18 m/18 m Diameter: 90 cm/90 cm Casing: perforated PVC Monitoring devices: Water meter + Digital rain gauge + 1 datalogger	
Zona 3, Ciudad de Guatemala	Water source: rain from roofs (1,624 m^2) Number of wells: 1 Depth: 19.4 m Diameter: 90 cm Casing: perforated PVC Monitoring devices: Water meter + Digital rain gauge	
Zona 15, Ciudad de Guatemala	Water source: rainfall from roofs (236 m^2) + runoff Number of wells: 1 Depth: 18 m Diameter: 90 cm Casing: perforated PVC Monitoring devices: Digital rain gauge	

as compared to the other sites, where some water is 'lost' to soil infiltration and evaporation.

The most relevant final product of this project is a technical guide that provides guidelines and advice for implementing MAR at the local level without leaving aside the technical details and important considerations. The result is positive and improves as the cases progress. The guide will specify the idea's conception, the construction process, monitoring, and maintenance of the works.

Another objective of the guide is to lay the foundation for formal regulation or a normative guide to be used at the municipal level to promote and facilitate this type of MAR by providing construction licenses within a specific range of intervened square meters for the infiltration well. Figure 18.4 shows a diagram of the design of a MAR scheme based on one infiltration well, such as applied in the pilot project.

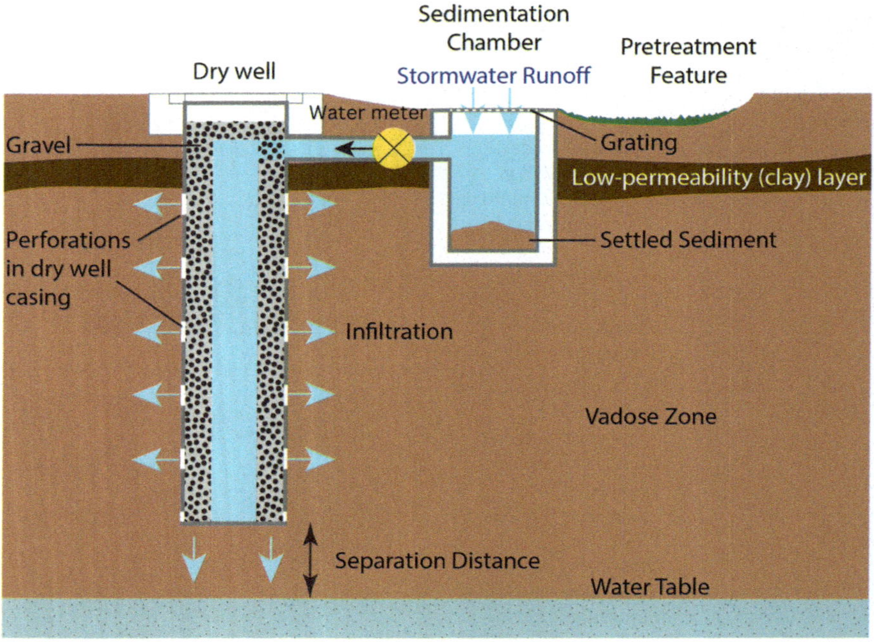

Fig. 18.4 Diagram of an infiltration well for managed aquifer recharge. *Source* Own elaboration based on CNR (2020)

Further evidence and knowledge will be leveraged through partnerships and learning from international experience and upscaled cases (Zheng et al. 2021).

Municipality of San Antonio Aguas Calientes—Successful MAR Experience

A case of success in using MAR is the municipality of San Antonio Aguas Calientes, located in the department of Sacatepéquez, 52 km from Guatemala City. The primary objective of implementing MAR was to reduce the impacts of flooding events. This municipality is home to 11,347 inhabitants (INE 2018), is located 1,520 m above sealevel, and has a territorial extension of 17 km^2 (SEGEPLAN 2010) (Table 18.5).

The municipality's urban area is characterized by mountains with steep slopes, making it vulnerable to landslides and flooding associated with extreme precipitation events. The last recorded event with disaster consequences was Hurricane Mitch in 1998.

Approximately since 2000, the municipal administration began the construction of more than 2,000 recharge works, like infiltration ditches on hillsides, dikes and dams, absorption wells, as well as the increased use of live barriers.

In 2005, when storm Stan occurred, the municipality did not detect any flood damage, despite the significant losses reported for the rest of the country. Nevertheless, the recharge wells were silted up and required maintenance.

Table 18.5 Economic and human losses in extreme events in Guatemala

Year	Events	Number of deaths	Number of people affected	Estimated damages (millions of USD)
1998	Mitch Hurricane	268	105,000	748
2005	Stan tropical storm	664	285,000	983
2008	Tropical storms, landslides, floods	74	180,000	2.5
2009	Landslides, floods, dry	36	73,080	14
2010	Agatha tropical storm	152	104,052	650
2011	Rina Hurricane	39	500,000	62
2011	Tropical Depression 12-E	51	254,000	5
2015	Landslides, floods, epidemic	358	–	–
2016	Tropical storm, landslides	10	445,628	–
2017	Tropical storm, landslides	20	–	–
2020	Tropical storms ETA and IOTA, landslides	61 (99 missing)	311,317	781

Source Own elaboration with information from Carrera (2019) and SEGEPLAN (2021)

It is important that the location of the recharge works on slopes using infiltration ditches or trenches, were carried out with local knowledge, ruling out the possibility of landslides, since excessive soil saturation could lead to landslides and therefore have repercussions in the form of new or other disasters.

The idea began as an experiment on a pilot scale. It has now become an established program to allocate municipal resources to ensure its maintenance. Although there was a change of administration in 2008–2012, the program continued because of its success and because there was no alternative storm drainage system to divert excess rainwater. The recharge works are registered in the land-use plan worked out for the municipality.

The location of the recharge works was empirically determined using the slope of the land as a guide because of the water runoff. Steep slopes increase the surface runoff velocity which reduces the water infiltration potential. Farmers were consulted for guidance on the location of the works in agricultural areas and on steep slopes surrounding the urban center. The work is done through manual excavation. In mountainous areas, absorption wells can reach a depth of 3 m. and in the urban center up to 15 m. The wells consist of perforated circular side walls of diameter between 80 and 90 cm. Each well has a cover and an upstream side de-sanding box to remove excess sand and gravel, covered by a grate to prevent the entry of oversized solid materials, such as branches and rubbish. By 2008, there were more than 70 wells of this type constructed in the urban center of the municipality.

The wells get silted up, so maintenance work is required annually to restore them to a depth of 10 m. As a result, a total of 62 wells were originally recovered by 2021. Eight of these wells are cased with perforated PVC pipe; the rest are maintained as a direct perforation in the soil with filtering media such as pebbles and river round rocks. The intention is to install PVC casings to all of them to prevent caving in.

The maintenance program for the recharge works starts every year in March–April before the first rains and ends in November. It begins with 12 people working full-time on daily maintenance and ends with a group of 8 people. Half of them are dedicated to cleaning and removing sand from the street grating points. Wells are cleaned manually using suspended buckets that are filled with the extracted sand and sediments. Currently, 50 wells are maintained, and it takes approximately 20 days to complete an entire cleaning cycle.

The municipal administration knows that even though the initial idea was the reduction of floods, these wells contribute to recharging the groundwater used for water supply. There are six abstraction wells in the area, supporting 95% water coverage in the city. These wells have been supplying the municipality for 20 years with no decrease in productivity. In addition, the recharge wells have saved the cost of removing accumulated sediments in the urban center from flooding. Their charge wells also alleviate park and cemetery cleaning as before their implementation, flooding resulted in accumulated sediments of about 80 cm annually.

The use of absorption or infiltration wells, especially on sloping land, could be widespread in different municipalities of Guatemala. Typically, this type of work is carried out empirically without understanding the direct relationship with groundwater recharge. It is important that the concept be socialized with this understanding, and the construction of the works to be more technical, thus, the risks of contamination with water that runs off the streets will be mitigated. In addition, an appropriate monitoring of the works must be carried out and the water contribution estimated, in this way, it will be possible to speak of "managed recharge" in all its scope.

18.3.3 Steps Toward a "Sponge City" in Guatemala

The private consulting sector in Guatemala offers solutions for the management and use of rainwater through three technologies:

- Filters for rainwater harvesting from roofs
- Water injection wells for water recharge
- Modular subsurface tank systems for harvesting, retention, and infiltration of rainfall and runoff (so-called SuDS—Sustainable Drainage Systems).

Emerging successful experiences in water harvesting and SuDS infiltration tanks are described below.

Quetzaltenango: sustainable urban drainage and infiltration systems

Quetzaltenango, the second most important city in Guatemala, has about 180,700 people (INE 2018). During the rainy season, floods become the main problem. Therefore, the municipal city has been part of the high-risk flood zones identified by the National Coordinator for Disaster Reduction (Zone 10 of the municipal capital has been part of the high flood risk zones identified by the Coordinadora Nacional para la Reducción de Desastres (CONRED 2015)) (Fig. 18.5).

To reduce the flooding problem, an urban drainage system (SuDS) has been designed consisting of five infiltration tank units of 14.4 m^3, built of modular units with polypropylene walls with a porosity of 95%. These tanks were named Envirotank by the company (Fig. 18.6). The infiltration area covers 144 m^2, allowing the retention and infiltration of up to 72 m^3/day below the tank.

The implementation of the tanks is along a 600 m long flood-prone street, embedded in the subsurface. Runoff is collected by the street surface and runs off to the sides of a pluvial pipeline connected to the tanks. Each tank has a geotextile layer

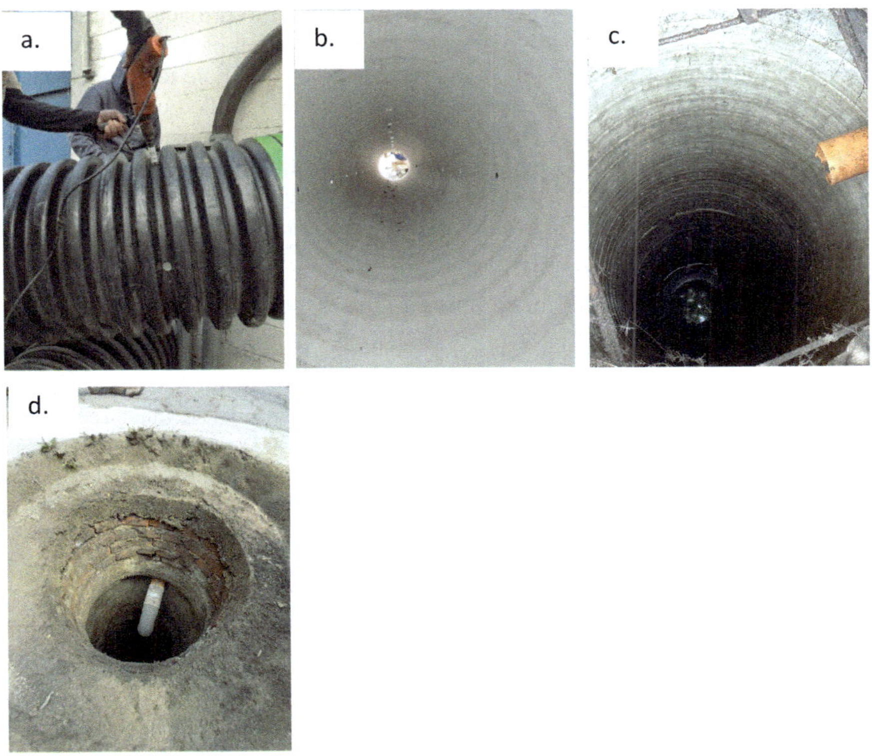

Fig. 18.5 **a** Detail of PVC pipe drilling. **b** A perforated PVC pipe that allows water absorption, before installation. **c** A completed infiltration well is actually in operation. **d** View of an infiltration well without PVC casing, river round rocks have been put at the base of the hole to prevent caving-in

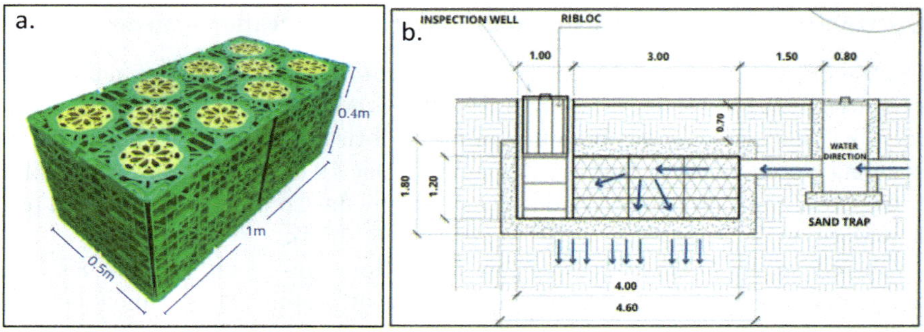

Fig. 18.6 Detail of a polypropylene infiltration tank module (Envirotank) and b) Schematic side view of one of the infiltration tanks designed for Zone 10, Quetzaltenango. *Source* Courtesy of Ramírez (2022)

and an outer filling (bottom and sides) of a mixture of 70% gravel and 30% sand, with a maximum grain size of 0.5 inches. The outer top of the tanks is filled with backfill material and placed with controlled compaction to provide strength. Each tank has an external sand trap through which the runoff water passes before entering the tank. The tanks also have a 90 cm diameter maintenance hole for inspection.

Fifteen of these modular polypropylene tanks will be installed one meter below the ground surface in the streets and avenues of the urbanized area. The tanks vary in volume between 35 and 75 m^3, reaching a total infiltration capacity of 725 m^3. An added benefit of these systems is that water is rapidly captured underground for subsequent infiltration, reducing the loss to evaporation from water standing on the ground.

Monterrico, Santa Rosa: sustainable urban drainage systems

This project, located at a hotel in Monterrico in Santa Rosa, implements five infiltration tanks of 32.4 m^3 (Fig. 18.7). The construction method is the same as described for Quetzaltenango, using modular polypropylene systems. These tanks receive the rainwater collected on the roofs of the hotel, allowing it to infiltrate directly into the ground, reaching the water table, which is very shallow. A total of 40,000 m^2 of water will be captured. Considering the annual rainfall of 1,300 to 1,500 mm, the infiltration potential is estimated to be between 44,800 and 50,000 m^3 per year. The infiltration scheme helps to replenish the local aquifer used by the tourism sector in the area.

As flooding becomes increasingly problematic in urban areas due to increasing excess runoff with climate change, technical options for stormwater management combined with groundwater recharge begin to be more valued, not only by the private sector, but also by the public sector. The operation of this type of solution will be more recognized by the general population, allowing an understanding of the benefits of rainwater management for managed recharge purposes.

Fig. 18.7 Drainage and infiltration system installation in Monterrico, Santa Rosa. *Source* Courtesy of Ramírez (2022)

18.4 Conclusions

In a country with relatively high rainfall such as Guatemala, but also with significant water scarcity due to geographic disparities, population growth, and poor service delivery, especially in larger urban and metropolitan areas, rainwater harvesting and managed aquifer recharge systems are a functional alternative to reliance on surface water, which is often contaminated. Challenges about costs and materials could be partly overcome by public–private partnerships and the sponsorship of materials by leading companies.

The concept of shallow recharge wells must be socialized and taught at the municipal level, imparting the understanding of how runoff rainwater can reach the aquifer for replenishment purposes. This should be done while improving the technical construction to reduce the risks of contamination. Also, the monitoring and estimation of volumetric recharge will transform empirical practice into more "managed" solutions.

Tested technical options for stormwater management combined with groundwater recharge are available in Guatemala and have begun to attract attention for various developments. In the near future, it is likely that these projects will gain greater visibility and be more valued for the water-regenerative approach through managed aquifer recharge.

References

Carrera JL (2019) Qué tan vulnerables somos Elementos para entender la vulnerabilidad de Guatemala [How vulnerable are we? Elements for understanding Guatemala's vulnerability]. In: Castellanos EEJ, Paiz-Estévez A, Escribá J, Rosales-Alconero M, Santizo A (eds) Primer reporte de evaluación del conocimiento sobre cambio climático en Guatemala [First assessment report on climate change knowledge in Guatemala]. Editorial Universitaria UVG, Guatemala, pp 64–85

Castañeda MD, Castañón y OA (2000) Lineamientos de Política Hídrica Nacional y Propuesta para elFortalecimiento del Marco Institucional y Legal del Sector Recursos Hídricos [National Water Policy Guidelines and Proposal for Strengthening the Institutional and Legal Framework of the Water Resources Sector]. Plan de Manejo Integrado de los Recursos Hídricos (PMIRH), Ministerio de Agricultura, Ganadería y Alimentación, Guatemala

Comisión Económica para América Latina y el Caribe (CEPAL), Fondo Nórdico de Desarrollo (NDF), Banco Interamericano de Desarrollo (BID), & Ministerio de Ambiente y Recursos Naturales (MARN) (2018) La economía del cambio climático en Guatemala [The economy of climate change in Guatemala) - Documento técnico (2018). Ciudad de México. https://repositorio.cepal.org/bitstream/handle/11362/43725/1/S1800650_es.pdf

Coordinación Nacional para la Reducción de Desastres (CONRED) (2015) Amenaza por deslizamientos e inundaciones [Landslide and flood hazards]. Quetzaltenango, Guatemala. https://conred.gob.gt/mapas/municipales_ameindes/QUETZALTENANGO/QUETZALTENANGO/QUETZALTENANGO%20901.pdf

Fundación para la Conservación del Agua de la Región Metropolitana de Guatemala (FUNCAGUA) (2022) Informe del estado del agua de la Región Metropolitana de Guatemala 2022: el agua nos une [Report on the State of Water in the Metropolitan Region of Guatemala 2022: Water brings us together]. Guatemala

Fundación para el desarrollo de Guatemala (FUNDESA) (2021) Índice de competitividad Global [Global Competitiveness Index]. https://fundesa.org.gt/content/files/publicaciones/Documento_Tecnico_ICG_2021.pdf

Gil-Rodas N, Calvo-Brenes G, Guerra A, Perdomo A (2021) Water quality assessment of six rivers of the Pacific side of Guatemala. Environ Earth Sci 80:196. https://doi.org/10.1007/s12665-021-09505-w

Global Water Partnership Centroamérica (GWP) (2015) Situación de los recursos hídricos en Centroamérica [Situation of water resources in Central America]. Tegucigalpa. https://www.gwp.org/globalassets/global/gwp-cam_files/srh_guatemala_2016.pdf

Hernández-Ulate, Aurora & Granados, Carlos & Delgado, Hannia & Rodríguez Herrera, Eduardo (2000) Cuencas internacionales: conflictos y cooperación en Centroamérica [International watersheds: conflicts and cooperation in Central America]..

Instituto Nacional de Estadística (INE) (2018) XII Censo Nacional de Población y VII Censo Nacional de Vivienda 2018 [XII National Population Census and VII National Housing Census]. https://www.censopoblacion.gt/explorador

Ministerio de Agricultura, Ganadería y Alimentación (MAGA) (2009) Evaluación del potencial de aguas subterráneas de la República de Guatemala a escala de reconocimiento (1:250,000), como apoyo al desarrollo del riego para la producción agrícola en comunidades de pequeños y medianos productores [Evaluation of the groundwater potential of the Republic of Guatemala at a recognition scale (1:250,000), to support the development of irrigation for agricultural production in communities of small and medium producers]. Guatemala. https://www.maga.gob.gt/download/aguas-subt.pdf

Ministerio de Agricultura, Ganadería y Alimentación (MAGA) (2009) Unidad de Planificación Geográfica y Gestión de Riesgo (UPGG). Mapa de cuencas hidrográficas a escala 1:50,000, República de Guatemala, Método de Pfafstetter (Primera aproximación) [Map of watersheds at a scale of 1:50,000, Republic of Guatemala, Pfafstetter Method (First approximation)]. Guatemala

Ministerio de Ambiente y Recursos Naturales (MARN) (2013) Informe Ambiental del Estado de Guatemala [Environmental Report of the State of Guatemala]. Guatemala

Nepomilueva D (2017) Water scarcity indexes: water availability to satisfy humanneeds. Helsinki Metropolia University of Applied Sciences. Helsinki, Finland. https://core.ac.uk/download/pdf/84797448.pdf

Ramírez P (2022) Envirotech experiences and projects in Guatemala. Personal Commun

Secretaría de Planificación y Programación de la Presidencia (SEGEPLAN) (2006) Política nacional de gestión integrada de los recursos hídricos y de la estrategia nacional de gestión integrada de los recursos hídricos [National integrated water resources management policy and the national integrated water resources management strategy]. Guatemala. https://cebem.org/revistaredesma/vol11/pdf/legislacion/engirh_guatemala.pdf

Secretaría de Planificación y Programación de la Presidencia (SEGEPLAN) (2010) Plan de Desarrollo Municipal de San Antonio Aguas Calientes. Dirección de planificación territorial y Consejo Municipal de Desarrollo [San Antonio Aguas Calientes Municipal Development Plan. Directorate of Territorial Planning and Municipal Development Council]. https://www.segeplan.gob.gt/nportal/index.php/departamento-de-sacatepequez/file/93-pdm-san-antonio-aguas-calientes

Secretaría de Planificación y Programación de la Presidencia] (SEGEPLAN) (2021) Más de Q6 mil millones en daños y pérdidas causaron las depresiones tropicales Eta e Iota en el país [Over Q6 billion in damages and losses caused by tropical depressions Eta and Iota in the country]. https://www.segeplan.gob.gt/nportal/index.php/sala-de-prensa/2016-01-26-18-14-30/noticias/2016-mas-de-q6-mil-millones-en-danos-y-perdidas-causaron-las-depresiones-tropicales-eta-e-iota-en-el-pais#:~:text=Guatemala%2C%2012%20de%20enero%20de,pa%C3%ADs%20ascienden%20a%20Q6%2C002%2C035%2C633

United States Corps of Engineers (2000) Evaluación de recursos de agua de Guatemala [Guatemala water resources assessment. http://www.sam.usace.army.mil/Portals/46/docs/military/engineering/docs/WRA/Guatemala/GuatemalaWRA Spanish.pdf

Universidad Rafael Landivar (URL) (2005) Situación del Recurso Hídrico en Guatemala [Report about Water Resourcesin Guatemala]. Guatemala. https://desastres.medicina.usac.edu.gt/documentos/docgt/pdf/spa/doc0135/doc0135.pdf

Universidad Rafael Landívar (URL) (2015) Balance hidrológico de las subcuencas de la República de Guatemala: Bases fundamentales para la gestión del agua con visión a largo plazo[Hydrological balance of watersheds in República de Guatemala: Fundamentals for longterm vision]. Instituto de investigación y protección sobre ambiente natural y sociedad (IARNA). Guatemala. https://www.url.edu.gt/publicacionesurl/FileCS.ashx?Id=40416

Universidad Rafael Landívar (URL) (2022) Mapa de disponibilidad hídrica per cápita. Estimación del estrés hídrico municipal. [Per capita Water Availability Map. Municipal Water stress Estimation]. Vicerrectoría de Investigación y Proyección (VRIP). Sistema de información estratégica (SIE). Guatemala. https://sie.url.edu.gt/mapa-de-disponibilidad-hidrica-pc/

World Bank Group (2018) Guatemala's water supply, sanitation and hygiene poverty diagnostic. Challenges and opportunities. The World Bank. Washington, DC: International Bank for Reconstruction and Development

Zheng Y, Ross A, Villholth KG, Dillon P (eds) (2021) Managing aquifer recharge: a showcase for resilience and sustainability. UNESCO, Paris. ISBN: 978-92-3-100488-9

Chapter 19
Scope of Induced Recharge to River Bank Aquifers in Bangladesh

Khairul Bashar

Abstract Bangladesh has a complex network of rivers, including the Ganges, Brahmaputra, Meghna, and their tributaries, which provide vital water resources for the country. Much of Bangladesh is an alluvial plain, which is a flat, low-lying area formed by the deposition of sediments, such as sand, silt, and clay by the rivers and their flood plains. The rivers are generally hydraulically well connected to the subsurface aquifers, made up of sands, in many parts of Bangladesh, which offers the scope to recharge the near bank aquifers artificially, through induced bank filtration (IBF) system. Infiltration galleries tapping the river bank aquifers can also be used to harness filtered river water for potable uses. IBF systems have already been setup on the bank of Padma River at Bheramara of Kushtia district and also on the bank of Jamuna River near Bangabandhu Jamuna Bridge of Sirajganj district which are inducing recharge of river water to aquifers. It has been concluded that the availability of groundwater for different societal uses can be increased by constructing pumping wells close to the rivers or other surface water bodies in Bangladesh after proper hydrogeological studies.

Keywords Bangladesh · Groundwater recharge · Induced bank filtration · Infiltration gallery · Induced recharge

19.1 Introduction

Bangladesh is a riverain country. The bottom reaches of the Himalayan Ranges' water sources lie within the borders of Bangladesh and serve as the primary sources of water for rivers that flow through such countries like China, Nepal, Bhutan, and India and eventually debouching into the Bay of Bengal in Bangladesh (Fig. 19.1).

K. Bashar (✉)
Department of Geological Sciences, Jahangirnagar University, Savar, Dhaka 1342, Bangladesh
e-mail: bashar@juniv.edu

D. Saha et al. (eds.), *Managed Groundwater Recharge and Rainwater Harvesting*, Water Resources Development and Management,
https://doi.org/10.1007/978-981-99-8757-3_18

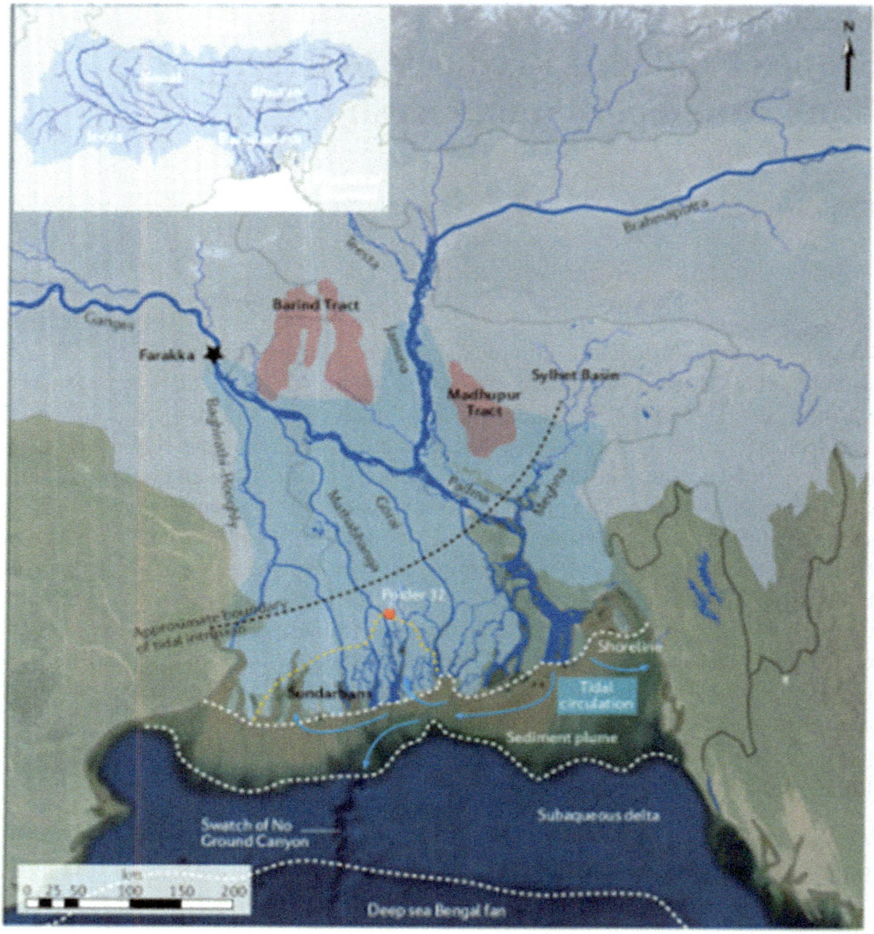

Fig. 19.1 Location of the Ganges–Brahmaputra–Meghna delta. Light blue shading shows the Ganges–Brahmaputra–Meghna (GBM) watershed, whereas blue shading shows outline of the delta. *Source* Paszkowski et al. (2021)

The physiography of Bangladesh is characterized by two distinctive features: a broad deltaic plain subject to frequent flooding, and a narrow hilly and terrace region crossed by swiftly flowing rivers. A highly irregular deltaic coastline of about 710 km exists in the south of Bangladesh which is crisscrossed by many rivers and streams flowing into the Bay of Bengal.

There are four major river systems in Bangladesh; (1) the Brahmaputra-Jamuna, (2) the Ganges-Padma, (3) the Surma-Meghna, and (4) the Chittagong Region River system. Brahmaputra is the 22nd longest (2,850 km) and the Ganges is the 30th longest (2,510 km) river in the world. The total number of rivers including tributaries is about 700 which have a total length of about 24,140 km (River—Banglapedia).

As located in the lower reaches, most of the major rivers are flood-prone. The average flood discharges of the Jamuna, Ganges, Padma (the main branches of the Ganges and Jamuna), and Upper Meghna Rivers are 70,000, 52,000, 95,000, and 13,700 m^3/s as measured at *Bahadurabad*, *Hardinge Bridge*, *Mawa*, and *Bhairab Bazar*, respectively. While, the average low flow discharges are 4250, 600, and 4800 m^3/s for the Jamuna, Ganges, and Padma Rivers (Sarker et al. 2003). The water flowing through the rivers is ultimately discharged to the Bay of Bengal.

The rivers generally flow on unconsolidated alluvial sediments comprising clay, silt, and sand, thus remain hydraulically well connected to the subsurface aquifer system (Mukherjee et al. 2015) as also seen in the adjoining parts of India. The mean sizes of the bed material in the Jamuna, Ganges, Padma, Upper Meghna, and Lower Meghna Rivers are 0.20, 0.15, 0.12, 0.14, and 0.09 mm, respectively (Sarker et al. 2003). The planform of the rivers varies from meandering to braiding over space and time, like the Jamuna is mainly braided, the Ganges is meandering, and the Padma is a wandering river. The Upper Meghna is anastomosing, and the Lower Meghna is anabranching (Sarker et al. 2003).

Bangladesh is heavily dependent on groundwater for its different societal needs. Rainfall is the main source of recharge. There is hardly any recharge from the rivers as they are effluent in nature. Using the IBF technique, it is possible to utilize groundwater and also to increase the recharge to the aquifers in areas close to the rivers. Currently, there are many wells operating in the vicinity of rivers or other surface water bodies and are inducing recharges artificially to the aquifers in different parts of Bangladesh. However, a planned construction of well fields along the banks of rivers may increase the total availability of fresh water for different uses in many parts of Bangladesh.

19.2 Groundwater–Surface Water Interaction

Groundwater and surface water (including rivers, lakes, reservoirs, wetlands, estuaries, etc.) are the two important components of the hydrological cycle that are interconnected and can interact with each other in various ways. This interaction takes place through the loss of surface water to groundwater, and also seepage of groundwater to surface water bodies, or a combination of both.

From its origin to the end, the river reaches can be influent or effluent depending on the water level of the river and the depth of groundwater level over the banks. The interaction of river with the groundwater system varies in space and time. When the groundwater level is higher than the river water level, the river gains water from the groundwater system and aquifers, thus the river is effluent. But when the groundwater level is lower than the river water level, river looses its water and the subsurface aquifer receives water from the river, calling it influent.

By artificially lowering the hydraulic head of the aquifer near or under a surface water body the recharge to the aquifer can be initiated. The term Managed Aquifer Recharge (MAR) describes the intentional recharge (and storage) of water into an aquifer for subsequent recovery or for environmental benefits (Dillon 2005). Induced bank filtration or infiltration galleries are types of MAR where the infiltration of surface water is induced by pumping from a nearby well (Jeffords 1945; Kazmann 1948; Rorabaugh 1951; USGS 1999).

19.3 Subsurface Intakes and Open Surface Water Intakes

Intakes are the structures used for admitting water from the surface sources (i.e., river, reservoir, or lake) and conveying it further to the treatment plant. Calm and still water conditions in the water bodies are needed, so that comparatively purer water may be collected from the source. Generally, an intake is a masonry or concrete structure with the aim of providing relatively clean water, free from pollution, sand, and objectionable floating material.

Intakes of water from lakes or rivers can be broadly classified as subsurface intakes and open surface water intakes. Induced Bank Filtration (IBF) System (or Riverbank Filtration, RBF) and Infiltration Gallery are the two widely used subsurface intake systems in the world.

Subsurface intake systems (also referred to as bank infiltration or infiltration wells) are usually preferred in Bangladesh because during the transport of water through the subsurface formation the water is getting filtered before its extraction. Thus the water obtained from such intake systems (for example, collector wells) usually contains lower concentrations of suspended solids and organic matter. They also yield water with less temperature variations, less biological activity, and lower dissolved oxygen content.

The local hydrogeological conditions determine the feasibility of the subsurface intake systems. If the hydraulic conductivity and the thickness of the subsurface porous medium is too low or if the depth of porous medium is very high then such intakes may not be effective.

A subsurface intake system typically consists of wells drilled near the river bank. The casing and the slotted pipes should be properly designed. The pump, risers, pump starters, chemical dosing systems and interconnecting pipes should also be properly planned. The wells can be either vertical or horizontal collector wells. Also, directionally drilled wells have been used where a well is drilled at an angle to the ground surface and under the body of water.

In contrast, open surface water intake systems require extensive treatment of water prior to supply for potable use. The quality of feed water obtained from an open intake is typically contained high suspended solids, and algae that can clog the intake system. In addition, water from an open intake is subjected to seasonal variations in temperature.

The waves and water currents in the river could be really large in Bangladesh and could exert high mechanical loads on the intake system components such as pipes, joints, anchors and other accessories. Because of these disadvantages, open surface water intake systems are preferred only when subsurface systems are not feasible.

19.4 Induced Bank Filtration System

Induced infiltration of river water by pumping wells can be located on or near the banks is used as a technique to withdraw water from a river for urban or industrial supply in the southeastern Asia including India (Saha and Roy 2018). The hydrogeological conditions of the area adjacent to the rivers in different parts of Bangladesh are favorable for setting up such type of intake.

19.4.1 Induced Infiltration Technique

In places, where aquifers are intersected by perennial streams, large groundwater supply systems can be developed by the installation of wells or other types of subsurface intakes, which are aligned parallel to the course of the stream. The strainers of such wells are set at a sufficient depth below the stream in order to develop adequate hydraulic gradients from the stream to the subsurface intake.

As groundwater is withdrawn, water levels are lowered down in the vicinity of the intake. As pumping continues, the cone of depression deepens and the area of interception expands. The water contributed from the river moves through the aquifer down gradient, towards the center of withdrawal when the piezometric head.

The interception and moving surface water by induced infiltration takes advantage of the aquifer materials, which behave like a slow sand filter provided by Nature as contrasted to the more widely used and costlier procedure of direct surface intake, which requires the construction of extensive filter beds.

As the surface water travels through the aquifer towards the RBF well, pathogens and suspended contaminants are removed or significantly reduced via a combination of physical, chemical, and biological processes (Ray et al. 2003) and can attain drinking water quality (Fig. 19.2). Bacterial pathogen removal efficiencies of >99.9% can be achieved and heavy metal concentrations are reduced (Kelly et al. 2006; Sandhu et al. 2011; Cady et al. 2013; Gillefalk et al. 2018).

In using a groundwater reservoir as a filter medium, careful consideration must be given to the proper location of the underground intakes in order to take full advantage of local hydrogeologic conditions and at the same time obtain adequate filtration. At first consideration, it may appear advisable to locate the subsurface intake as close to the stream bed as possible, because this would develop the maximum gradient and result in peak performance.

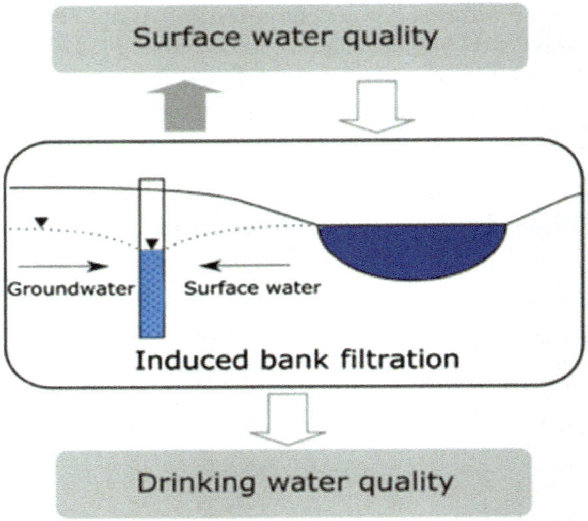

Fig. 19.2 Improvement water quality by Induced Bank Filtration Technique (Gillefalk et al. 2018)

In periods of flood when the stream is turbid, the method of direct-surface intake requires continual watchfulness on the part of the filter plant operator, but where a subsurface intake is used, much of this problem is eliminated. Where chlorination or water softening is necessary, operation schedules can be planned from weeks to months in advance.

Bank infiltration schemes commonly consist of a well gallery or a line of abstraction wells parallel to the bank of a surface water body. River Bank Filtration (RBF), has been used in Europe for over 100 years (Schmidt et al. 2003). Examples of bank filtration sites can be found in South Korea (Maeng and Lee 2019), Dresden, and Budapest (Schmidt et al. 2003; Ray et al. 2003). In India, RBF sites were developed in Uttarkhand, Jammu, Jharkhand, Andhra Pradesh, and Bihar (Sandhu et al. 2019). The quality of both surface water and RBF water was generally good with respect to most inorganic parameters and organic parameters when compared to Indian and World Health Organization drinking water standards (Sandhu et al. 2019).

Table 19.1 gives of list of cities in different parts of the world where the Induced Bank Filtration system was implemented successfully.

19.5 Induced Infiltration of River Water in Bangladesh

There are two examples of setting up planned induced infiltration of river water in Bangladesh. The first one is on the bank of the Padma River in Bheramara of Kushtia district which was set up in 2009. The second one is on the bank of Jamuna (Brahmaputra River) near the Bangabandhu Jamuna bridge in the Sirajganj district

Table 19.1 Lists of cities all over the world, which use Induced Bank Filtration (IBF) include the percentage of Induced Bank Filtration in different countries/cities and the source of surface water bodies. The "symbol" means same as above, * indicates that info is missing, – that info is not applicable

Country	City	Percentage of drinking water provided by IBF	Source water bodies	River discharge, lake
Austria	*	*	River Enns	65–206 m^3/s
”	Innsbruck	*	River Inn	730 m^3/s
”	Vienna, Linz	*	Danube	1900 m^3/s
Bulgaria	*	*	*	–
Finland	Kuopio	*	Lake Kallavesi	4730 million m^3
”	*	*	Not mentioned	*
France	Paris region	*	Seine River	450 m^3/s
Germany	–	9 to 16	–	–
”	Berlin	60	–	–
”	”	–	Lake Müggelsee	36 million m^3
”	”	–	Lake Tegel	26 million m^3
”	”	–	Lake Wannsee	15 million m^3
”	Radeburg	–	Radeburg Reservoir	0.35 km^2 Max depth: 3 m
”	Düsseldorf	~100	River Rhine	2300 m^3/s
”	Frankfurt am Main	*	River Rhine	2300 m^3/s
”	”	*	Lower River Main	193 m^3/s
”	Torgau and Mockritz	*	Elbe River	700 m^3/s
Hungary	–	45	–	–
”	Budapest	–	Danube	6460 m^3/s
”	*	–	Rivers Raba, Drava, Ipoly, Sajo, Hernád	17, 500, 21, 67, 27 m^3/s
”	*	–	Sajo, Hernád	
Italy	Lucca, Pisa, Livorno	(300,000 inhabitants)	River Serchio Lake Mazais	46 m^3/s
Country	City	Percentage of Drinking Water Provided by IBF	Source Water Bodies	River Discharge, Lake
Latvia	Riga	*	Baltezers Lake LielaisBaltezers	10 million m^3 18 million m^3
The Netherlands	–	5	–	–
”	Remmerden	*	River Rhine	2300 m^3/s

(continued)

Table 19.1 (continued)

Country	City	Percentage of drinking water provided by IBF	Source water bodies	River discharge, lake
,,	Zwijndrecht	*	River Rhine	2300 m^3/s
,,	Roosteren	*	River Meuse	276 m^3/s
,,	Roermond	*	Gravel pit lake De Lange Vlieter	1.2 km^2 Max depth: 35 m
Norway	Hemne	*	Lake Rovatnet	8 km^2
,,	,,	*	River Buga	*
Poland	Pozna´n	*	River Warta	60 m^3/s
Romania	Iasi	*	Moldova River	143 m^3/s
Slovak Republic	–	50	–	–
Slovenia	Maribor	–	Drava River	50 m^3/s
Switzerland	–	10–30	–	–
			River Thur	40–50 m^3/s
UK	*	*	StreamsWissey, Rhee and Pang	1.9, 1.25, 0.64 m^3/s
Canada	*	*	Lake A and B (artificial)	*
USA	Jeffersonville	*	OhioRiver	3512m^3/s
,,	SantaRosa	*	RussianRiver	66 m^3/s
,,	Cincinnati	*	GreatMiamiRiver	109 m^3/s
,,	Columbus	*	Scioto/ BigWalnutCreek	6 m^3/s
,,	Galesburg	*	MississippiRiver	16,792 m^3/s
,,	Independence Kansas City Parkville	*	Missouri River	2158 m^3/s
,,	Jacksonville	*	Illinois River	659 m^3/s
,,	Kalama	*	KalamaRiver	40 m^3/s
,,	Kennewick	*	ColumbiaRiver	7500 m^3/s
,,	Lincoln	*	PlatteRiver	203 m^3/s
,,	Mt. CarmelTerreHaute	*	WabashRiver	837 m^3/s
,,	Sacramento	*	Sacramento River	660 m^3/s
,,	Cape Cod	*	Ashumet Pond	6 million m^3
Brazil	*	*	Beberibe River	0.3–0.4 m water depth
,,	*	*	Lake Lagoa do Peri	36 million m^3
China	Matan	96	Yellow River	1839 m^3/s
,,	Baisha Town	82.1	Yangtze River	31,100 m^3/s

(continued)

Table 19.1 (continued)

Country	City	Percentage of drinking water provided by IBF	Source water bodies	River discharge, lake
"	Jiuwutan	82.6	Yellow River	1839 m^3/s
"	Qingpu district	70–80	Taipu River	300 m^3/s
"	Xuzhou	>80	Kui River	*
"	Chengdu	80	Yinma River	30 m^3/s
India	–	*	–	–
"	Delhi	*	Yamuna River	100–1300 m^3/s
"	Satpuli	*	East Nayar River	–
"	Srinagar	*	River Alaknanda	507 m^3/s
"	Haridwar	*	River Ganga	1455 m^3/s
"	Nainital	*	Lake Nainital	6 million m^3
Malaysia	Kuala Kangsar	*	Sungai Perak (river)	57 m^3/s
South Korea	*	*	Nakdong River	37–3462 m^3/s
Thailand	Chiang Mai	*	Ping River	287 m^3/s
Egypt	–	0.1 (increasing)	Upper Nile	1548 m^3/s
"	Sidfa	*	Nile	2830 m^3/s
"	Aswan	*	"	"

Gillefalk et al. (2018)

of Bangladesh, which was constructed in 2018 (Fig. 19.3). Narayanganj City Corporation has also planned to set up an IBF system on the bank of Dhaleswari River in the south of Narayanganj in 2023. However, there are numerous pumping wells in different parts of Bangladesh, which were set on the bank of rivers or other surface water bodies and are inducing recharge to aquifers under exploitation. For example, a large number of tube wells constructed on the banks of the Sitalakhya River to supply potable water for Narayanganj city, are inducing recharge of river water to the subsurface aquifers.

19.5.1 Induced Infiltration of Padma River Water at Bheramara of Kushtia

Bheramara Combined Cycle Power Station is situated about 400 m east of the Padma River bank and about 600 m north of the Pump Station of G.K. (Ganges–Kobadak) Irrigation Project. The site is also situated about 1.8 km south from Hardinge Bridge (railway)/Lalon Shah Bridge (road) and about 350 m east from the Bheramara bypass road toward Kushtiacity (Fig. 19.3). The power station needs 1300 m^3/h (31.2 MLD) of water for cooling purposes.

In 2009, a number of tube wells, parallel to the Padma River bank, were sunk to supply water to the Bheramara Power Plant (https://openjicareport.jica.go.jp/pdf/11925369_04.pdf).

Fig. 19.3 Location of the two planned IBF systems in Bangladesh

The project area at Bheramara is underlain by an extensive unconsolidated and potential aquifer system, which in some parts is covered by 20–30 m thick clay and silty clay aquitard. The Padma River cuts the top aquitard layer. The thickness of sand (aquifer) is more than 100 m throughout the area as none of the well-encountered bottoms of the aquifer. In general, the sequence of sands gets coarser with depth. Analysis of pumping test data of the project site provided aquifer transmissivity of 4456 m^2/d, horizontal hydraulic conductivity of 42.5 m/d, and storativity of 0.007 (Bashar 2008).

As the Padma River is well connected to the subsurface aquifer, it was intended to set up an induced bank filtration system on the one km bank of the Padma. The minimum elevation of water level in the study area varies from 6 to 8.5 m in dry season and the maximum varies from 11 to 14 m in wet season. In the dry season, groundwater moves toward the river from the aquifers, while in the wet season, surface water moves from the river into the riverbank aquifers. Pre- and post-monsoon fluctuation of water level is higher in the narrow zone adjacent to the river Padma.

A conceptual model was developed taking into consideration of the hydrostratigraphy of the study area using MODFLOW, based upon the current level of information and understanding of the hydro-meteorological, hydrological, and hydro-geological patterns prevailing in the model area. The model area of about 16.5 × 22.7 km^2 was divided into a coarse grid of 80 rows and 100 columns. The model grid consists of 8000 active cells, spaced at about 206 m along rows and about 227 m along the columns, covering an area of about 374 km^2. The model parameters such as hydraulic conductivity and storage coefficient were defined for each of the 8000 active cells. In addition, the time variant parameters such as recharge, abstraction, and boundary conditions were also specified for monthly time steps. The constructed model simulates groundwater flow and takes into account all principal processes that drive the model like recharge, abstractions, and groundwater gradients that have been observed in the study area.

The model showed that the Padma River water infiltrates into the aquifer in both wet and dry seasons during pumping from the wells at the rate of 1300 m^3/h. The model predicted that there would be no significant declination of groundwater level in the project area or in the surrounding areas because pumping is neither in post-monsoon or pre-monsoon periods even after 20 years. Figures 19.4 and 19.5 showed the predicted groundwater level in the model area after 20 years for both dry and wet seasons, respectively.

The study reveals that the aquifers in Bheramara would be suitable for supplying 31,200 m^3/day of water continuously without any significant lowering of groundwater level and any environmental degradation. Padma River contributes significantly in recharging the aquifer each year.

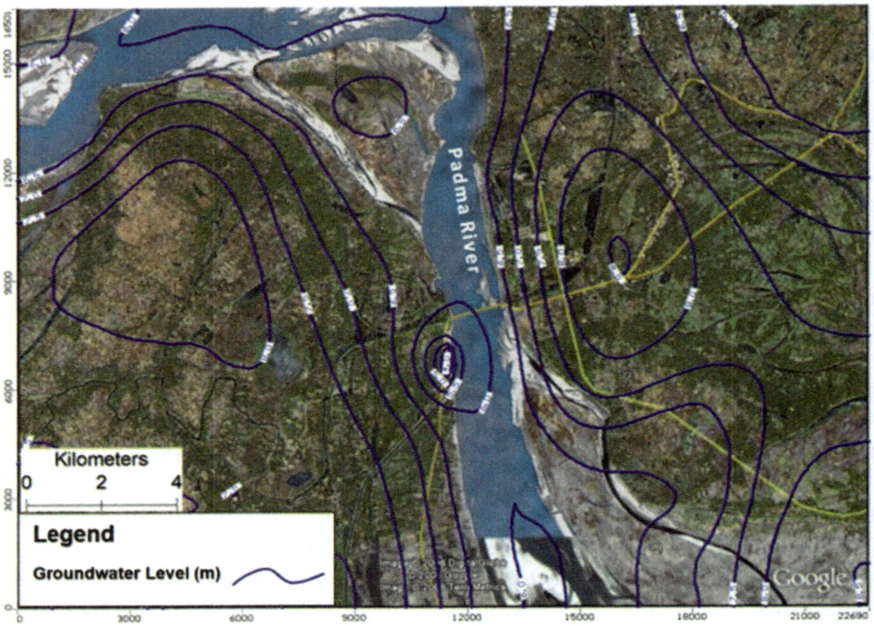

Fig. 19.4 Model predicted dry season (March) water table contour map after 20 years of pumping at 1300 m^3/h using tube wells at the bank of the river Padma

19.5.2 Induced Infiltration of Jamuna River Water at Sirajganj

Semb Corp North-West Power Company Ltd. Bangladesh, (400 ± 10% Dual-Fuel Combined Cycle Power Plant, Sirajganj Unit-4) generates about 400 MW of electricity at Sirajganj near the Bangabandhu Jamuna Bridge. This plant requires about 600 m^3/h (14.4 MLD) of water for operating the plant. Four tube wells sunk on the bank of the river Jamuna are currently supplying the required 600 m^3/h of water for operating the plant (Bashar 2018).

Figure 19.6 represents the schematic diagram of the induced infiltration of Jamuna River water caused by the shallow pumping wells. Four tube wells were constructed on the '*char*' land on the bank of the river Jamuna. The capacity of each well is about 300 m^3/h. Figure 19.7 shows one such tube well, constructed on the '*char*' of Jamuna River bank.

The Jamuna is a braided stream characterized by a network of interlacing channels with numerous sandbars enclosed in between them (Best et al. 2007; Haque et al. 2022). During low flows, the river becomes a multiple channel stream with sand bars in between and the channels shift back and forth between the main stream banks. The sandbars, locally known as '*chars*', however, occupy a semi-permanent position.

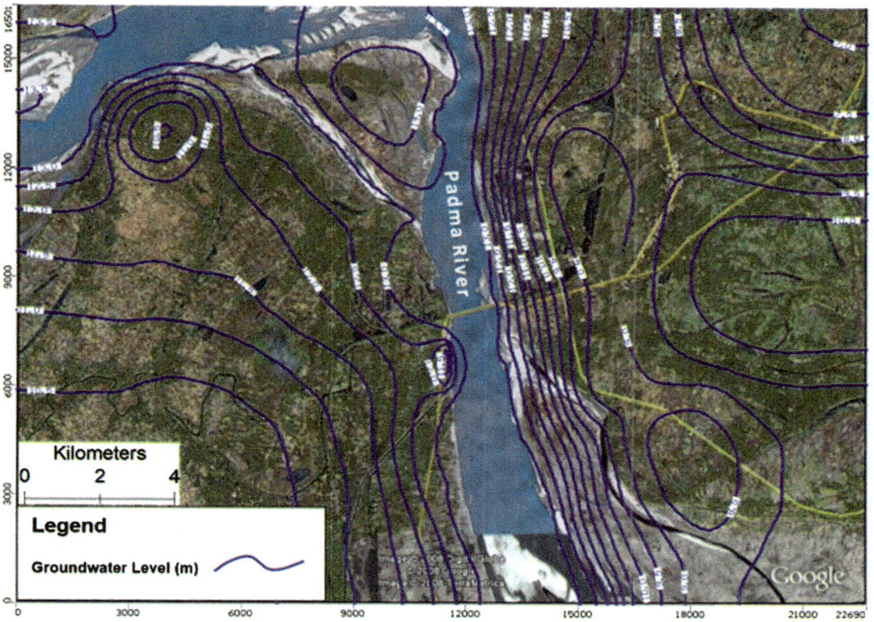

Fig. 19.5 Model predicted wet season (September) water table contour map after 20 years of pumping at 1300 m^3/h using tube wells at the bank of the river Padma

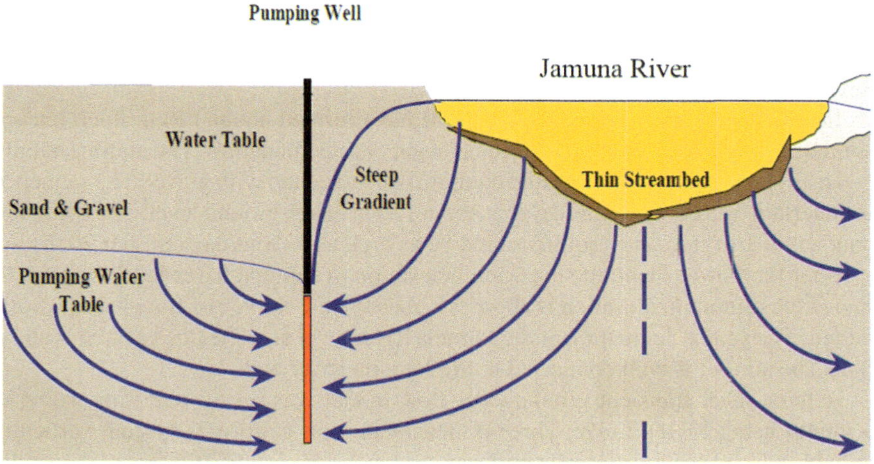

Fig. 19.6 Schematic diagram of the induced infiltration of Jamuna River water by shallow pumping wells

Fig. 19.7 Subsurface River Water Intake constructed on the Char land at the Bank of Jamuna River near Jamuna Bridge with a discharge rate of about 300 m^3/h

The river deposits them in one year very often to destroy and redeposit them in the next rainy season.

In the Project area, the Holocene sands have formed about 120 m thick unconsolidated sand and gravel aquifer which is unconfined in nature. The aquifer mainly consists of silty sand, fine sand, and medium to coarse sand with gravel. The sequence is underlain by a clay aquitard layer at about 120 m below ground level. The hydraulic conductivity of the sandy aquifer could be as high as 62.8 m/day (Bashar 2018).

Near the Power Plant, the maximum elevation of Jamuna River level is about 14 m and the minimum elevation is about 5.5 m above the Survey of Bangladesh (SoB) datum. The yearly fluctuation of the Jamuna River level near the Power Plant is about 8 m. The aquifer is well connected to the Jamuna River bed.

A three-layer transient groundwater flow model was set up with time steps of a month using MODFLOW. Three-dimensional, non-uniformly spaced, vertically variable spaced finite-difference grid was used in the model. The bottom of the model was taken −120 m. The model grid extends 10 km in the east–west direction and 10 km in the north–south direction covering an area of 100 km^2. The finite-difference grid was discretized in 97 columns and 97 rows. Fine grids have been introduced in areas where the gradient of groundwater level is high. In the coarse grid part, the cell dimension is about 0.01563 km^2 whereas, in the fine grid part, the cell dimension is about 0.0039 km^2.

The model parameters, such as hydraulic conductivity and storability, need to be defined for each active cell in the model. The model simulates groundwater flow and considers various time-varying parameters, including recharge, abstraction, and boundary conditions, on a monthly basis. It accounts for important processes that affect groundwater flow in the study area, such as recharge, abstraction, and observed groundwater gradients.

Figure 19.8 shows the modeled groundwater level of the dry season (March) after 20 years of pumping in the whole model area. Although the River Jamuna is effluent in the dry season over most parts of the model area. Due to the lowering of the head at the river bank near the Project site, it becomes influent due to the pumping wells.

The maximum elevation of the water level occurs in August or September. In the wet season, the Jamuna River recharges the river bank aquifers. The pumping wells along the Jamuna River bank get plenty of water infiltrated from the river to

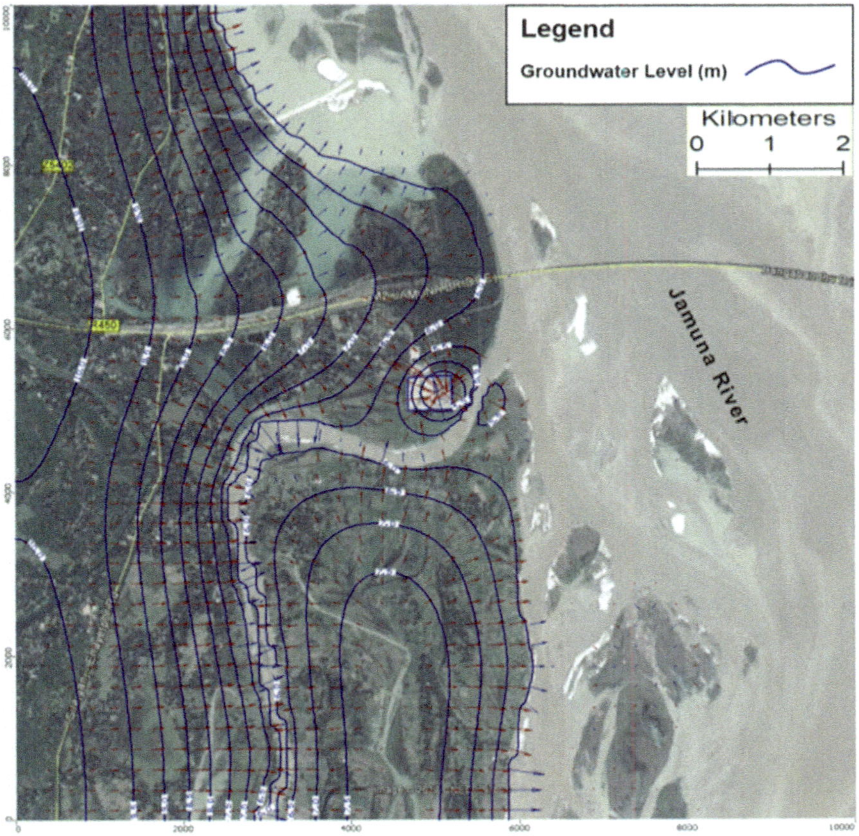

Fig. 19.8 Model predicted dry season (March) water table contour map after 20 years of pumping at 600 m^3/h using tube wells at the bank of the river Jamuna

the shallow aquifer. The aquifer is fully recharged in the wet season. Figure 19.9 shows the modeled groundwater level contour map of wet season after 20 years of pumping. The model clearly indicates that the shallow aquifer at the Jamuna River bank has the potential to supply water at the rate of 600 m^3/h for the next 20 years. The proportion of river water varies between 64 and 97% of the total water abstracted on a monthly scale by the tube wells of the project area. The model estimated that yearly about 83.9% of the total water abstracted by the tube wells along the bank of the Jamuna River will be infiltrated from the river.

Close water table contours at the river bank indicate a steep gradient of water level and high flow of water into the aquifer from the river. The length and direction of the velocity vectors at the river bank near the project site also indicate the high flow velocity of water in the shallow aquifer from the riverside (Figs. 19.8 and 19.9).

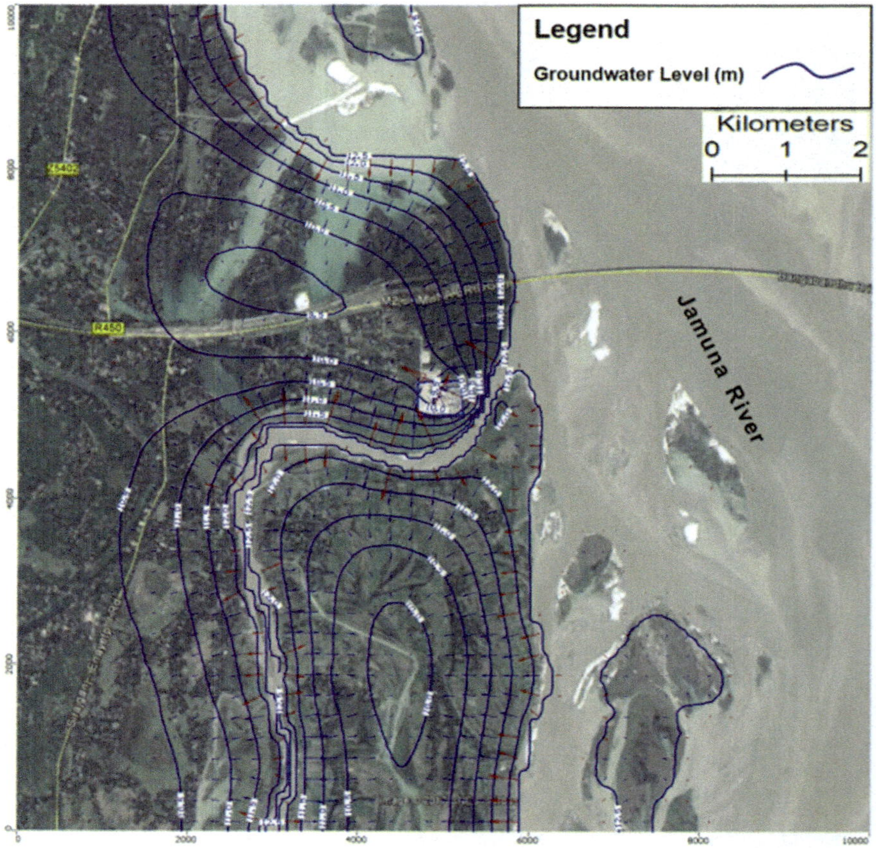

Fig. 19.9 Model predicted wet season (September) water table contour map after 20 years of pumping at 600 m^3/h using tube wells at the bank of the river Jamuna

Besides these two examples of planned Induced Infiltration cases, there are many examples of unplanned ones in Bangladesh. Just one example is given.

19.5.3 Induced Infiltration of Sitalakhya River Water at Narayanganj

Narayanganj city is located in the east-central part of Bangladesh. Narayanganj City Corporation (NCC) is situated along both the banks of the Sitalakhya River up to its confluence with the Dhaleswari River in the south (Fig. 19.10). Sitalakhya River is a distributary of the Brahmaputra (Jamuna) River. The development of well field on both sides of the River Sitalakhya at Narayanganj can be an example of unplanned induced bank filtration of river water.

The aquifers are made up of unconsolidated to poorly consolidated clastic sediments (i.e., sand and gravels) of Recent to Tertiary age. The DupiTila Sandstone Formation of the Plio-Pleistocene age forms the main aquifer from which most of the potable water is currently being supplied to Narayanganj city.

About 36 production tube wells (PTWs) were constructed on or near the bank by the Dhaka Water Supply and Sewerage Authority (DWASA) in the last 22 years. In addition, different industries and private households also installed a large number of tube wells for harnessing water for urban and industrial uses (Fig. 19.10).

Increasing extraction of groundwater in Narayanganj City has led to a progressive decline of the groundwater level over the last 20 years in most of the area, indicating abstraction is greater than recharge. Currently, the groundwater level in the Narayanganj City Corporation area is more than 30 m below ground level and the Sitalakhya River is flowing on an unsaturated zone, which means the influent nature of the River. This resulted in the increase of actual groundwater recharge to the aquifers in Narayanganj area induced by the large-scale groundwater development in the vicinity of the river.

The Sitalakhya River remains navigable around the year but remains generally polluted. DOE (2014) found that the dissolved oxygen content (DO) of water varies from 0 to a maximum of 5.6 mg/L, Biological Oxygen Demand (BOD) varies from 0 to 32 mg/L and Chemical Oxygen Demand (COD) varies between 4 and 141 mg/L.

Chloride concentration of the Sitalakhya River in 2014 was below the Environmental Quality Standard (EQS) (600 mg/L) (ECR 1997) for wastewater after treatment from industrial units. Chloride concentration varied between 3.1 and 44 mg/L. EC of Sitalakhya River at different locations was mostly within the EQS (1200 μS/cm) for treated wastewater from industrial units.

The polluted water of the Sitalakhya River is sufficiently decontaminated as it passes through the more than 100 m thick, predominantly sandy sediments (lie beneath the river bed) to the production wells. The supply water from the DWASA,

Fig. 19.10 Tube wells of DWASA constructed along the banks of the Sitalakhya within the Narayanganj City Corporation area are inducing infiltration of river water to the subsurface aquifers

which is getting water through induced infiltration of the Sitalakhya River, is found to be fully compliant with the Bangladesh drinking water standard (ECR 1997).

19.5.4 Proposed Induced Bank Filtration System on Dhaleswari River in Narayanganj

In 2022, a hydrogeological study was carried out on the bank of Dhaleswari River in the south of Narayanganj in order to assess the feasibility of setting up an IBF system for supplying 50 MLD potable water to Narayanganj City Corporation (Bashar 2023).

Both Sitalakhya and Dhaleswari rivers encircle the study area in the east, west, and south directions (Fig. 19.10). Meghna River is about 3 km away from the study area. Dhaleshwari river is about 290 km long having an average depth of 37.19 m and a maximum depth of 80.79 m (Real et al. 2017). Both the Dhaleswari and Sitalakhya rivers are perennial and potential source of renewable surface water for ensuring a long-term supply of potable water to Narayanganj city.

The study found that on the bank of the Dhaleswari River, there is only one single aquifer in the subsurface that extends up to a depth of 250 m or more below ground level. The River is hydraulically well-connected to the aquifer

A four-layer transient groundwater flow model was set up with time steps of a month using MODFLOW. The bottom of the model was taken −300m. The model grid extended 38 km in the east–west direction and 32 km in the north–south direction covering an area of about 1216 km^2. The finite-difference grid was discretized in 76 columns and 64 rows. The cell dimension is 0.25 km^2 for the whole model area.

The model assigns different hydrogeological parameters to each active cell. The model also includes various time-variant features, such as river stage, recharge, and evapotranspiration boundaries, as well as domestic, municipal, and irrigation pumping wells, which were specified for monthly time steps. These parameters and boundaries are used to simulate the movement of water and the interactions between the subsurface and surface water. The Sitalakhya and the Dhaleswari Rivers have a distinct influence on the water balance of the study area (Figs. 19.10 and 19.11).

Figure 19.11 shows the model-predicted groundwater level of the dry season of 2040 near the proposed well field at southern Narayanganj for pumping water at the rate of 62 MLD. It can be mentioned here that 12 MLD water will be required for local consumptions and 50 MLD water is expected to be exported to the Narayanganj City Corporation. The model predicted that an elongated cone of depression will be developed occupying a vast area below the bottom of the perennial rivers to induce large-scale infiltration of river water into the sub-surface aquifer. Figure 19.11 also shows that substantial water will flow from the eastern side where there is abundant water in the aquifers under the Meghna River. The minimum elevation of groundwater level (GWL) in the central part of the cone of depression could be about −35 m (Fig. 19.11).

It is very unlikely that there would be any permanent depletion of the groundwater or surface water level in the area in the long run. Narayanganj City Corporation has already started the implementation of setting up of the proposed IBF system on the bank of Dhaleswari to ensure the supply of about 50 MLD of water for potable uses.

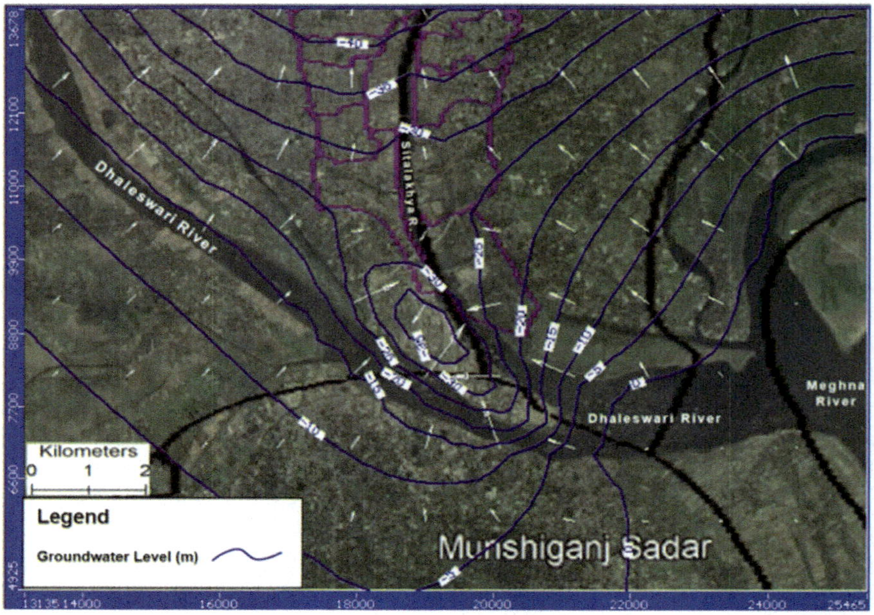

Fig. 19.11 Model predicted groundwater table contour map of the southern Narayanganj city area for the dry season of 2040 for pumping water at the rate of 62 MLD from the proposed IBF system

19.6 Summary and Conclusions

Surface water intake systems to meet societal needs can broadly be classified as subsurface intakes and open surface water intakes. Subsurface intake systems (e.g., bank infiltration or infiltration wells) are usually preferred because the transport of water through the aquifer material acts as a filter medium. During bank filtration, water quality improvement (treatment) of induced surface water is commonly the main objective.

In contrast, open surface water intake systems require extensive treatment of water prior to supply for potable use. The quality of feed water obtained from an open intake typically contains high suspended solids and algae that can clog the intake system. In addition, water from an open intake is subjected to seasonal variations in temperature. Because of these disadvantages, open surface water intake systems are preferred only when subsurface systems are not feasible.

The technique of induced infiltration of river water by pumping wells on its bank can replace the conventional open surface water intakes in Bangladesh. The water is not turbid and has low or no micro-biological contaminations. Moreover, water can be withdrawn from the aquifer even in the dry season when the flow in the river is at its lowest level and direct withdrawal of water from the river is cumbersome.

Bangladesh is a riverine country drained by about 700 rivers and their tributaries. In addition to that, there are a lot of natural lakes and surface water bodies. Most of the rivers flow through the unconsolidated alluvium comprising clay, silt, and sand with occasional gravels in Bangladesh. By inducing recharge to the subsurface aquifers using river or lake water it is possible to increase the availability of potable water in different parts of Bangladesh at low cost.

The advantage is that, the rivers are generally hydraulically well connected with the underlying high potential aquifer system in Bangladesh. Depending on the hydraulic head difference between the river and the aquifer river may recharge the aquifer or get discharge water from the aquifer under normal conditions. During pumping by wells located over banks, the hydraulic head of the aquifer is lowered and the recharge to the aquifer is increased. Such induced bank filtration is a type of Managed Aquifer Recharge (MAR) where the infiltration of surface water is induced by pumping from a nearby well.

Two examples of planned induced bank infiltration of river water in Bangladesh are discussed. The first one is on the bank of the Padma River which was set up in 2009 for extracting 1300 m^3/h (31.2 MLD), while the second one is on the bank of Jamuna (Brahmaputra River) constructed in 2018. However, in Bangladesh, a large number of pumping wells have been installed on the banks of rivers; resulting in unplanned induced bank filtration systems. An example of such case at Narayanganj on the banks of Sitalakhya River is discussed.

The groundwater model (MODFLOW) at Bheramara of Kushtia district showed that the sandy aquifers on the bank of the Padma River are substantially recharged by the river water, induced by pumping wells, in both dry and wet seasons. The model also predicted that there will be no significant decline in groundwater levels in the project or surrounding areas in the next 20 years. Similarly, at Sirajganj about 600 m^3/h (14.4 MLD) is withdrawn by four tube wells on the bank of the river Jamuna. Here, groundwater model predicted no significant decline in water level after 20 years of pumping. The proportion of river water varies between 64 and 97% of the total water abstracted monthly by the pumping tube wells from the riverbank aquifer.

There are numerous pumping wells in different parts of Bangladesh, which were set on the bank of rivers or other surface water bodies and are inducing recharge to river bank aquifers. Massive extraction of groundwater by tube wells installed on or near the banks of the Sitalakhya River in Narayanganj city is inducing recharge of river water to the subsurface aquifers.

The feasibility study through groundwater modeling, for setting up IBF system on the bank of Dhaleswari River in the south of Narayanganj city found that an elongated cone of depression will be developed below the bottom of the Dhaleswari and Sitalakhya rivers to induce large-scale infiltration of river water into the sub-surface aquifer for pumping water at the rate of 62 MLD. Narayanganj City Corporation has already started to set up an IBF system on the bank of the Dhaleswari River to supply 50 MLD water to the city.

The technique of induced infiltration of river water by pumping wells can be a cheaper and better way for withdrawing water from the rivers in all seasons in comparison to the direct withdrawal of water from river by centrifugal pumps in

Bangladesh. The technique is not only cheaper to install but also cheaper and easier for operation and maintenance. It is suggested that a proper feasibility study should be undertaken to assess the suitability of setting up induced bank filtration systems in different parts of Bangladesh in the future.

19.7 Recommendations

Water is an essential resource in Bangladesh, and it plays a critical role in the country's agriculture, industry, and daily life. Efforts are being made to improve the management of water resources in Bangladesh, including the development of infrastructure for water supply and sanitation, and the implementation of policies to promote water conservation and efficiency.

The widespread occurrence of Quaternary age unconsolidated shallow sandy aquifers in Bangladesh which are hydraulically well connected with a large number of perennial rivers, provides the opportunity to set up induced bank filtration systems, which can be a reliable source of water for different uses. The researchers and planners should conduct detailed studies to identify suitable areas to set up IBF systems for ensuring long-term water supply for urban and industrial uses.

Acknowledgements This article is based on the knowledge the author gained while working for Combined Cycle Power Plant, Bheramara, Kushtia, Bangladesh; Sembcorp Design and Construction Pte Ltd. (Singapore) and Urban Infrastructure Improvement Preparatory Facility (Narayanganj City Corporation Component) project. It should be noted that this article has been written independently and only reflects the views of the author. I thank the reviewer for his valuable comments, which helped improve the article.

References

Agriculture and Agri-food Canada Diversion of Water from Surface-Water Sources Through Infiltration Galleries http://www.pfra.ca/doc/Water%20Supply%20Engineering/InfiltrationGalleriesFinal.pdf

Bashar K (2008) Groundwater Modelling of the proposed site of bheramara power station and surrounding areas, Japan International Cooperation Agency (JICA) In "Feasibility Study of Bheramara Combined Cycle Power Plant, Bheramara, Kushtia" Microsoft Word - 00_Contents of FR Main Report.doc (jica.go.jp)

Bashar K (2018) Supplementary Hydrogeological Study for Determining the Source of Water Abstracted from Shallow Aquifer Along the Jamuna River Bank Using Groundwater Modelling Technique. Unpublished Report, submitted to SembCorp North-West Power Generation Company Ltd.

Bashar K (2023) Feasibility study for setting up Induced Bank Filtration (IBF) system on the bank of Dhaleswari-Sitalakhya Rivers in southern Narayanganj, Bangladesh.Unpublished Report, submitted to'Urban Infrastructure Improvement Preparatory Facility (Narayanganj City Corporation Component)

Best JL, Ashworth PJ, Sarker MH, Roden JE (2007) The Brahmaputra-Jamuna River, Bangladesh. In: Large rivers: geomorphology and management. Wiley, New York, pp 395–433

Cady P, Boving TB, Choudri BS, Cording A, Patil K, Reddy V (2013) Attenuation of bacteria at a riverbank filtration site in rural India. Water Environ Res 2013(85):2164–2174
Dillon P (2005) Future management of aquifer recharge. Hydrogeol J 13:313–316
DoE (2014) River Water Quality Report 2014. Natural Resource Management Section Department of Environment (DoE) Govt. of Bangladesh
ECR (1997) The Environment Conservation Rules 1997, Department of Environment, Ministry of Environment and Forest, Govt. of the People's Republic of Bangladesh
Gillefalk M, Massmann G, Nützmann G, Hilt S (2018) Potential impacts of induced bank filtration on surface water quality: a conceptual framework for future research. Water 10:1240. https://doi.org/10.3390/w10091240
Haque M, Shishir S, Mazumder A, Iqbal M (2022) Change detection of Jamuna River and its impact on the local settlements. Phys Geography 1–21. https://doi.org/10.1080/02723646.2022.2026075
Jeffords (1945) Recharge to water-bearing formations along the Ohio Valley. Am Water Works Assoc J 37:144–145
Kazmann (1948) River infiltration as a source of groundwater supply. Am Soc Civil Eng Trans 113:404–420
Kelly BP, Rydlund J, Paul H (2006) Water-quality changes caused by riverbank filtration between the Missouri River and Three Pumping Wells of the Independence, Missouri, Well Field 2003–2005, p 48. Available online: http://purl.access.gpo.gov/GPO/LPS75695 .
Maeng SK, Lee K-H (2019) Riverbank filtration for the water supply on the Nakdong River, South Korea. Water 11:129. https://doi.org/10.3390/w11010129
Mukherjee A, Saha D, Harvey CF, Taylor RG, Ahmed KM, Bhanja SN (2015) Groundwater systems of the IndianSub-Continent. J Hydrol Regional Stud. https://doi.org/10.1016/j.ejrh.2015.03.005
Paszkowski A, Goodbred S, Borgomeo E, Khan SMA, Hall JW (2021) Geomorphic change in the Ganges–Brahmaputra–Meghna delta. Nat Rev Earth Environ 2:763–780. https://doi.org/10.1038/s43017-021-00213-4
Ray C, Melin G, Linsky RB (2003) Riverbank filtration. Improving source water quality. Kluwer Academic Publishers, Dordrecht, The Netherlands
Real M, Khanam N, Mia M, Nasreen M (2017) Assessment of water quality and microbial load of Dhaleshwari River Tangail, Bangladesh. Adv Microbiol 7:523–533. https://doi.org/10.4236/aim.2017.76041
Rorabaugh (1951) Stream-bed percolation in development of water supplies. Union Geol Geophys Internat Assoc Internat.d'HydrologieScientifique, Assemblee Gen de Bruxelles 1951, Extract du Tome II 165–174
Saha D, Roy R (2018) Groundwater resources of India: potential, challenges and management. In: Sikdar P (ed) Groundwater development and management, issues and challenges in South Asia. Springer, Berlin. https://doi.org/10.1007/978-3-319-75115-3_2
Sandhu C, Grischek T, Kumar P, Ray C (2011) Potential for riverbank filtration in India. Clean Technol Environ Policy 2011(13):295–316
Sandhu C, Grischek T, Börnick H, Feller J, Sharma SK (2019) A water quality appraisal of some existing and potential riverbank filtration sites in India. Water 2019(11):215. https://doi.org/10.3390/w11020215
Sarker MH, Huque I, Alam M, Koudstaal R (2003) Rivers, chars and char dwellers of Bangladesh. Int J River Basin Manag 1:61–80
Schmidt CK, Lange FT, Kuehn HJB (2003) Schmidt, Experiences with riverbank filtration and infiltration in Germany
https://www.researchgate.net/publication/267779083_Experiences_
USGS (1999) Induced infiltration from the rockaway river and water chemistry in a stratified-drift aquifer at Dover, New Jersey, with a section on modeling ground-water flow in the Rockaway River Valley. In: Dysart JE, Rheaume SJ, Kontis AL (eds) Water-resources investigations report, pp 96–4068

Index

D. Saha et al. (eds.), *Managed Groundwater Recharge and Rainwater Harvesting*, Water Resources Development and Management,
https://doi.org/10.1007/978-981-99-8757-3

The manufacturer's authorised representative in the EU is Springer Nature Customer Service Centre GmbH, Europaplatz 3, 69115 Heidelberg, Germany. If you have any concerns regarding our products, please contact ProductSafety@springernature.com

Printed and bound by CPI Group (UK) Ltd, Croydon, CR0 4YY
01/10/2026
02225866-0016